A PRIMER OF POPULATION GENETICS

A PRIMER OF
POPULATION

DANIEL L. HARTL WASHINGTON UNIVERSITY SCHOOL OF MEDICINE

GENETICS

SECOND EDITION

SINAUER ASSOCIATES, INC. • PUBLISHERS
SUNDERLAND, MASSACHUSETTS 01375

To Christopher

A PRIMER OF POPULATION GENETICS, SECOND EDITION

 For information address Sinauer Associates, Inc., Sunderland, Mass., 01375, USA.

Library of Congress Cataloging-in-Publication Data

Hartl, Daniel L.
A primer of population genetics.

Includes bibliographies and index.
1. Population genetics. 2. Quantitative genetics.
3. Population genetics—Problems, exercises, etc.
I. Title.
QH455.H36 1987 575.1′5 87-20530
ISBN 0-87893-301-8 (pbk.)

THE COVER
A dense population of dwarf monkey flowers (*Mimnius nanus*) growing in the high chaparral of northern California; white *Collinsia torreyi* (blue lips) are interspersed with the bright yellow *Mimnius*. Photograph by L. Egede-Nissen/BPS.

Printed in U.S.A.

5 4 3 2 1

CONTENTS

2 CAUSES OF EVOLUTION 69

3 MOLECULAR POPULATION GENETICS 143

4 QUANTITATIVE GENETICS 215

PREFACE

Population genetics combines observation with theory for the purpose of understanding genetic changes that occur within and among populations. Population genetics occupies a special place in biology because it cuts across so many diverse disciplines, among them molecular biology, genetics, ecology, evolutionary biology, systematics, natural history, plant breeding, animal breeding, many areas of conservation and wildlife management, human genetics, sociology, anthropology, mathematics, and statistics. Because the principles of population genetics have such wide-ranging applications, it seems desirable to have them briefly summarized in a form accessible to a wide audience. The *Primer* is my attempt at such a summary.

The intended audience for the *Primer* consists of students who seek a working knowledge of population genetics, without technical details, for use in other types of studies. Focus has been centered on the basic elements of population genetics and their application to real-life situations. No mathematics beyond simple algebra is included, and all principles are illustrated with numerical examples, using actual data wherever possible.

The *Primer* is divided into four parts. The overall theme is the origin, maintenance, and significance of genetic variation. Chapter 1 focuses on the detection and measurement of genetic variation as influenced by particular mating systems, such as random mating or inbreeding, and it includes sections on DNA restriction fragment length polymorphisms and their uses. Chapter 2 deals with the evolutionary effects of mutation, migration, selection, and small population size. The chapter also discusses how the major evolutionary forces have been synthesized by Sewall Wright into his shifting balance theory of evolution, and it includes sections on interdeme selection and kin selection.

Molecular population genetics, which represents the exciting hybrid of molecular biology and population genetics, is the subject of Chapter 3, which is new in this Second Edition. Topics include the neutral theory of molecular evolution; rates and patterns of molecular evolution; molecular clocks; evolution in multigene families through duplication, divergence, gene conversion and other mechanisms; multigene superfamilies; and SINEs and LINEs. Topics related to transposable elements include insertion sequences and composite transposons in bacteria, P-elements and hybrid dysgenesis in *Drosophila*, and endogenous retroviruses and their horizontal transfer in mammals.

Chapter 4 considers multifactorial traits that are influenced by many genes as well as by the effects of environment. Although particular attention is devoted to methods of artificial selection that are used for the genetic improvement of crop plants and domesticated animals, Chapter 4 also discusses how the principles can be applied to traits in humans and natural populations of other species. There is a new approach to the treatment of stabilizing selection, a discussion of the current controversy between the relative importance of stabilizing selection and antagonistic pleiotropy in maintaining genetic variation in natural populations, and some previously unpublished approximations that simplify equations used in the quantitative genetics of threshold traits.

Each chapter contains about 25 problems for solution, which have been chosen to illustrate the application of important principles. Three special features have been incorporated in an attempt to make the problems interesting as well as useful. First, almost all of the problems use authentic data collected in the field or in laboratory experiments. Second, each problem is inserted into the main text immediately following the principle it is designed to illustrate. Third, each problem is provided with its solution worked out in full.

Casual thumbing through the book will reveal symbols included in angle braces like $\langle p \rangle$ and $\langle \mu \rangle$. These are used consistently to distinguish *parameters* like p and μ from their *estimated values* $\langle p \rangle$ and $\langle \mu \rangle$. Rather then adopt standard embellishments like circumflexes, tildes, or overbars, which are easily confused with embellishments used for equilibrium values and other purposes, I resolved to use one consistent symbolism for estimates that can be implemented with all standard word processors. I am pleased with how well the symbolism works, as are the reviewers. I recommend it, not only for textbooks because of its unambiguousness, but also for scientific papers. It has unanticipated advantages, for example, in distinguishing clearly between $\overline{w}$ as the parametric value of the average fitness of a population, and $\langle \overline{w} \rangle$ as an estimate of $\overline{w}$ based on some set of observations, or in requiring that students learn only one symbol μ for the parametric mean of a normal distribution, with an estimate symbolized as μ rather than by the entirely unrelated symbol $\overline{x}$.

Each chapter is provided with a chapter summary at the end, along with numerous suggestions for further reading. In the interest of brevity, citations within the text have been pared to a minimum, and the literature cited can be found at the back of the book. I regret that, because of the paring, many original investigators do not receive overt credit for their important contributions.

The process of revising the *Primer* was begun by Joan Zurheidt, who converted hard copy of the first edition into ASCII files. I then went over every element of the book thoroughly in order to update where necessary and to improve the presentation where possible. The entire manuscript was read carefully by Andrew G. Clark, and Chapter 3 was thoroughly reviewed by Clark, J. Bruce Walsh, and Bruce S. Weir, to whom I am indebted for their many suggested improvements. Tendai Mutangadura-Mhlanga and Hope Hollocher made many helpful comments. As usual, Andrew Sinauer was very patient and supportive during the revision process, and Joseph Vesely did another outstanding job in production. Carol Wigg deserves special credit for coding the files for electronic typesetting.

I remain indebted to my colleagues Daniel E. Dykhuizen and Stanley A. Sawyer for discussing population genetics with me on almost a daily basis and for keeping my thinking straight. My laboratory is blessed with superb doctoral and postdoctoral students, memorable personalities all, who have created a very special atmosphere of excitement and friendly cooperation. For this I am grateful to Glenn Bryan, Antony Dean, Robert DuBose, Pat McEwen, Danny Garza, Louis Green, Rob DeSalle, James Jacobson, Ann MacPeek, Kyoko Maruyama, Meetha Medhora, Tendai Mutangadura-Mhlanga, and Nancy Scavarda. Thanks to the Genetics office staff, Carol Jones, Kay Webb, and Becky Meininger, for helping me keep my sanity, such as it is, amid a blizzard of nitty details. I owe continuing thanks to my wife Christine for her patience during the numerous occasions when my mind was preoccupied with untangling a sentence, phrasing a problem, or understanding an equation.

Daniel L. Hartl

GENETIC VARIATION 1

Population genetics is the study of Mendel's laws and other genetic principles as they apply to entire populations of organisms. The genetic principles underlying population genetics are, for the most part, simple and straightforward, but it may be helpful to preface the discussion with a few key definitions.

Gene is a general term meaning, loosely, the physical entity transmitted from parent to offspring during the reproductive process that influences hereditary traits. Genes influence human traits such as hair color, eye color, skin color, height, weight, and various aspects of behavior—although most of these traits are also variably influenced by environment. Genes also determine the makeup of proteins such as hemoglobin, which carries oxygen in the red blood cells, or insulin, which is important in maintaining glucose balance in the blood. Genes can exist in different forms or states. For example, a gene for hemoglobin may exist in a normal form, or in any one of a number of different forms resulting in hemoglobin molecules that are abnormal. All alternative forms of a gene, including the normal form, are called **alleles**.

From a biochemical point of view, a gene corresponds to a specific sequence of constituents (called nucleotides) along a molecule of DNA (deoxy-

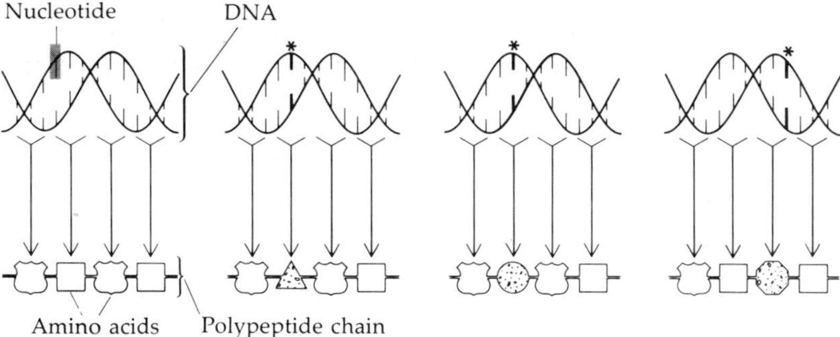

Figure 1. Alleles are alternative forms of a gene. DNA is composed of two intertwined strands, each consisting of a long linear sequence of nucelotides. Genes are fundamental units of genetic information that correspond chemically to the sequence of nucleotides in a segment of DNA. A typical gene consists of hundreds of thousands of nucleotides, only a few of which are shown here. The arrows show how the genetic information in a portion of the nucleotide sequence of DNA specifies the amino acid sequence in a portion of a polypeptide (or protein), each group of three adjacent nucleotides corresponding to one amino acid. Substitutions of one nucleotide for another (at positions indicated by the asterisks and heavy lines) in the DNA can lead to substitutions of one amino acid for another in the polypeptide (indicated by stippling).

ribonucleic acid); DNA is thus the genetic material. Different sequences of nucleotides that may occur in a gene represent alleles (Figure 1). Most genes code for polypeptide chains that form proteins. Each adjacent group of three nucleotides constitutes a coding group or **codon**, which specifies the corresponding amino acid subunit in the polypeptide chain.

The essentials of gene expression in the cells of higher organisms (eukaryotes) are outlined in Figure 2. The coding regions of the DNA in a gene (CR1 and CR2), which code for amino acids, are often interrupted by one or more noncoding regions known as intervening sequences (IVS) or **introns**. The sequence of nucleotides present in the introns is faithfully copied into the nucleotides in an RNA molecule during the first, or transcription, step in gene expression. In the second, or processing, step of gene expression, the introns are removed by splicing (cutting and rejoining). RNA processing results in a messenger RNA molecule, in which the coding regions, called **exons**, have been made contiguous. The messenger RNA molecule undergoes translation on ribosomes in the cytoplasm to produce a polypeptide.

Within a living cell, genes are arranged in linear order along microscopic threadlike bodies called **chromosomes**. A typical chromosome may consist of several thousand genes. The position of a gene along a chromosome is called a **locus**. In most higher organisms, each cell contains two copies of each type of chromosome, one inherited from the mother through the egg

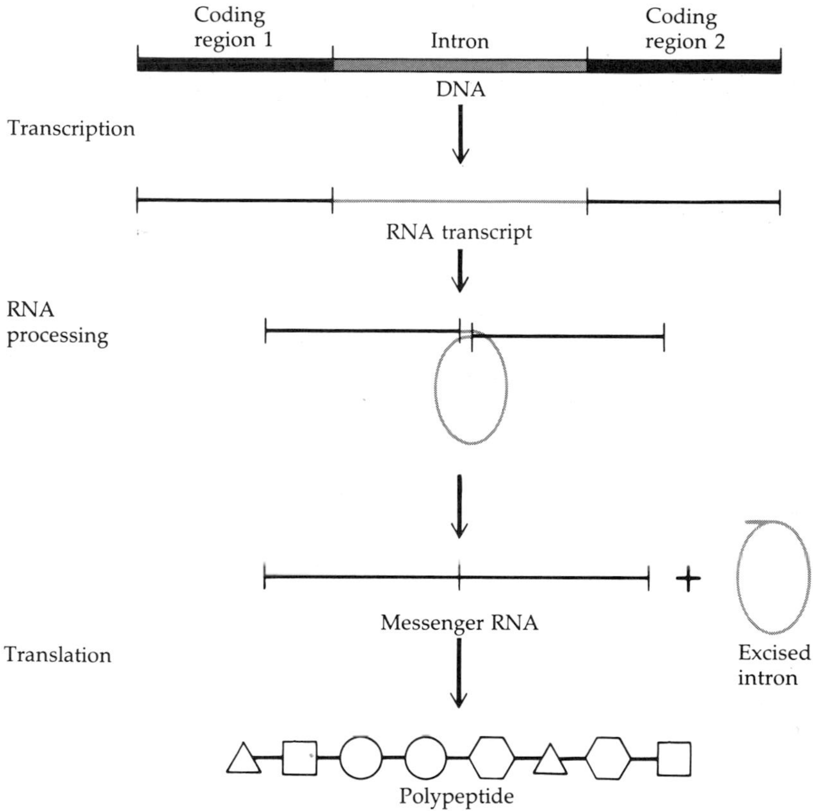

Figure 2. Processes in gene expression in eukaryotic cells. DNA regions coding for the amino acids in a single polypeptide can be interrupted by noncoding regions (intervening sequences or introns). When the DNA is copied into RNA in transcription, both coding and noncoding regions are transcribed. However, the introns are removed from the transcript during processing, and in the messenger RNA which is translated into the polypeptide, the coding regions are contiguous.

and one inherited from the father through the sperm. At any locus every normal individual contains exactly two alleles—one at each corresponding position in the maternal and paternal chromosome. If the two alleles at a locus are chemically identical (in the sense of having the same nucleotide sequence along the DNA), the individual is said to be **homozygous** at the locus under consideration; if the two alleles at a locus are chemically different, the individual is said to be **heterozygous** at the locus. The term "gene" is a general term usually used in the sense of "locus."

Geneticists make a fundamental distinction between the genetic constitution of an individual and the individual's physical or biochemical attributes. The genetic constitution of an individual is called its **genotype**; genotype thus refers to the specific alleles contained in an individual at all loci that affect the trait in question. For example, if a trait is influenced by two genes, each with two alleles, then there are nine possible genotypes, as follows:

AA BB	*AA Bb*	*AA bb*
Aa BB	*Aa Bb*	*Aa bb*
aa BB	*aa Bb*	*aa bb*

where *A* and *a* refer to the alleles of the first gene and *B* and *b* refer to the alleles of the second gene. In some cases when the genes are *linked* (located in the same chromosome), it is necessary to distinguish between the genotypes *A B*/*a b* and *A b*/*a B*, yielding ten possible genotypes. Whether each genotype has a single, unique expression of the trait depends on the manner in which the genes interact during development. For example, suppose that the trait in question is degree of pigmentation and that the amount of pigment in an individual is determined by the total number of *A* and *B* alleles present, each of which adds a single unit of pigmentation to the phenotype. Then, as shown in Table 1, there are only five possible levels of pigmentation, and genotypes *aa BB*, *Aa Bb*, and *AA bb* all have the same level.

The physical expression of a genotype is called the **phenotype**. In the preceding pigment example, we would say that genotypes *aa BB*, *Aa Bb*, and *AA bb* all have the same phenotype, namely, two units of pigmentation. The distinction between the genetic constitution of an individual (genotype) and the physical or biochemical attributes of the individual (phenotype) is particularly important in cases in which the environment can affect the trait; in such cases, two individuals having the same genotype can nevertheless have different phenotypes because of differences in the environment.

Table 1. A model of additive gene action.[a]

Genotype	Amount of pigmentation[b]
AA BB	4
Aa BB, *AA Bb*	3
aa BB, *Aa Bb*, *AA bb*	2
Aa bb, *aa Bb*	1
aa bb	0

[a]At left are shown the nine possible genotypes with two loci and two alleles at each. At right is shown the amount of pigmentation expected in each genotype when it is assumed that each allele designated by an uppercase letter is responsible for producing a certain amount of pigment.
[b]Measured as the increase in pigmentation over that found in *aa bb* genotypes.

PHENOTYPIC DIVERSITY AND GENETIC VARIATION

One of the universal attributes of natural populations is phenotypic diversity. For most traits, many differing phenotypes can be found among the individuals in any population. Phenotypic diversity in many traits is impressive even with the most casual observation. Among humans, for example, there is diversity with respect to height, weight, body conformation, hair color and texture, skin color, eye color, and many other physical and psychological attributes or skills. Population genetics must deal with this phenotypic diversity, and especially with that portion of the diversity which is caused by differences in genotypes among individuals. In particular, the field of population genetics has set for itself the tasks of determining how much genetic variation exists in natural populations, and of explaining this variation in terms of its origin, maintenance, and evolutionary importance. Genetic variation, in the form of multiple alleles of many genes, exists in most natural populations; this conclusion can be documented by any number of methods, the most important of which are discussed in this chapter. On the other hand, all methods are limited to the study of a certain number of genes, usually a small number of genes in comparision to the total number in the organism. Extrapolation of the results for a small number of genes to the entire genome is therefore questionable, and because of the uncertainty as to whether the genes studied are truly representative of the genome, we still do not know, quantitatively, how much genetic variation is present in natural populations.

Whatever total amount of genetic variation may exist, sufficient variation has already been found in genes amenable to study that, for most populations, no two individuals (barring identical twins or other multiple identical births) could be expected to have the same genotype for all genes. Thus, it becomes important to describe how genetic variation in natural populations is organized into genotypes—to determine, for example, whether alleles of different genes are associated at random.

Protein Electrophoresis

One of the most widely used procedures for revealing genetic variation in enzymes and other proteins is **electrophoresis**, a technique that came into widespread use in the late 1960s. A typical laboratory setup for enzyme electrophoresis is illustrated schematically in Figure 3. In a typical population study using electrophoresis, a small sample of tissue is ground up or a small amount of blood serum is drawn from each of a number of individuals in the population to be surveyed. Each sample is placed in a small slot near the edge of a rectangular, fairly thin slab of a jelly-like material (usually starch,

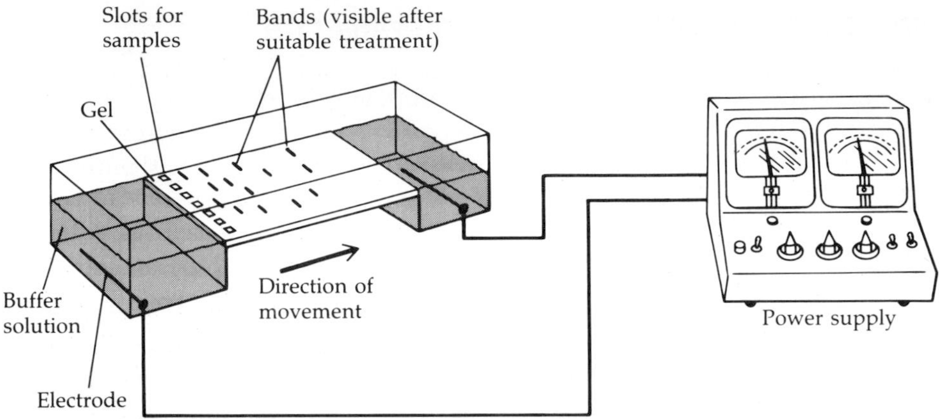

Figure 3. One type of laboratory apparatus for electrophoresis. The procedure is widely used to separate protein or DNA molecules. In conventional gels, DNA fragments smaller than about 20 kb migrate approximately in proportion to the logarithm of their molecular weights.

polyacrylamide, or agarose) located inside a plastic box. At each end of the box are chambers containing electrodes and buffered solution, and in performing the actual electrophoresis, an electric current is applied across the gel for several hours. Molecules in the samples, including the proteins of interest, move through the gel in response to the electric field. After the electrophoresis is finished, the gel is stained for a particular enzyme by soaking it in a solution containing a substrate for the enzyme along with a dye that precipitates where the enzyme-catalyzed reaction occurs. A dark band thus appears in the gel, revealing the position of the enzyme.

If an enzyme in an individual has an amino acid substitution that leads to a difference in the overall ionic charge of the molecule, then the enzyme will have a somewhat altered **electrophoretic mobility**, which is to say that the enzyme moves at a different rate. The change in mobility occurs because enzymes of the same size and shape move at a rate determined largely by the ratio of the number of positively charged amino acids (primarily lysine, arginine, and histidine) to the number of negatively charged ones (principally aspartic acid and glutamic acid). Electrophoresis can therefore be used to detect mutations that result in differences in electrophoretic mobility of the corresponding enzymes.

One possible result of an electrophoresis experiment is shown in the hypothetical gel in Figure 4, in which all individuals in the sample manifest an enzyme with the same electrophoretic mobility. Another kind of result is shown in Figure 5, in which some individuals are homozygous for an allele *F* associated with a rapidly migrating enzyme (*fast*), others are homozygous

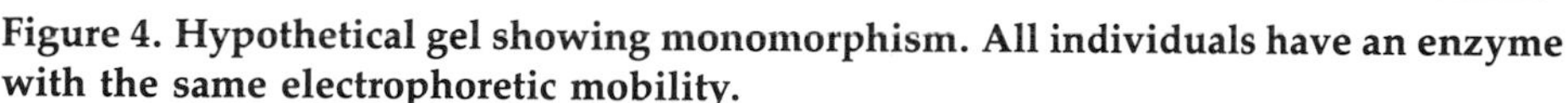

Figure 4. Hypothetical gel showing monomorphism. All individuals have an enzyme with the same electrophoretic mobility.

for an allele *S* associated with an enzyme that migrates more slowly (*slow*), and still others are heterozygous (*F*/*S*) for the alleles. Two enzyme bands appear in heterozygotes because the enzyme itself is **monomeric**—it consists of a single polypeptide chain—and heterozygous individuals produce two kinds of polypeptide chains, one corresponding to each allele. When the enzyme in question is **multimeric**, say a dimer (consisting of two polypeptide chains), then the heterozygote will produce three types of enzymes that differ in their electrophoretic mobility: rapidly migrating (*fast* + *fast*) and slowly migrating (*slow* + *slow*) **homodimers**, which are like those produced in the corresponding homozygous genotypes, and a *fast* + *slow* **heterodimer**, which typically is found to have intermediate electrophoretic mobility.

Allele Frequency

Enzymes that differ in electrophoretic mobility as a result of allelic differences at a single locus are called **allozymes**. Thus, allozyme variation in a population is an indication of genetic variation, and such genetic variation is very common. Extensive allozyme variation has been found in almost all natural populations studied by electrophoresis, including organisms such as the bacterium *Escherichia coli*, plants, *Drosophila*, mice, and humans.

In order to compare different genes and different populations, it is necessary to have some convenient quantitative measure of genetic variation. Genetic variation can be quantified using the concept of allele frequency. The

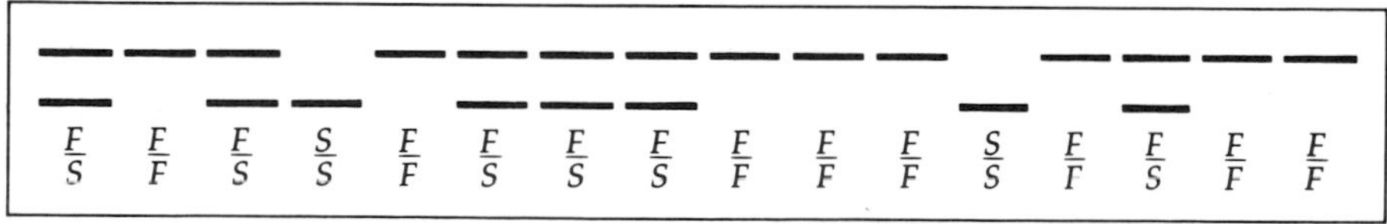

Figure 5. Hypothetical gel showing allozyme polymorphism. Eight individuals are homozygous for an allele (*F*) that codes for a rapidly migrating enzyme; two individuals are homozygous for a different allele (*S*) that codes for a slowly migrating enzyme; and six individuals are heterozygous (*F*/*S*) and therefore exhibit enzyme bands corresponding to both alleles.

allele frequency of a prescribed allele among a group of individuals is simply the proportion of all alleles at the locus that are of the prescribed type. The frequency of any prescribed allele in a sample is therefore equal to twice the number of homozygous genotypes with that allele (because each homozygote carries two copies of the allele) plus the number of heterozygous genotypes with that allele (because each heterozygote carries one copy) divided by two times the total number of individuals in the sample (because each individual carries two alleles at the locus). Among the small sample of 16 individuals in the hypothetical gel shown in Figure 4, for example, there are 8 *F*/*F* homozygotes, 6 *F*/*S* heterozygotes, and 2 *S*/*S* homozygotes: the allele frequency of *F* is therefore $(2 \times 8 + 6)/(2 \times 16) = 0.6875$, whereas the allele frequency of *S* is $(2 \times 2 + 6)/(2 \times 16) = 0.3125$. (Detailed calculations are shown in Table 2.) Note that the sum of all allele frequencies, in this case those for *F* and *S*, must equal 1. The allele frequency in a sample of individuals from a population is only an estimate of the true allele frequency in the whole population, of course, but the estimate will usually be close to the true frequency if the sample is sufficiently large; for this reason, allele frequency estimates should be based on samples of 100 or more individuals whenever possible. Since the term *gene* is sometimes used synonymously with *allele*, the expression *gene frequency* is sometimes used synonymously with *allele frequency* if the particular allele being referred to is clear from the context.

Formally, suppose that among n individuals sampled from a population the numbers of *AA*, *Aa* and *aa* genotypes are (*AA*), (*Aa*), and (*aa*), respectively. Following convention, we let p and q represent the allele frequencies of *A* and *a*, respectively, with $p + q = 1$. The estimate $\langle p \rangle$ of the allele frequency p in the population that was sampled is

$$\langle p \rangle = [2(AA) + (Aa)]/2n \qquad \textbf{(1.1)}$$

and the estimated sampling variance is

$$\langle \mathrm{Var}(\langle p \rangle) \rangle = \langle p \rangle (1 - \langle p \rangle)/2n \qquad \textbf{(1.2)}$$

Equations 1.1 and 1.2 make use of several important concepts in statistics. Quantities used in describing populations are **parameters**. Although the exact values of parameters are usually unknown, their values can be estimated using samples from the population. In this book, whenever it is necessary to distinguish parameters from their estimates, we use unembellished symbols for parameters, for example p for the (unknown) frequency of an allele in a specified population, and the same symbol in angular brackets for the estimated value, in this example $\langle p \rangle$. The variance of an estimate is used for judging the reliability of the estimate. Since the variance is also estimated from data in the sample, the estimated variance is $\langle \mathrm{Var}(\langle p \rangle) \rangle$. The square root of the variance of an estimate is known as the **standard error** of the estimate.

Table 2. Calculation of allele frequencies.

	GENOTYPE			
	F/F	F/S	S/S	Total
No. of individuals	8	6	2	16
No. of *F* alleles	16	6	0	22
No. of *S* alleles	0	6	4	10
No. of *F* + *S* alleles	16	12	4	32
Allele frequency of *F* = 22/32 = 0.6875				
Allele frequency of *S* = 10/32 = 0.3125				

If the sampling and estimation were repeated many times using the same population, then approximately 68 percent of the estimates would fall within plus or minus one standard error of the true value of the parameter, approximately 95 percent would fall within two standard errors, and approximately 99.7 percent would fall within three standard errors. These intervals are known as the 68, 95, and 99.7 percent **confidence intervals**, respectively. To put the matter the other way around, 32 percent of the estimates would differ from the true value by more than one standard error, 5 percent by more than two, and only 0.3 percent by more than three. These approximations all assume that the repeated estimates conform to the familiar, bell-shaped *normal distribution*.

The estimate in Equation 1.2 is the sampling variance of a binomial distribution, which occurs in such familiar probability contexts as tossing coins or drawing balls from urns. The estimate assumes that the $2n$ observed genes are independent samples, as are the successive tosses of a coin.

PROBLEM 1

Using the data in Table 2, calculate the standard error of $\langle p \rangle$ and the 68, 95, and 99.7 percent confidence intervals. To appreciate the effect of sample size on the precision of an estimate, suppose the observed numbers of each of the three genotypes in Table 2 were increased by a factor of 10, resulting in a total sample size of 160 individuals. Carry out the same calculations as above for the enlarged sample size.

ANSWER Here $\langle p \rangle = 0.6875$ with $2n = 32$. Using Equation 1.2, the estimated variance of $\langle p \rangle$ equals $(0.6875)(0.3125)/32 = 0.0067$, and the standard error is the square root of this, or $(0.006714)^{1/2} = 0.0819$. One, two and three multiples of the standard error equal 0.0819, 0.1638, and 0.2457, respectively; thus the 68 percent confidence interval is 0.6056 to 0.7694 (that is, 0.6875 − 0.0819 to 0.6875 + 0.0819), the 95 percent confidence interval is 0.524 to 0.851, and the 99.7 percent confidence interval is 0.442 to 0.933. For comparison, exact intervals have been calculated using the binomial distribution for repeated samples of 32 genes from a population with p = 0.6875, and 66 percent of the estimates lie in the range 0.625 to 0.750, 97 percent in the range 0.531 to 0.844, and 99.7 percent in the range 0.469 to 0.938. Thus, the approximation is reasonably accurate in this case despite the relatively small sample size.

With sample size increased by a factor of 10, the standard error becomes $[(0.6875)(0.3125)/320]^{1/2} = 0.0259$ and the 68, 95 and 99.7 percent confidence intervals are now (0.662, 0.713), (0.636, 0.739), and (0.610, 0.765), respectively. (The intervals are smaller because of the larger sample size.) The exact results are that 69 percent of the samples lie in the range (0.662 to 0.713), 95 percent in the range (0.641 to 0.738), and 99.7 percent in the range (0.612 to 0.762). The approximations are much better with the larger sample size.

PROBLEM 2 Levin (1978) has studied allozyme polymorphisms in the wild annual plant *Phlox drummondii*. For the alcohol dehydrogenase gene (alleles Adh^a and Adh^b), one population in Texas had the following genotype frequencies: Adh^a/Adh^a (0.04), Adh^a/Adh^b (0.32), and Adh^b/Adh^b (0.64). Estimate the allele frequencies of Adh^a ($\langle p \rangle$) and Adh^b ($\langle q \rangle$).

ANSWER From Equation 1.1, $\langle p \rangle = 0.04 + 0.32/2 = 0.20$; and $\langle q \rangle = 0.64 + 0.32/2 = 0.80$. Without knowledge of the size of the sample (it was $n = 35$), the variance cannot be estimated.

PROBLEM 3 Mukai, Watanabe, and Yamaguchi (1974) captured 660 fertilized females of *Drosophila melanogaster* from a natural population in Raleigh, North Carolina. These females were used to found a larger experimental population. After about five months (ten generations), 489 third chromosomes in the population were examined for allozymes coding for the enzymes esterase-6 (alleles $E6^F$ and $E6^S$), esterase-C (alleles EC^F and EC^S), and octanol dehydrogenase (alleles Odh^F and Odh^S). The order of the genes in the third chromosome is known to be *E6–EC–Odh*. Results were as follows:

$E6^F\ EC^F\ Odh^F$	152	$E6^S\ EC^F\ Odh^F$	264
$E6^F\ EC^F\ Odh^S$	7	$E6^S\ EC^F\ Odh^S$	13
$E6^F\ EC^S\ Odh^F$	15	$E6^S\ EC^S\ Odh^F$	29
$E6^F\ EC^S\ Odh^S$	1	$E6^S\ EC^S\ Odh^S$	8

Estimate the allele frequencies and their standard errors for $E6^F$ and $E6^S$, for EC^F and EC^S, and for Odh^F and Odh^S. In this case, the estimate of the allele frequency in the population is equal to the allele frequency in the sample, and the estimate of the variance in Equation 1.2 is still valid.

ANSWER For esterase-6, there were 175 $E6^F$ and 314 $E6^S$ alleles, yielding $\langle p \rangle = 175/489 = 0.3579$ for $E6^F$ and $\langle q \rangle = 1 - \langle p \rangle = 0.642$ for $E6^S$. The standard error equals $[(0.3579)(0.6421)/489]^{1/2} = 0.0217$, and it is identical for both $\langle p \rangle$ and $\langle q \rangle$. The estimate and standard error are usually written as $\langle p \rangle = 0.3579 \pm 0.0217$. For the other alleles the estimates and standard errors are 0.892 ± 0.014 for EC^F and 0.108 ± 0.014 for EC^S; and 0.941 ± 0.011 for Odh^F and 0.059 ± 0.011 for Odh^S.

Polymorphism

The concept of allele frequency underlies the idea of polymorphism. A **polymorphic** gene is a gene for which the most common allele has a frequency of less than 0.95 (some authors prefer a less stringent cutoff of 0.99). Conversely, a **monomorphic** gene is one that is not polymorphic. The cutoff at 0.95 (or 0.99) in the definition of polymorphism is arbitrary, but it serves to focus attention on those genes in which allelic variation is common. In any large population, rare alleles occur in virtually every gene. **Rare alleles** are alleles with frequencies of less than 0.005, and between one and two individuals per thousand are heterozygous for rare alleles at any locus, as judged by an electrophoretic survey of 43 enzyme genes in 250,000 Europeans (Harris and Hopkinson 1972). Many rare alleles are deleterious and are presumably maintained in populations by recurrent mutation, as will be discussed in Chapter 2. The definition of polymorphism is an attempt to focus on genes which have alleles with frequencies too high to be explained solely by recurrent mutation to harmful alleles. With the 0.95 definition of polymorphism given above, at least 10 percent of the population is heterozygous for the most common allele, for reasons that will become clear when we discuss the Hardy–Weinberg principle.

Allozyme Polymorphisms

Polymorphism for alleles that determine allozymes is extremely widespread. This is shown in Figure 6, which summarizes the results of electrophoretic surveys of 14 to 71 (mostly around 20) genes in populations of 243 species. Each point in the figure (except for the one pertaining to humans) gives the type of organism studied and the number of species examined. The vertical axis, labeled "$\langle P \rangle$," refers to the estimated proportion of genes that are polymorphic by the 0.95 criterion; and the horizontal axis, labeled "$\langle H \rangle$," refers to the estimated proportion of genes expected to be heterozygous in an average individual. The average heterozygosity H is estimated by totaling the number of heterozygous individuals for each gene, dividing this by the total number of individuals in the sample, and averaging over all genes. For example, the heterozygosity estimated for the one gene in the small sample in Figure 5 is $\langle H \rangle = 6/16 = 0.375$. To illustrate further the estimation of P and H, we may use data of Harris (1966), who studied ten enzyme genes in the English population. Of these ten genes, three were found to be polymorphic; based on this sample of ten, the estimated proportion of polymorphic genes in the genome is $\langle P \rangle = 3/10 = 0.30$. The observed proportion of heterozygous individuals for each of the seven monomorphic genes was, of course, 0, and the proportion of heterozygous individuals for the three

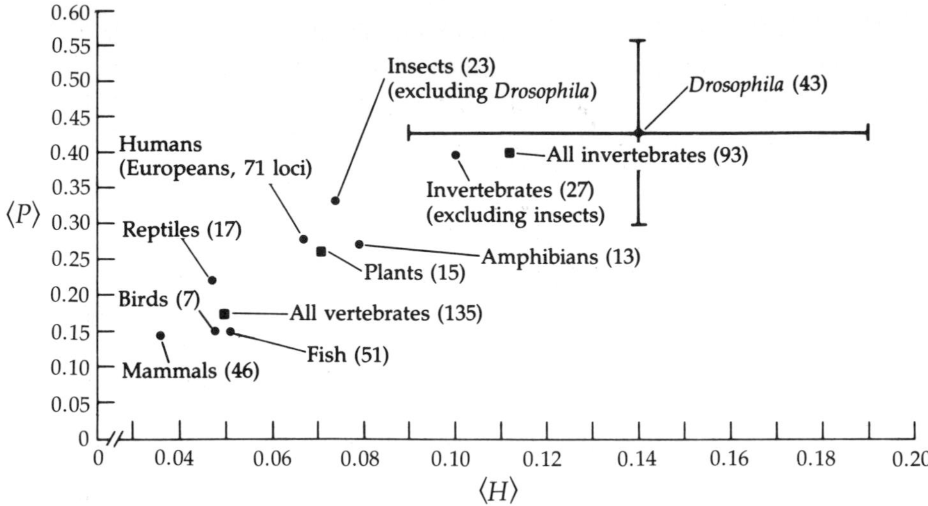

Figure 6. Estimated levels of heterozygosity (⟨*H*⟩) and proportion of polymorphic genes (⟨*P*⟩) derived from allozyme studies of various groups of plants and animals. The number of species studied is shown in parentheses beside each point. Squares denote averages for plants, invertebrates, and vertebrates. The bars across the *Drosophila* point show the range of *H* and *P* within which about 70 percent of the *Drosophila* species fall. Other groups would have similarly large bars. (Data from Nevo 1978.)

polymorphic genes was 0.509 (for red-cell acid phosphatase), 0.385 (for phosphoglucomutase), and 0.095 (for adenylate kinase); thus, using this sample, the average heterozygosity in the population is estimated as $\langle H \rangle = (0.509 + 0.385 + 0.095 + 7 \times 0)/10 = 0.099$.

The vertical and horizontal bars on the point corresponding to *Drosophila* indicate substantial variability found in *P* and *H* among species in a group. The bars indicate the limits of *P* and *H* within which 68 percent of the individual species are expected to fall. Thus, among *Drosophila* species, 68 percent are expected to have a proportion of polymorphic genes in the range 0.30 to 0.56 and an average heterozygosity in the range 0.09 to 0.19. Such bars could be attached to each point, but their lengths would be comparable to those for *Drosophila*.

Figure 6 admits of no simple summary because of the immense variability in polymorphism and heterozygosity found within each group of organisms, as indicated by the length of the variability bars corresponding to *Drosophila*. On the whole, there is a positive relationship between amount of polymorphism and degree of heterozygosity, which is as expected because the greater the fraction of polymorphic genes in a population, the more genes that are expected to be heterozygous in an average individual. The overall mean (average) of $\langle P \rangle$ in Figure 6 is 0.26 ± 0.15, and the mean of $\langle H \rangle$ is 0.07 ± 0.05. Vertebrates have the lowest average amount of genetic variation among the groups in Figure 6, plants come next, and invertebrates have the highest. *Drosophila* is the most genetically variable group of higher organisms so far studied, and mammals the least variable. Humans are fairly typical of large mammals, and an extensive electrophoretic survey of 104 genes in a sample including all major human races gave estimates of $\langle P \rangle = 0.32$ and $\langle H \rangle = 0.06$ (Harris et al. 1977). The one obvious conclusion that can be reached from Figure 6 is that allozyme polymorphisms are widespread among higher organisms. Genetic variation is even more prevalent among some prokaryotes. For example, natural isolates of the mammalian intestinal bacterium *Escherichia coli* exhibit levels of genetic polymorphism two or three times greater than occur in species of eukaryotes (Selander and Levin 1980; Whittam et al. 1983).

Although genetic polymorphisms are widespread, they are not universal. For example, both major subspecies of the cheetah (*Acinonynx jubatus*) are virtually monomorphic. A survey of 49 enzymes among 30 individuals from the East African subspecies (*A. jubatus raineyi*) yielded only two polymorphic genes and estimates of $\langle P \rangle = 0.04$ and $\langle H \rangle = 0.01$; among 98 individuals from the South African subspecies (*A. jubatus jubatus*), the estimates were $\langle P \rangle = 0.02$ and $\langle H \rangle = 0.0004$ (O'Brien et al. 1987). Most unusual was the finding of skin graft acceptance between unrelated cheetahs from the South African subspecies. Graft acceptance means that the cheetah population is monomorphic for the major histocompatibility locus, which is abundantly polymorphic in other mammals. Apparently the cheetah, which was worldwide in its range at one time but presently numbers less than 20,000 animals, underwent at least two severe constrictions in population number resulting in the loss of most of its genetic variability (O'Brien et al. 1987).

How Representative Are Allozymes?

The validity of estimates of polymorphism based on electrophoresis is open to question. The amount of polymorphism may be underestimated because conventional electrophoresis fails to detect many amino acid substitutions. For example, in a study of 14 myoglobin proteins from 13 species of cetaceans (whales, dolphins, and porpoises) and the horse, only between six and eight were distinguished by conventional electrophoretic methods (McLellan and Inouye 1986). However, the resolving power of electrophoresis is enhanced by running a sequence of gels at different pH values (**sequential electrophoresis**), because proteins not separated at one pH may be separated at another (Singh et al. 1976). Using five pH conditions, McLellan and Inouye (1986) were able to distinguish 13 of the 14 different myoglobins. Some amino acid substitutions that are difficult to resolve electrophoretically can be detected because they render the enzyme sensitive to high temperatures. Use of these techniques increased the number of identified alleles at the xanthine dehydrogenase gene in *Drosophila pseudoobscura* from 6 to 37 and increased estimates of average heterozygosity from 0.44 to 0.73 (Singh et al. 1976). Bonhomme and Selander (1978) were able to identify about twice as many alleles in the house mouse (*Mus musculus*) as had been previously identified using conventional electrophoresis. On the other hand, while the more precise techniques reveal additional alleles of genes known to be polymorphic, thus increasing estimates of H, genes classified as monomorphic by means of routine electrophoresis tend to remain monomorphic even with more critical techniques, so estimates of P remain much the same as before.

Electrophoretic surveys might also overestimate the amount of polymorphism because the enzymes typically surveyed are those found in relatively high concentration in tissues or body fluids; such enzymes are often called "Group II enzymes" in order to distinguish them from more substrate-specific "Group I enzymes" involved in processes such as energy transformation. Group II enzymes are hardly a random sample of proteins because they do not include Group I enzymes or most structural or regulatory molecules. In *Drosophila*, among ten Group II enzymes studied by conventional electrophoresis, estimates of $\langle P \rangle$ and $\langle H \rangle$ were 0.70 and 0.24, respectively, but among eleven Group I enzymes, the estimates were 0.27 and 0.04, respectively (Gillespie and Langley 1974). In a two-dimensional electrophoretic study of the 54 proteins found in greatest concentration in *Drosophila*, Leigh Brown and Langley (1979) obtained estimates of $\langle P \rangle = 0.11$ and $\langle H \rangle = 0.04$, which are very much smaller than those usually quoted for *Drosophila*. However, the two-dimensional technique has even less resolving power than conventional electrophoresis (McLellan and Inouye 1986).

In summary, sequential electrophoresis is presently the best method for distinguishing among protein molecules, but it is nevertheless difficult to extrapolate from electrophoretic surveys of enzymes to the entire genome because the enzymes may not be representative.

DNA Restriction-Fragment Polymorphisms

The ultimate resolution of genetic variation occurs at the level of DNA sequences, the study and interpretation of which is included in Chapter 3. However, polymorphisms in DNA sequences can be detected by the use of **restriction enzymes** that cut DNA molecules at sites of particular nucleotide sequences (the **restriction sites** of the enzymes). Several restriction enzymes and their restriction sites are shown in Figure 7. For example, the enzyme *Alu*I cuts DNA at sites containing the four-nucleotide sequence AGCT, and

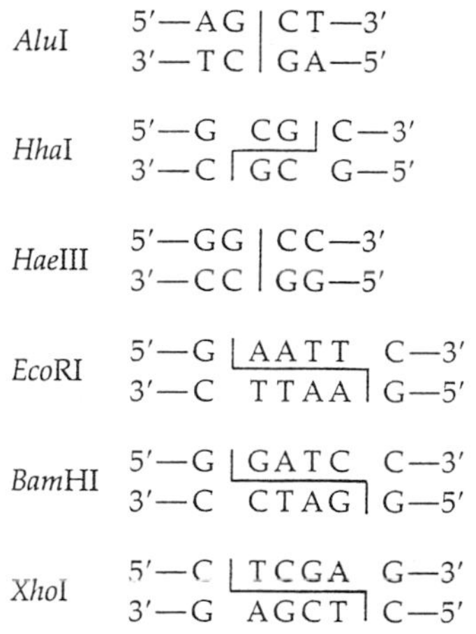

AluI
5′—AG | CT—3′
3′—TC | GA—5′
HhaI
5′—G CG | C—3′
3′—C | GC G—5′
HaeIII
5′—GG | CC—3′
3′—CC | GG—5′
EcoRI
5′—G | AATT C—3′
3′—C TTAA | G—5′
BamHI
5′—G | GATC C—3′
3′—C CTAG | G—5′
XhoI
5′—C | TCGA G—3′
3′—G AGCT | C—5′

Figure 7. Restriction enzymes cleave DNA molecules at sites of specific, short nucleotide sequences. More than 500 different restriction enzymes are commercially available. They are essential tools in DNA analysis and gene cloning. In these examples, the short vertical lines show the positions of cleavage in the DNA backbones.

*Eco*RI cuts at the six-nucleotide sequence GAATTC. Most restriction enzymes used in population studies have either four- or six-nucleotide restriction sites.

Like enzymes, DNA fragments can be separated by electrophoresis. Unlike enzymes, which can be identified by making use of their enzymatic activity, the identification of DNA fragments requires special procedures, the most widely used of which is the Southern blot (Southern 1975), illustrated in Figure 8. The key element is the use of a **probe** DNA molecule containing

Gel
Filter
DNA restriction fragments
(a) Blot
(b) Hybridize filter with radioactive probe
(c) Photographic film exposed to filter

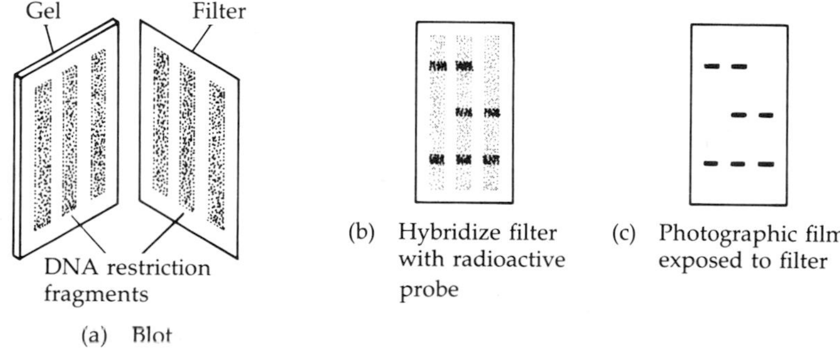

Figure 8. Southern transfer procedure (Southern blot). (a) DNA fragments separated by electrophoresis are transferred and chemically attached to a filter. (b) The filter is mixed with radioactive probe DNA, which sticks to homologous DNA molecules in the filter. (c) After washing, the filter is exposed to photographic film, which develops dark bands caused by radioactive emissions from the probe.

the nucleotide sequences of interest, which is obtained from a gene that has been cloned into, for example, bacterial cells. In the Southern procedure, DNA restriction fragments that have been separated by electrophoresis are blotted onto a nitrocellulose or nylon filter, where subsequent chemical treatment attaches them (Figure 8a). The filter is heated in order to separate the two complementary DNA strands making up each molecule, and bathed in a solution containing probe DNA that has been rendered radioactive (Figure 8b). As the solution cools, the probe DNA strands form double-stranded molecules with their complementary counterparts on the filter, and careful washing removes all of the probe DNA which has remained unpaired. The filter is sandwiched with photographic film, where radioactive disintegrations from the bound probe result in visible bands (Figure 8c).

Polymorphisms resulting in the presence or absence of restriction sites can be identified because they change the length of characteristic restriction fragments (Figure 9). The upper part of each panel shows the location of restriction sites in the DNA molecules in a diploid individual, and the *a*-type molecule contains one additional restriction site not present in the *A*-type molecule. The lower part of the figure demonstrates that, with suitable probe DNA, all three genotypes can be distinguished by their pattern of restriction fragments. Polymorphisms in restriction fragments are **restriction fragment length polymorphisms** (abbreviated RFLPs, pronounced *rifleps*). RFLPs need only cloned DNA fragments to be revealed, are distributed throughout the genome, and are widespread in virtually all organisms. RFLPs have assumed major importance in human population genetics as genetic markers for mapping the genome and as genetic markers of disease.

The amount of polymorphism at the nucleotide level can be estimated from restriction-site data. For example, suppose a study includes n homologous DNA molecules from different individuals in a population or in different species, which are cut with a restriction enzyme and found to contain some number m of cleavage sites (sites at which cleavage occurs in at least one molecule in the sample). For cleavage site i, let c_i represent the number of molecules in the sample which are cleaved. Thus, for $i = 1, 2, \ldots, m$, the particular value of c_i may range from 1 to n. The quantity $c = \Sigma c_i$ equals the total number of cuts at all cleavage sites in the sample. Estimates of the proportion of polymorphic nucleotide sites in the sample P_{nuc} and the heterozygosity per nucleotide site in the sample H_{nuc} have been determined by Engels (1981) as

$$\langle P_{\text{nuc}} \rangle = \frac{c - n(m - k)}{jc} \tag{1.3}$$

$$\langle H_{\text{nuc}} \rangle = \frac{nc - \Sigma c_i^2}{jc(n - 1)} \tag{1.4}$$

where k is the number of cleavage sites which are absent in at least one member of the sample, and j is the number of nucleotides in the restriction site (usually either 4 or 6). The subscript "nuc" indicates that the polymorphism and heterozygosity are estimated at the level of individual nucleotides. When several enzymes are used, m, k, and c_i refer to the total number of cleavage sites, and j is redefined as the average length of the restriction sites, weighted by the total number of cuts made by each enzyme (see Engels 1981 for further detail and estimates of the standard errors). The estimators 1.3 and 1.4 both assume that a given restriction site contains no more than one nucleotide polymorphism.

PROBLEM 4

Cross and Birley (1986) studied restriction-site variation in the region of the alcohol dehydrogenase gene of *Drosophila melanogaster* in a population descended from animals trapped at a Dutch fruit market in Groningen. The region contained a total of 23 sites for five restriction enzymes, each having a six-base restriction site, and 16 of the sites were cut in all individuals in the sample. The data below document the presence (+) or absence (−) of the seven sites found to be polymorphic in a sample of 10 chromosomes. Use Equations 1.3 and 1.4 to estimate the proportion of polymorphic nucleotides and the heterozygosity of nucleotide sites in the sample.

***Bam*H1**	***Hind*III**	***Pst*I**	***Xho*I**	***Pst*I**	***Eco*RI**	***Eco*RI**
+	−	−	+	+		−
+	−	−	−	−	+	+
−	−	+	−	−	+	−
−	+	−	+	−	+	+
−	+	−	+	−	+	+
−	+	−	+	−	+	+
−	+	−	+	−	+	+
−	+	−	+	−	+	+
−	+	−	+	−	+	+
−	−	−	+	−	+	−

ANSWER For this sample, $n = 10$, $m = 16 + 7 = 23$, $k = 7$, and $j = 6$. Taking the sites from left to right, $c_1 = 2$, $c_2 = 6$, $c_3 = 1$, $c_4 = 8$, $c_5 = 1$, $c_6 = 9$, $c_7 = 7$, and there are 16 additional sites for which $c_i = 10$. Thus, $c = \Sigma c_i = 34 + 160 = 194$ and $\Sigma c_i^2 = 236 + 1600 = 1836$. From Equation 1.3, $\langle P_{nuc} \rangle = [194 - 10(23 - 7)]/(6)(194) = 0.029$ and $\langle H_{nuc} \rangle = [(10)(194) - 1836]/(6)(194)(9) = 0.010$. (The actual sample of 34 chromosomes gave almost the same estimates.)

Visible Genetic Variation in Natural Populations

Most of the genetic variation due to allozymes or RFLPs is **hidden genetic variation** which is not apparent at the phenotypic level. However, genetic

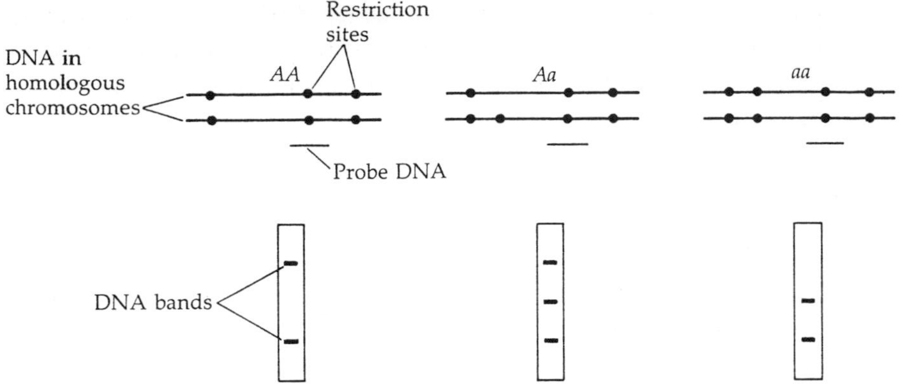

Figure 9. Restriction fragment length polymorphisms (RFLPs, pronounced "rifleps") result from the presence or absence of particular restriction sites in DNA. In this example, the DNA molecule designated *A* contains three restriction sites, and the one designated *a* contains four. Genotypes *AA*, *Aa* and *aa* each give a different pattern of bands in Southern transfer experiments using the indicated probe DNA.

variation that does affect phenotype is also widespread in most natural populations. This generalization is supported by the results of several kinds of observations, particularly by the generally deleterious effects of close inbreeding (discussed later in this chapter) and by the success of artificial selection in altering almost any morphological or physiological characteristic (discussed in Chapter 4). In most cases, genetic variation that affects phenotype within the normal range is not traceable to the effects of alleles or individual genes. Most of the traits are **multifactorial**, meaning that they are influenced by "many factors," typically several or many genes acting in concert with environmental factors. Among humans, for example, the obvious differences between normal people in hair color, eye color, skin color, stature, and other such traits are not usually traceable to individual genes. Even abnormal phenotypes outside of the normal range are difficult and sometimes impossible to trace to single genes. In humans, about five percent of live-born babies have some serious physical or mental disability, yet many of these conditions have no known hereditary basis (lack of oxygen at the time of birth can be an environmental cause of mental retardation, for example), or their mode of inheritance is **polygenic** (resulting from the combined effects of several or many genes) and therefore complex. Similarly, about five percent of the individuals in natural populations of *Drosophila melanogaster* have phenotypic abnormalities, many of which resemble those found in special laboratory strains containing particular mutant genes. Yet breeding tests indicate that no more than one-third of phenotypically abnormal flies owe their abnormality to individual mutant genes—the rest owe their phenotypic abnormality to more complex genetic causes or to environmental accidents.

Populations

The word "population" has so far been used in an informal, intuitive sense to refer to a group of organisms belonging to the same species, but further discussion and clarification of the concept is appropriate at this time. In population genetics, the word "population" does not usually refer to an entire species. It refers instead to a group of individuals of the same species living within a sufficiently restricted geographical area that any member can potentially mate with any other member (provided, of course, that they are of the opposite sex). Precise definition of such a unit is difficult and varies from species to species because of the almost universal presence of some sort of **geographical structure** in species—some typically nonrandom pattern in the spatial distribution of organisms. Members of a species are rarely distributed homogeneously in space: there is almost always some sort of clumping or aggregation, some schooling, flocking, herding, or colony formation. Population subdivision is often caused by environmental patchiness: areas of favorable habitat intermixed with unfavorable areas. Such environmental patchiness is obvious in the case of, for example, terrestrial organisms on islands in an archipelago, but patchiness is a common feature of most habitats—freshwater lakes have shallow and deep areas, meadows have marshy and dry areas, forests have sunny and shady areas. Population subdivision can also be caused by social behavior, as when wolves form packs. Even the human population is clumped or aggregated—into towns and cities, away from deserts and mountains.

In population genetics, focus is on the local interbreeding units of possibly large, geographically structured populations, because it is within such local units that systematic changes in allele frequencies occur that ultimately result in the evolution of adaptive characteristics. Such local interbreeding units are often called **local populations** or **demes**, and these are the fundamental units of population genetics, usually referred to simply as **populations**. For example, in Figure 6, the point pertaining to humans is actually the average value of $\langle P \rangle$ and $\langle H \rangle$ among local populations. The other points are averages of a similar sort for the other groups. In this book, we use the word "population" to mean "local population"—the actual, evolving unit of a species—unless a broader meaning is clear from the context. Local populations are also sometimes referred to as **Mendelian populations** or **subpopulations**.

Uses of Genetic Polymorphisms

Whether studied through allozymes or restriction fragment length polymorphisms, naturally occurring genetic variation is a boon in many applications. Genetic variation provides a set of built-in markers for the genetic study of organisms in their native habitats, including organisms for which domestication or laboratory rearing is not feasible, and for which conventional genetic manipulation is impossible.

Studies of genetic variation are useful in inferring the ancestral history of a group of organisms. In some cases, merely the amount of genetic variation is revealing. For example, the extreme genetic uniformity found in the cheetah, in which allozyme polymorphisms are two to five times less frequent than in other large cats, supports the hypothesis that the species underwent at least one very severe constriction in population number at some time in the geologically recent past, probably no later than 10,000 to 12,000 years ago (O'Brien et al. 1987). At the other end of the spectrum, the finding that genetic variation in *Escherichia coli* is even more widespread than in eukaryotes (Milkman 1973, Selander and Levin 1980) gave convincing disproof to the traditional prejudice that species of haploid asexual organisms must be genetically uniform.

Genetic polymorphisms are useful in investigating the genetic relationships among subpopulations in a species. The principle is that alleles are shared among subpopulations because of migration, and the similarity in allele frequencies can be used to estimate the rate of migration (Chapter 2). Shared alleles also result from common ancestry. For example, the Ainu people of northern Japan have numerous Caucasoid-like features, including their facial features, light skin, and hairy bodies, yet their genetic polymorphisms clearly show them to be more closely related to other Mongoloid groups. For example, they possess the *D(Chi)* allele of transferrin protein and the Di^a allele of the Diego blood group, both of which are virtually restricted to Mongoloid populations, while lacking several alleles that are polymorphic in Caucasoids (Watanabe et al. 1975).

From a practical point of view, genetic polymorphisms are useful in human populations as genetic markers which may be linked in chromosomes with harmful genes that cause disease. In kinships with a family history of the disease, the genetic markers can be monitored to determine which members of the kinship are carriers of the harmful gene and for early diagnosis of individuals likely to be affected. RFLPs that are linked to disease genes have great potential as probes for identifying recombinant DNA clones containing the defective genes, which may enable the function of the defective genes to be pinpointed and lead to the development of effective treatments.

The natural mating systems of plants and animals can be investigated through studies of the frequencies in which alternative genotypes occur in a population. For example, certain patterns of genotype frequencies are expected when a population undergoes mating which is random relative to the alleles of interest. Genotype frequencies are also used to estimate the amount of self-fertilization that occurs in populations of monoecious plants or hermaphroditic animals. The influence of the mating system on the frequencies of occurrence of alternative genotypes is the subject of the next section.

ORGANIZATION OF GENETIC VARIATION

The purpose of this section is to determine what effect the matings that occur in a population have on the relative frequencies of the genotypes that are formed from the alleles of a polymorphic gene. The frequency of a specified genotype in a population is the **genotype frequency**, and the frequencies of the various possible types of matings determines the mathematical relation between the allele frequencies and the genotype frequencies. Such mathematical relations are usually inferred from models of a population.

Models

A **model** is an intentional simplification of a complex situation designed to eliminate extraneous detail in order to focus attention on the essentials of the situation. In population genetics, we must contend with factors such as population size, patterns of mating, geographical distribution of individuals, mutation, migration, and natural selection. Although we wish ultimately to understand the combined effects of all these factors and more, the factors are so numerous and interact in such complex ways that they cannot usually be grasped all at once. Simpler situations are therefore devised, situations in which a few identifiable factors are the most important ones and others can be neglected.

One kind of model frequently used in population genetics is the **mathematical model**, which is a set of hypotheses that specifies the mathematical relations between measured or measurable quantities (the parameters) in a system or process. Mathematical models can be extremely useful: They express concisely the hypothesized quantitative relationships between parameters; they reveal which parameters are the most important ones in a system and thereby suggest critical experiments or observations; they serve as guides to the collection, organization, and interpretation of observed data; and they make quantitative predictions about the behavior of a system that can, within limits, be confirmed or shown to be false. The validity of a model must, of course, be tested by determining whether the hypotheses on which it is based and the predictions that grow out of it are consistent with observations. Mathematical models are always simpler than the actual situation they are designed to elucidate. Many features of the actual system are intentionally left out of the model, because to include every aspect of the system would make the model too complex and unwieldy. Construction of a model always involves compromise between realism and manageability; a completely realistic model is likely to be too complex to handle mathematically, and a model that is mathematically simple may be so unrealistic as to be useless. Ideally, a model should include all essential features of the system and exclude all nonessential ones. How good or useful a model is often depends on how closely this ideal is approximated. In short, a model is a sort of metaphor or analogy. Like all analogies, it is valid only within certain limits and, when pushed beyond these limits, becomes misleading or even absurd.

One of the most important mathematical models in population genetics is known as the **nonoverlapping generation model** because the model pertains to populations with a very simple sort of life history in which the individuals in each generation die before the members of the next generation are born. The nonoverlapping generation model is diagrammed in Figure 10. The model applies literally only to certain short-lived insects or to annual plants that have short growing seasons. In such plants, all members of any generation germinate at about the same time, mature together, shed their pollen and are fertilized almost simultaneously, and die immediately after producing the new generation. This sort of hypothetical population, with its simple life history, is used in population genetics as a first approximation to populations that have more complex life histories. Although at first glance it seems oversimplified, the nonoverlapping generation model turns out to be a surprisingly good approximation. As will be seen shortly, calculations of expected genotype frequencies based on the model are adequate for most purposes, and they are often satisfactory approximations even for populations with long and complex life histories such as in humans.

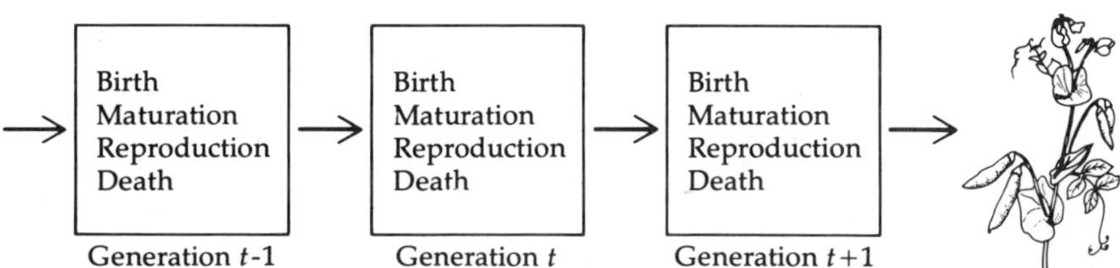

Figure 10. Nonoverlapping generation model. The life history of the organism is assumed to be like that of an annual plant (or any short-lived organism), and the generations are assumed to be separated in time (discrete generations). Although the model is a simple one, it provides a useful and surprisingly accurate approximation to populations with more complex life histories.

Random Mating

Prediction of genotype frequencies of a gene from knowledge of allele frequencies is quite straightforward in the nonoverlapping generation model, but there are still a few minor complications to be discussed. Genotype frequencies are determined in part by mating patterns. For example, plant species that regularly undergo self-fertilization are expected to have very low levels of heterozygosity, for reasons that will become clear when self-fertilization is discussed later in this chapter. One of the simplest and most important mating patterns is **random mating**, in which mating takes place at random with respect to the genotypes under consideration. With random mating, the chance that an individual mates with another having a prescribed genotype is equal to the frequency of that genotype in the population. For example, suppose that in some population the genotypes *AA*, *Aa*, and *aa* are in the proportions 0.16, 0.48, and 0.36, respectively; if mating is random, *AA* males mate with *AA*, *Aa*, and *aa* females in the proportions 0.16, 0.48, and 0.36, respectively, and these same proportions apply to the mates of *Aa* and *aa* males. In this section, we focus on random mating; effects of departures from random mating are taken up in later sections of this chapter.

It is important to keep in mind that mating can be random with respect to some traits but, at the same time, nonrandom with respect to others. In humans, for example, mating seems to be random with respect to blood groups, allozyme phenotypes, restriction fragment length polymorphisms, and many other characteristics, but mating is nonrandom with respect to other traits such as skin color and height. Genotype frequencies are also influenced by various evolutionary forces—among them, mutation, migration, and natural selection. For the moment, these evolutionary forces will be assumed to be absent or at least negligibly small in magnitude (their effects will be discussed in Chapter 2). Additionally, genotype frequencies are affected by chance statistical fluctuations that occur in all small populations (also a subject of Chapter 2). For now, imagine that each local population is sufficiently large that small-population effects can be neglected. Although small-population effects do occur unless the population is infinite in size, the magnitude of the effects is sufficiently small that they can usually be neglected if population size is 500 or more.

The Hardy–Weinberg Principle

The assumptions made so far in developing the model for predicting genotype frequencies may be summarized as:

1. The organism in question is diploid.
2. Reproduction is sexual.

3. Generations are nonoverlapping.
4. Mating is random.
5. Population size is very large.
6. Migration is negligible.
7. Mutation can be ignored.
8. Natural selection does not affect the alleles under consideration.

Collectively, these assumptions constitute the Hardy–Weinberg model, the name given to genotype frequencies with random mating after their discovery in the early 1900s by G. H. Hardy and W. Weinberg. Once the model is carefully stated, as in the preceding list, the genotype frequencies for a gene with two alleles can be worked out very easily, as shown in Table 3. In Table 3, we assume that genotype frequencies of *AA*, *Aa*, and *aa* in the parental generation are *P*, *Q*, and *R*, respectively, where $P + Q + R = 1$. The allele frequencies of *A* and *a* can be calculated as in Table 2—namely, the allele frequency of *A* is *P* times 2 (because each *AA* genotype contains two *A* alleles) plus *Q* (because each *Aa* genotype contains one *A* allele) divided by 2 (because each diploid genotype contains two alleles at the locus). In symbols, the allele frequency *p* of *A* is given by

$$p = (2P + Q)/2 = P + Q/2 \tag{1.5}$$

and the frequency *q* of *a* is given by

$$q = (2R + Q)/2 = R + Q/2 \tag{1.6}$$

Table 3. Demonstration of the Hardy–Weinberg principle.

Mating	Frequency of mating	OFFSPRING GENOTYPE FREQUENCIES *AA*	*Aa*	*aa*
AA × *AA*	P^2	1	0	0
AA × *Aa*	$2PQ$	½	½	0
AA × *aa*	$2PR$	0	1	0
Aa × *Aa*	Q^2	¼	½	¼
Aa × *aa*	$2QR$	0	½	½
aa × *aa*	R^2	0	0	1
	Totals (next generation)	P'	Q'	R'

where $P' = P^2 + 2PQ/2 + Q^2/4 = (P + Q/2)^2 = p^2$

$Q' = 2PQ/2 + 2PR + Q^2/2 + 2QR/2 = 2(P + Q/2)(R + Q/2) = 2pq$

$R' = Q^2/4 + 2QR/2 + R^2 = (R + Q/2)^2 = q^2$

Note that $p + q = P + Q + R = 1$, which is a consequence of the fact that the gene has only two alleles.

With two alleles of a gene, there are six possible types of matings. When mating is random, these mating types occur in proportion to the genotypic frequencies in the population. For example, the mating *AA* × *AA* occurs only when an *AA* male mates with an *AA* female, and this occurs a proportion $P \times P$ (or P^2) of the time; similarly, an *AA* × *Aa* mating occurs when an *AA* male mates with an *Aa* female (proportion $P \times Q$), or when an *AA* female mates with an *Aa* male (proportion $P \times Q$)—so the overall proportion of *AA* × *Aa* matings is $PQ + PQ = 2PQ$. The frequencies of these and the other types of matings are given in the second column of Table 3.

The offspring genotypes produced by the matings are given in the last three columns of Table 3. The offspring frequencies follow from Mendel's law of segregation, which states that an *Aa* heterozygote produces an equal number of *A*-bearing and *a*-bearing gametes (**gamete** is a useful word meaning sperm or egg). *AA* genotypes produce only *A*-bearing gametes and *aa* genotypes produce only *a*-bearing gametes. Thus, a mating of *AA* with *aa* produces all *Aa* offspring, a mating of *AA* with *Aa* produces 1/2 *AA* and 1/2 *Aa* offspring, a mating of *Aa* with *Aa* produces 1/4 *AA*, 1/2 *Aa*, and 1/4 *aa* offspring, and so on.

The genotype frequencies of *AA*, *Aa*, and *aa* after one generation of random mating are denoted in Table 3 as P', Q', and R', respectively. These numbers are calculated as the sum of the cross-products shown at the bottom of the table. The genotype frequencies simplify very nicely to p^2, $2pq$, and q^2, which is a result known as the **Hardy–Weinberg principle**; that is, with random mating for a gene with two alleles *A* and *a* at respective frequencies *p* and *q*, the genotype frequencies of *AA*, *Aa*, and *aa* are

$$\begin{aligned} AA&: \ p^2 \\ Aa&: \ 2pq \\ aa&: \ q^2 \end{aligned} \qquad (1.7)$$

PROBLEM 5

In the *Drosophila* population cage of Mukai, Watanabe, and Yamaguchi discussed in Problem 3, the genotype frequencies of the esterase-6 gene were in Hardy–Weinberg proportions, i.e., the genotype frequencies of $E6^F/E6^F$ (hereafter called *FF*), $E6^F/E6^S$ (*FS*), and $E6^S/E6^S$ (*SS*) were $(0.3579)^2 = 0.1281$, $2(0.3579)(0.6421) = 0.4596$, and $(0.6421)^2 = 0.4123$, respectively. Assuming that all of the assumptions of the Hardy–Weinberg model hold, particularly those pertaining to random mating in a large population with no mutation, selection, or migration, make a table of mating frequencies similar to Table 3 for the esterase-6 alleles. Then calculate the genotype frequencies expected in the next generation along with the corresponding allele frequencies.

ANSWER Mating frequencies are as follows: *FF* × *FF* (0.0164); *FF* × *FS* (0.1177);

FF × *SS* (0.1056); *FS* × *FS* (0.2112); *FS* × *SS* (0.3790); and *SS* × *SS* (0.1700). Genotype frequencies among the offspring are, for *FF*, 0.0164 + 0.1177/2 + 0.2112/4 = 0.1281; for *FS*, 0.1177/2 + 0.1056 + 0.2112/2 + 0.3790/2 = 0.4596; for *SS*, 0.2112/4 + 0.3790/2 + 0.1700 = 0.4123; note that these are the same as in the parental generation. The allele frequencies of *F* and *S* are again 0.3579 and 0.6421, respectively.

PROBLEM 6 A four-base cleavage site for the restriction enzyme *Ban*I, which is located within a large intron of the larval transcript of the alcohol dehydrogenase gene in *Drosophila melanogaster*, was found to be present in 29 of 60 chromosomes isolated from a Raleigh, North Carolina, Farmer's Market (Kreitman and Aguadé 1986). Letting *B* and *b* represent the presence or absence of the *Ban*I site in a chromosome, respectively, and assuming Hardy–Weinberg genotype frequencies, calculate the expected frequencies of the genotypes *BB*, *Bb*, and *bb*. Apply Equations 1.3 and 1.4 to estimate the level of nucleotide polymorphism among the four nucleotides in the *Ban*I site.

ANSWER $\langle p \rangle = 29/60 = 0.483$ and $\langle q \rangle = 1 - \langle p \rangle = 0.517$. Expected genotype frequencies are $\langle p \rangle^2 = 0.234$ (*BB*), $2\langle p \rangle\langle q \rangle = 0.499$ (*Bb*), and $\langle q \rangle^2 = 0.267$. For application of Equations 1.3 and 1.4, $n = 60$, $m = 1$, $k = 1$, $j = 4$, $c_1 = c = 29$, and $\Sigma c_i^2 = (29)^2 = 841$. Thus, $\langle P_{nuc} \rangle = [29 - 60(1 - 1)]/(4)(20) = 0.25$. The reason for $\langle P_{nuc} \rangle = 1/4$ is that Equation 1.3 assumes the *Ban*I polymorphism results from a single nucleotide polymorphism at one of the four positions in the restriction site. Likewise, $\langle H_{nuc} \rangle = [(60)(29) - 841]/[(4)(29)(50)] = 0.131$. Note that this is approximately the frequency of *Bb* heterozygotes divided by four ($2\langle p \rangle\langle q \rangle/4 = 0.124$), which occurs because Equation 1.4 assumes that no more than one nucleotide in the restriction site can be heterozygous. The agreement is not exact because Equation 1.4 also corrects for sampling error.

Table 3 illustrates an important principle of population genetics, which is that *random mating of individuals is usually equivalent to random union of gametes*. Because of this equivalence, we can cross-multiply the allele frequencies along the margins in the sort of square shown in Figure 11. As shown at the right of the figure, the cross multiplication leads to the relationships $P' = p^2$, $Q' = 2pq$, and $R' = q^2$, which are exactly the same as reached by the more laborious method in Table 3. The cross multiplication in such a square works because it is just a systematic way of going through all the possibilities of gamete combination. The probability that a sperm or egg carries *A* is p; the probability that a sperm or egg carries *a* is q. With random combination of gametes, the chance that an *A*-bearing sperm fertilizes an *A*-bearing egg is $p \times p = p^2$; therefore, this is the frequency of *AA* genotypes. The probability that an *A*-bearing sperm fertilizes an *a*-bearing egg is $p \times q = pq$; and the probability that an *a*-bearing sperm fertilizes an *A*-bearing egg is $q \times p = qp$. Altogether, the frequency of *Aa* heterozygotes is $pq + qp = 2pq$. Finally, the probability that an *a*-bearing sperm fertilizes an *a*-bearing egg is $q \times q = q^2$,

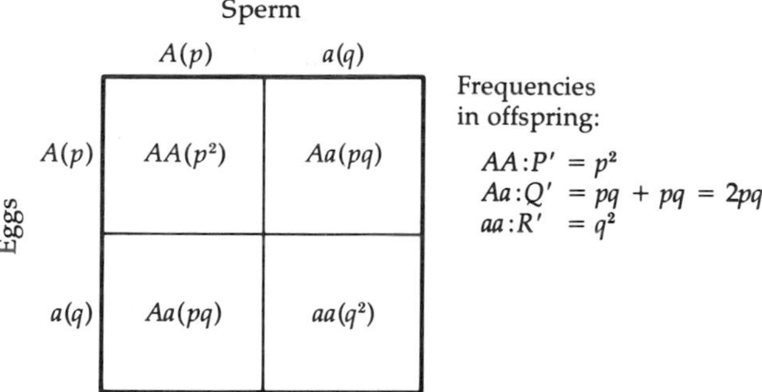

Figure 11. Cross-multiplication square showing Hardy–Weinberg frequencies resulting from random mating with two alleles.

which is the genotype frequency of *aa*. Note that $p^2 + 2pq + q^2 = (p + q)^2 = (1)^2 = 1$, thus accounting for all the offspring.

It is perhaps a little hard to believe that so simple a result from such a simple model can be so important, but the Hardy–Weinberg principle provides the foundation for many investigations in population genetics. One of the most important implications of the Hardy–Weinberg principle emerges when we calculate the allele frequencies of *A* and *a* in the next generation from the formulas for P', Q', and R' in Table 3 and Figure 11. Using Equation 1.5, the allele frequency of *A* equals

$$(2P' + Q')/2 = (2p^2 + 2pq)/2 = p(p + q) = p$$

Likewise, using Equation 1.6, the allele frequency of *a* equals

$$(2R' + Q')/2 = (2q^2 + 2pq)/2 = q(q + p) = q$$

Thus, the allele frequencies in the next generation are exactly the same as they were the generation before. With random mating, the allele frequencies remain the same generation after generation. In any generation, therefore, the genotype frequencies are p^2, $2pq$, and q^2 for *AA*, *Aa*, and *aa*, respectively—as given in Equation 1.7—and these frequencies constitute what is often called the **Hardy–Weinberg equilibrium** for two alleles. The constancy of allele frequency (and therefore of the genotypic composition of the population) is the single most important implication of the Hardy–Weinberg principle. The constancy of allele frequencies implies that, in the absence of specific evolutionary forces to change allele frequency (assumptions 5, 6, 7, and 8 in the preceding list), the mechanism of Mendelian inheritance, by itself, will keep the allele frequencies constant and thus preserve genetic variation. A second item of interest is that the Hardy–Weinberg frequencies are attained in just one generation of random mating if the allele frequencies are the same in males and females. This, however, is true only with non-overlapping generations; in populations with more complex life histories, the Hardy–Weinberg frequencies are attained gradually over a period of several generations.

It is important to note here that the Hardy–Weinberg principle is not very sensitive to certain kinds of departures from the assumptions numbered 3 to 8 in the earlier list, particularly numbers 5 to 8 (i.e., very large population size with no migration, mutation, or selection). That is, the mere fact that observed genotype frequencies may happen to fit Hardy–Weinberg proportions cannot be taken as evidence that the assumptions in the list are valid. Indeed, the relative insensitivity of the Hardy–Weinberg frequencies to departures from assumptions 5 to 8 makes the formulation even more useful, as it implies that the Hardy–Weinberg frequencies are often valid to a first approximation even when one or more of the assumptions is violated.

The Hardy–Weinberg Principle in Operation

Use of the Hardy–Weinberg principle can be illustrated with actual data pertaining to the MN blood group phenotypes among a sample of 1000 Britishers (Race and Sanger 1975). The MN blood groups are due to the alleles of a single gene. The two alleles are designated *M* and *N*, and each allele results in the production of molecules on the surface of red blood cells that can be differentiated by means of appropriate chemical reagents. The substances corresponding to the *M* and *N* alleles are designated M and N, respectively. Individuals of genotype *MM* produce only the M substance, those of genotype *NN* produce only the N substance, and those of genotype *MN* produce both the M and the N substances.

Among the 1000 individuals there were 298 M, 489 MN, and 213 N phenotypes. Because in this system each genotype is associated with a unique phenotype, we know immediately that the sample contained 298 *MM* genotypes, 489 *MN* genotypes, and 213 *NN* genotypes. To determine whether these genotype frequencies are in accord with Hardy–Weinberg frequencies, the allele frequencies of *M* and *N* must first be estimated. The allele frequency of *M*, which we will designate as p, is estimated from Equation 1.1 as $\langle p \rangle = (596 + 489)/2000 = 0.5425$, and the allele frequency of *N*, designated q, is estimated as $\langle q \rangle = (426 + 489)/2000 = 0.4575$. (Note again that $p + q = 1$.) Were the population in Hardy–Weinberg equilibrium, we would expect the genotype frequencies of *MM*, *MN*, and *NN* to be p^2, $2pq$, and q^2, respectively, which can be estimated as $\langle p \rangle^2 = (0.5425)^2 = 0.2943$, $2\langle p \rangle\langle q \rangle = 2(0.5425)(0.4575) = 0.4964$, and $\langle q \rangle^2 = (0.4575)^2 = 0.2093$, respectively. Since the total number of individuals studied is 1000, the expected numbers of the *MM*, *MN*, and *NN* genotypes are $(0.2943)(1000) = 294.3$, $(0.4964)(1000) = 496.4$, and $(0.2093)(1000) = 209.3$, respectively.

At this point, it is convenient to tabulate the data into three columns, the first giving the genotypes, the second giving the observed numbers, and the third giving the expected numbers:

MM	298	294.3
MN	489	496.4
NN	213	209.3

With the data so arrayed, it is evident that the fit between the observed numbers and the expected numbers, though not perfect because of chance statistical fluctuations in the number of the various genotypes that may be included in any given sample, is nevertheless very close. To verify this conclusion, we will apply a standard statistical test to the data in order to assess quantitatively the closeness of fit. A test commonly employed in population genetics is called the **chi-square test**. It is based on the value of a number, called χ^2, calculated from the data as

$$\chi^2 = \sum \frac{(\text{obs} - \text{exp})^2}{\text{exp}} \qquad (1.8)$$

where "obs" refers to the observed number in any genotypic class, "exp" refers to the expected number in that same genotypic class, and the Σ sign denotes that the values are to be summed over all genotypic classes. In the case at hand,

$$\begin{aligned}\chi^2 = &(298 - 294.3)^2/294.3 + \\ &(498 - 496.4)^2/496.4 + \\ &(213 - 209.3)^2/209.3 = 0.222\end{aligned}$$

To be completely rigorous, some statisticians prefer use of the symbol X^2 for the realized value of the test statistic defined Equation 1.8, in order to distinguish between the test statistic and the true χ^2 distribution itself. The distinction should certainly be kept in mind, but we will not recognize it formally with different symbols.

Associated with any χ^2 value is a second number called the **degrees of freedom** for that χ^2. In general, the number of degrees of freedom associated with a χ^2 equals the number of classes of data (in this case, 3) minus one, minus the number of parameters estimated from the data (in this case, one, because the parameter p was estimated from the data), so the number of degrees of freedom for our chi-square value is $3 - 1 - 1 = 1$. (Note: a degree of freedom is not deducted for estimating q because of the relation $q = 1 - p$; that is, once p has been estimated, the estimate of q is automatically fixed, so we deduct just the one degree of freedom corresponding to p.)

Calculation of χ^2 and its associated degrees of freedom is carried out in order to obtain a number for assessing goodness of fit; the number is determined from Figure 12. To use the chart, find the value of χ^2 along the

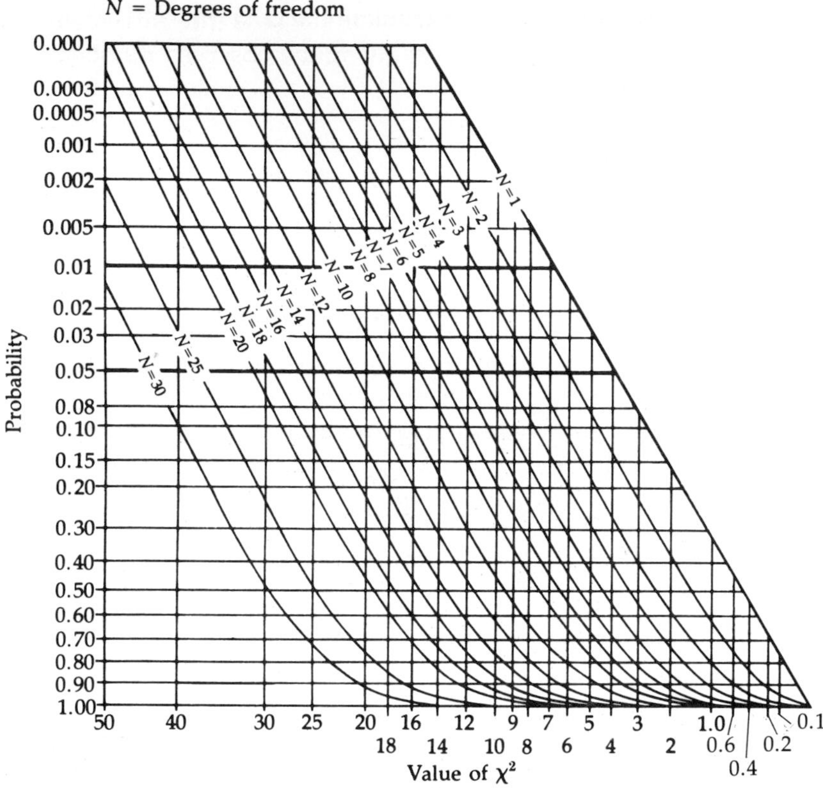

Figure 12. Graph of χ^2. To use the graph, find the value of χ^2 along the abscissa (the horizontal axis), then read the probability value corresponding to the appropriate number of degrees of freedom. (From Crow 1945.)

horizontal axis; then move vertically from this value until the proper degrees-of-freedom line is intersected; then move horizontally from this point of intersection to the vertical axis and read the corresponding probability value. In our case, with $\chi^2 = 0.222$ and one degree of freedom, the corresponding probability value is about 0.67. This probability has the following interpretation: It is the probability that chance alone could produce a deviation between the observed and expected values at least as great as the deviation actually obtained. Thus, if the probability is large, it means that chance alone could account for the deviation, and it strengthens our confidence in the validity of the model used to obtain the expectations—in this case, the Hardy–Weinberg model. On the other hand, if the probability associated with the χ^2 is small, it means that chance alone is not likely to lead to a deviation as large as actually obtained, and it undermines our confidence in the validity of the model. Where exactly the cutoff should be between a "large" probability and a "small" one is, of course, not obvious, but there is an established guideline to follow: If the probability is less than 0.05, then the goodness of fit is considered sufficiently poor that the model is judged invalid for the data; alternatively, if the probability is greater than 0.05, the fit is considered sufficiently close that the model is not rejected.

Because the probability in the MN example is 0.67, which is greater than 0.05, we have no reason to reject the hypothesis that the genotype frequencies are in Hardy–Weinberg proportions for this gene.

PROBLEM 7

Another human blood group, related to the MN system, is also controlled by two alleles (*S* and *s*), and three distinct phenotypes can be identified by means of the appropriate reagents. Among the same 1000 Britishers who gave the MN data above, the numbers found of each genotype in the *Ss* system were: 99 *SS*, 418 *Ss*, and 483 *ss*. Estimate the allele frequency of *S* (p) and *s* (q) and carry out a χ^2 test of goodness of fit between the observed genotype frequencies and their Hardy–Weinberg expectations. Is there any reason to reject the hypothesis of Hardy–Weinberg proportions for this gene?

ANSWER $\langle p \rangle = 0.308$ and $\langle q \rangle = 0.692$. Expected numbers of *SS*, *Ss*, and *ss* are 94.86, 426.27, and 478.86, respectively. $\chi^2 = 0.377$ with one degree of freedom; the associated probability from Figure 12 is about 0.55, so there is no reason to reject the hypothesis.

Frequency of Heterozygotes

One of the uses of the Hardy–Weinberg principle is in the calculation of genotype frequencies in cases when dominance occurs. **Dominance** refers to situations in which one allele has such a strong phenotypic effect in heterozygotes that it conceals the presence of the weaker allele. The strong allele is said to be **dominant**, the weak one **recessive**. When dominance occurs, only two phenotypic classes can be distinguished—the dominant phenotype (which occurs in the dominant-allele homozygotes and the heterozygotes) and the recessive phenotype (which occurs in the recessive-allele homozygotes). An example is a constituent of the surface of human red blood cells coded by the *D* allele in the Rh blood group system. Genotypes *DD* and *Dd* are Rh^+ (Rh positive), whereas *dd* is Rh^- (Rh negative). Thus, *D* is dominant to *d*. The *D* allele codes for the Rh^+ antigen on the surface of human red blood cells. Since *DD* and *Dd* genotypes both produce the antigen, both result in the Rh^+ phenotype, even though the *Dd* genotype might be expected to produce only half as much antigen as *DD* because it contains only one *D* allele.

Among American Caucasians, the frequency of Rh^+ is about 85.8 percent and the frequency of Rh^- is about 14.2 percent (Mourant et al. 1976). Although we might wish to use these data to calculate the frequencies of the individual *DD, Dd,* and *dd* genotypes, the genotype frequencies cannot normally be estimated from the frequencies of dominant and recessive phenotypes without making some assumptions about the mating system. However, assuming that mating is random for this gene and that the genotype frequencies satisfy the Hardy–Weinberg principle, then the genotype frequencies can be estimated. Letting p stand for the allele frequency of D, and q stand for the allele frequency of d, the allele frequencies can be estimated by noting that, with random mating, the frequency of homozygous recessives, in this case 0.142, should equal q^2. So $\langle q\rangle^2 = 0.142$, or $\langle q\rangle = (0.142)^{1/2} = 0.3768$. More generally, if R is the frequency of homozygous recessive genotypes found among a sample of n individuals, then $\langle q\rangle$ and its sampling variance are estimated as

$$\langle q\rangle = R^{1/2}$$
$$\langle Var(\langle q\rangle)\rangle = (1 - R)/(4n) \quad \textbf{(1.9)}$$

With $\langle q\rangle$ estimated from Equation 1.9 as $\langle q\rangle = 0.3768$, then $\langle p\rangle = 1 - \langle q\rangle = 1 - 0.3768 = 0.6232$, and the frequencies of *DD, Dd,* and *dd* are expected to be $\langle p\rangle^2 = (0.6232)^2 = 0.3884$, $2\langle p\rangle\langle q\rangle = 2(0.6232)(0.3768) = 0.4696$, and $\langle q\rangle^2 = (0.3768)^2 = 0.1420$, respectively. The proportion of Rh^+ individuals that are actually heterozygous is therefore 0.4696/(0.4696 + 0.3884) = 54.7 percent.

Unfortunately, in cases when dominance occurs, there is no chance for a goodness-of-fit test to Hardy–Weinberg proportions because there are 0 degrees of freedom [i.e., 2 (for two classes of data) − 1 − 1 (for estimating the frequency of the recessive allele) = 0]. This lack of degrees of freedom is the reason why the calculated frequencies of Rh^+ and Rh^- (0.3884 + 0.4696 = 0.858 and 0.142, respectively) fit the observed frequencies exactly.

PROBLEM 8 The Basque people, who live in the Pyrenees mountains between France and Spain, have one of the highest frequencies of the *d* allele in the Rh system so far reported. In one study of 400 Basques, 230 were found to be Rh^+ and 170 Rh^- (Mourant et al. 1976). Estimate the frequencies of the *D* and *d* alleles, the genotype frequencies, and the proportion of Rh^+ individuals who are actually heterozygous *Dd*. What is the standard error of the estimate $\langle q\rangle$?

ANSWER $\langle q\rangle = (170/400)^{1/2} = 0.65$, $\langle p\rangle = 0.35$, and the estimated genotype frequencies of *DD*, *Dd*, and *dd* are 0.121, 0.454 and 0.425, respectively. The proportion of *Dd* among Rh^+ individuals is 0.454/(0.121 + 0.454) = 79 percent. The standard error of $\langle q\rangle$ equals $[(1 - 0.425)/(1600)]^{1/2} = 0.02$.

The Hardy–Weinberg principle also finds application in studies of industrial melanism, one of the most famous and best studied cases of evolution

Figure 13. Melanic and nonmelanic moths, showing camouflage of light moths on light background and dark moths on dark. (Photograph by H. B. D. Kettlewell.)

in action (Kettlewell 1973). **Industrial melanism** refers to the evolution of black (melanic) color patterns in several species of moths that accompanied progressive pollution of the environment by coal soot during the industrial revolution. (The various color forms of the moths are known as **morphs**.) Industrial melanism occurs in Great Britain, West Germany, Eastern Europe, the United States, and in other heavily industrialized areas. The species involved in industrial melanism are large moths that fly by night and rest in a sort of cataleptic state by day, often on the trunks of trees, using their cryptic black-and-white mottled color pattern for concealment from visually cued predators such as hedge sparrows, redstarts, and robins (Figure 13). Of nearly 800 species of large moths in the British Isles, where industrial melanism has been most intensively studied, about 100 species are industrial melanics (Bishop and Cook 1975). The best known of these are the peppered moth (*Biston betularia*) and the scalloped hazel moth (*Gonodontis bidentata*). In most instances, the melanic color pattern has been found to be due to a single dominant allele.

PROBLEM 9

In one study of a heavily polluted area near Birmingham, England, Kettlewell (1956) observed a frequency of 87 percent melanic *Biston betularia*. Calculate the frequency of the dominant allele leading to melanism in this population and the frequency of melanics that are heterozygous.

ANSWER Observed frequency of homozygous recessives is $R = 0.13$, so frequency of recessive allele is estimated as $\langle q \rangle = (0.13)^{1/2} = 0.36$, and $\langle p \rangle = 0.64$. Frequencies of dominant homozygotes, heterozygotes, and recessive homozygotes are estimated as 0.41, 0.46, and 0.13, respectively. Proportion of melanics that are heterozygous is $0.46/0.87 = 52.9$ percent.

One of the important implications of the Hardy–Weinberg principle pertains to rare recessive alleles: *When a recessive allele is rare, most individuals who carry the allele are heterozygous.* This principle follows immediately from the genotype frequencies with random mating. If the frequency of the recessive allele is denoted q, then, with random mating, there are $2pq$ heterozygotes and q^2 recessive homozygotes. The ratio of heterozygote to homozygote frequencies is $2pq/q^2 = 2p/q$. For $q = 0.10$, this ratio is 18 (i.e., there are 18 times as many heterozygotes as recessive homozygotes), for $q = 0.01$ the ratio is 198, and for $q = 0.001$ the ratio is 1998. Thus, as a recessive allele becomes more rare, the greater is the proportion of heterozygous carriers relative to homozygous recessives.

To take a specific example, the disease cystic fibrosis, a severe condition with abnormal glandular secretions in individuals homozygous for an autosomal recessive gene, affects about 1 in 1700 newborn Caucasians. For this allele, $\langle q \rangle = (1/1700)^{1/2} = 0.024$. The frequency of heterozygotes is therefore estimated as $2(0.024)(1 - 0.024) = 0.047$, or about 1 in 21. That is, although only 1 person in 1700 is actually affected with cystic fibrosis, 1 person in 21 is a heterozygous carrier of the harmful allele.

PROBLEM 10 Phenylketonuria is a severe form of mental retardation caused by a homozygous recessive allele. The condition affects about 1 in 10,000 newborn Caucasians. Estimate the frequency of heterozygotes for this allele.

ANSWER 0.0198. About 1 person in 50 carries the allele.

SPECIAL CASES OF RANDOM MATING

In this section we extend the Hardy–Weinberg principle to cases of multiple alleles, to genes located in the X chromosome, and to pairs of genes linked in the same chromosome.

Three or More Alleles

Genotype frequencies under random mating for genes with three alleles are shown in Figure 14. Here it is convenient to label the alleles as A_1, A_2, and

Sperm

Eggs	$A_1(p_1)$	$A_2(p_2)$	$A_3(p_3)$
$A_1(p_1)$	$A_1A_1(p_1^2)$	$A_1A_2(p_1p_2)$	$A_1A_3(p_1p_3)$
$A_2(p_2)$	$A_1A_2(p_1p_2)$	$A_2A_2(p_2^2)$	$A_2A_3(p_2p_3)$
$A_3(p_3)$	$A_1A_3(p_1p_3)$	$A_2A_3(p_2p_3)$	$A_3A_3(p_3^2)$

Frequencies in offspring:

A_1A_1: p_1^2
A_1A_2: $p_1p_2 + p_1p_2 = 2p_1p_2$
A_1A_3: $p_1p_3 + p_1p_3 = 2p_1p_3$
A_2A_2: p_2^2
A_2A_3: $p_2p_3 + p_2p_3 = 2p_2p_3$
A_3A_3: p_3^2

Figure 14. Cross-multiplication square showing Hardy–Weinberg frequencies for three autosomal alleles.

A_3 and the corresponding allele frequencies as p_1, p_2, and p_3. Since there are only three alleles, $p_1 + p_2 + p_3 = 1$. The genotype frequencies of A_1A_1, A_1A_2, A_2A_2, A_1A_3, A_2A_3, and A_3A_3 are p_1^2, $2p_1p_2$, p_2^2, $2p_1p_3$, $2p_2p_3$, and p_3^2, respectively, which can be obtained by expanding $(p_1A_1 + p_2A_2 + p_3A_3)^2$, a calculation that the cross-multiplication square does automatically.

Use of Figure 14 can be illustrated with the familiar ABO blood groups in humans. This system is due to three alleles that are designated I^O, I^A, and I^B. Genotypes I^AI^A and I^AI^O have blood type A; genotypes I^BI^B and I^BI^O have blood type B; genotype I^OI^O has blood type O; and genotype I^AI^B has blood type AB. In one test of 6313 Caucasians in Iowa City, the numbers of individuals with blood types A, B, O, and AB were found to be 2625, 570, 2892, and 226, respectively (Mourant et al. 1976). The best estimates of allele frequency in this case are $\langle p_1 \rangle = 0.2593$ (for I^A), $\langle p_2 \rangle = 0.0625$ (for I^B), and $\langle p_3 \rangle = 0.6755$ (for I^O). (Estimation of allele frequencies for the ABO blood groups is complicated because of dominance. For methods see Cavalli-Sforza and Bodmer, 1971, and Vogel and Motulsky, 1986.) The expected numbers of the four blood types are therefore

$$\text{A: } [(0.2593)^2 + 2(0.2593)(0.6755)](6313) = 2636.0$$
$$\text{B: } [(0.0652)^2 + 2(0.0652)(0.6755)](6313) = 582.9$$
$$\text{O: } (0.6755)^2(6313) = 2880.6$$
$$\text{AB: } 2(0.2593)(0.0652)(6313) = 213.5$$

Comparison of observed with expected numbers is thus

A	2625	2636.0
B	570	582.9
O	2892	2880.6
AB	226	213.5

Chi-square for goodness of fit to Hardy–Weinberg proportions is 1.11. There is one degree of freedom for this test: 4 (to start with) − 1 − 1 (for estimating $\langle p_1 \rangle$ from the data) − 1 (for estimating $\langle p_2 \rangle$ from the data) = 1; again, a degree of freedom is not deducted for estimating $\langle p_3 \rangle$ because $\langle p_3 \rangle = 1 - \langle p_1 \rangle - \langle p_2 \rangle$. For a χ^2 of 1.11 with one degree of freedom, the associated probability from Figure 12 is about 0.30, so the Iowa City population gives no evidence against Hardy–Weinberg proportions for this gene.

PROBLEM 11 In a sample of 1617 Spanish Basques, the numbers of A, B, O, and AB blood types observed were 724, 110, 763, and 20, respectively (Mourant et al. 1976). Best estimates of allele frequency are $\langle p_1 \rangle = 0.2661$ (for I^A), $\langle p_2 \rangle = 0.0411$ (for I^B), and $\langle p_3 \rangle = 0.6928$ (for I^O). Calculate the expected numbers of the four phenotypes and carry out a χ^2 test for goodness of fit to the Hardy–Weinberg expectations.

ANSWER Expected numbers of A, B, O, and AB are 710.7, 94.8, 776.1, and 35.4, respectively. χ^2 equals 9.61 with one degree of freedom, for which the corresponding probability is 0.0025. Since a deviation as large or larger than that observed could be expected by chance in only 0.0025 samples (i.e., about 1 in 400), there is very good reason to think that this population is not in Hardy–Weinberg proportions for this gene. The reason for the discrepancy is not known. One likely possibility is migration into the population by individuals having allele frequencies significantly different from those among the Basques themselves.

PROBLEM 12 Among many aboriginal American Indian tribes, the allele frequency of I^B is extremely low. Mourant et al. (1976) report that among 600 Papago Indians of Arizona there were 37 A and 563 O blood types. What are the best estimates of the allele frequencies of I^A, I^B, and I^O in this population, and what are the expected genotype frequencies assuming random mating?

ANSWER There are no I^B alleles among the sample, so the best estimate of p_2 is 0. Thus there are only three genotypes: I^AI^A, I^AI^O (both of blood type A), and I^OI^O (blood type O), and these are expected to occur in the proportions p_1^2, $2p_1p_3$, and p_3^2, respectively. Because $\langle p_2 \rangle = 0$ in this case, $\langle p_1 \rangle + \langle p_3 \rangle = 1$. The best estimate of $\langle p_3 \rangle$ is thus obtained from Equation 1.9 as $(563/600)^{1/2} = 0.9687$ and that of $\langle p_1 \rangle$ as $1 - \langle p_3 \rangle = 0.0313$. Expected genotype frequencies are $(0.0313)^2 = 0.0010$ for I^AI^A, $2(0.0313)(0.9687) = 0.0606$ for I^AI^O, and $(0.9687)^2 = 0.9384$ for I^OI^O.

In general, if there are n alleles ($A_1, A_2, \ldots, A_n$) with frequencies $p_1, p_2, \ldots, p_n$ (with $p_1 + p_2 + \ldots + p_n = 1$), then the genotype frequencies under random mating are

$$\begin{array}{l} p_i^2 \text{ for } A_iA_i \text{ homozygotes} \\ 2p_ip_j \text{ for } A_iA_j \text{ heterozygotes} \end{array} \qquad \textbf{(1.10)}$$

Equation 1.10 may be applied to data from Prakash (1977), who studied allozyme polymorphisms in *Drosophila persimilis* in California. In one sample of 108 individuals from the Fish Creek population, he observed four alleles of the xanthine dehydrogenase (*Xdh*) gene. We may call these alleles *Xdh*−1, *Xdh*−2, *Xdh*−3, and *Xdh*−4, and their respective frequencies were estimated as $\langle p_1 \rangle = 0.08$, $\langle p_2 \rangle = 0.21$, $\langle p_3 \rangle = 0.62$, and $\langle p_4 \rangle = 0.09$. With four alleles, there are four possible homozygotes (e.g., *Xdh*−1/*Xdh*−1) and six possible heterozygotes (e.g., *Xdh*−1/*Xdh*−2). In a random-mating population, the frequency of any homozygous genotype is expected to be the square of the corresponding allele frequency (e.g., the frequency of *Xdh*−1/*Xdh*−1 is expected to be $\langle p_1 \rangle^2$), and the frequency of any heterozygous genotype is expected to be two times the product of the corresponding allele frequencies (e.g., the frequency of *Xdh*−1/*Xdh*−2 is expected to be $2\langle p_1 \rangle\langle p_2 \rangle$). The Hardy–Weinberg frequencies for all ten possible genotypes can be obtained by multiplying out the expression (0.08 *Xdh*−1 + 0.21 *Xdh*−2 + 0.62 *Xdh*−3 + 0.09 *Xdh*−4)^2.

PROBLEM 13

In a Texas population of *Phlox cuspidata*, Levin (1978) found four alleles of the alcohol dehydrogenase (*Adh*) gene. The alleles may be designated *Adh*−1, *Adh*−2, *Adh*−3, and *Adh*−4, and their respective frequencies were estimated as 0.11, 0.84, 0.01, and 0.04. What are the expected Hardy–Weinberg proportions of the ten genotypes?

ANSWER *Adh*−1/*Adh*−1: $(0.11)^2 = 0.0121$; *Adh*−1/*Adh*−2: $2(0.11)(0.84) = 0.1848$; *Adh*−2/*Adh*−2 = $(0.84)^2 = 0.7056$; *Adh*−1/*Adh*−3 = $2(0.11)(0.01) = 0.0022$; *Adh*−2/*Adh*−3 = $2(0.84)(0.01) = 0.0168$; *Adh*−3/*Adh*−3 = $(0.01)^2 = 0.0001$; *Adh*−1/*Adh*−4 = $2(0.11)(0.04) = 0.0088$; *Adh*−2/*Adh*−4 = $2(0.84)(0.04) = 0.0672$; *Adh*−3/*Adh*−4 = $2(0.01)(0.04) = 0.0008$; *Adh*−4/*Adh*−4 = $(0.04)^2 = 0.0016$. It should be pointed out that the observed genotype frequencies were nowhere near these Hardy–Weinberg expectations, for the reason that *Phlox cuspidata* undergoes a substantial frequency of self-fertilization (about 78 percent), and this violates the random-mating assumption. How to deal with such departures from random mating will be discussed later in this chapter.

X-Linked Genes

There is one important exception to the rule that individuals contain two alleles of every gene, which has to do with the special chromosomes that determine sex. In mammals and many insects, females have two copies of a sex-determining chromosome designated the X chromosome, whereas males have one copy of the X chromosome and one copy of a different sex-determining chromosome called the Y chromosome. The X and Y chromosomes segregate during the formation of sperm, so half the sperm from a male carry the X chromosome and half carry the Y chromosome. Although the Y chromosome carries very few genes other than those involved in the determination of sex and male fertility, the X chromosome carries as full a complement of genes as any other chromosome. Genes on the X chromosome are called **X-linked genes**, and the important consequence of X linkage is that a recessive allele on the X chromosome in a male is expressed phenotypically because the Y chromosome lacks any compensating allele. For X-linked genes with two alleles, therefore, there are three female genotypes (*AA*, *Aa*, and *aa*, say) but only two male genotypes (*A* and *a*).

The consequences of random mating with two X-linked alleles are shown in Figure 15, where the alleles are denoted X^A and X^a. Note that in females, which have two X chromosomes, the genotype frequencies are as given by the Hardy–Weinberg principle in Equation 1.7; in males, which have only one X chromosome, the genotype frequencies are simply equal to the allele frequencies. The calculations in Figure 15 are valid only if the allele frequencies are identical in eggs and sperm. When they differ, approximate equality of allele frequencies in the sexes is usually attained for X-linked genes in a period of ten or so generations of random mating.

One of the important features of random mating for X-linked genes is that conditions due to a recessive allele will be more common in males than in females. In Problem 14, for example, the proportion of Xg(a−) males is 346/1013 = 34 percent, whereas the proportion of Xg(a−) females is only

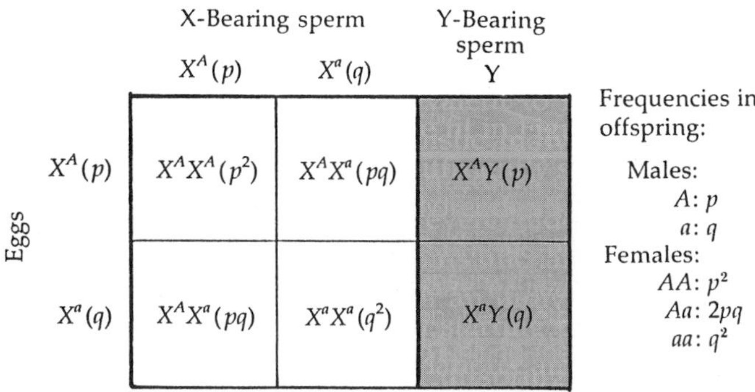

Figure 15. Consequences of random mating with X-linked genes. Genotype frequencies in *females* equal the Hardy-Weinberg frequencies, and genotype frequencies in *males* equal the allele frequencies.

PROBLEM 14

The human Xg blood group is controlled by an X-linked gene with two alleles, designated Xg^a and Xg. Two phenotypes can be distinguished by means of the appropriate antisera, Xg(a+) and Xg(a−). Xg^a is dominant to Xg, so females of genotype Xg^a/Xg^a and Xg^a/Xg have blood type Xg(a+), whereas females of genotype Xg/Xg are phenotypically Xg(a−). Males of genotype Xg^a have blood type Xg(a+); those of genotype Xg have blood type Xg(a−). Among a sample of 2082 British (Race and Sanger 1975), there were 967 Xg(a+) females, 667 Xg(a+) males, 102 Xg(a−) females, and 346 Xg(a−) males; the best estimates of allele frequency are $\langle p \rangle = 0.675$ (for Xg^a) and $\langle q \rangle = 0.325$ (for Xg). Calculate the expected numbers in the four phenotypic classes, assuming random-mating proportions, and carry out a χ^2 test for goodness of fit. (The number of degrees of freedom in this case is 1; there are four degrees of freedom to start with, but one must be deducted for using the observed number of males in calculating the expectations for males, one must be deducted for using the observed number of females in calculating their expectations, and one more must be deducted for estimating p from the data.)

ANSWER Expected numbers of Xg(a+) and Xg(a−) males are $(0.675)(1013) = 683.8$ and $(0.325)(1013) = 329.2$, respectively. Expected numbers of Xg(a+) and Xg(a−) females are $[(0.675)^2 + 2(0.675)(0.325)](1069) = 956.1$ and $(0.325)^2(1069) = 112.9$, respectively. The χ^2 is 2.45 and, as noted above, it has one degree of freedom. The associated probability is about 0.12 (Figure 12), so there is no reason to reject the hypothesis of random-mating proportions.

102/1069 = 10 percent. The sex difference occurs because q (which equals the proportion of males with the recessive phenotype) will always be greater than q^2 (which is the proportion of females with the recessive phenotype). Indeed, the discrepancy grows larger as the recessive allele becomes more rare. For example, with the X-linked "green" type of color blindness, $q = 0.05$ in Western Europeans, so the ratio of affected males to affected females is $q/q^2 = 1/q = 1/0.05 = 20$; for the X-linked "red" type of color blindness, by contrast, $q = 0.01$, so in this case, the ratio of affected males to affected females is $1/0.01 = 100$.

PROBLEM 15

Policansky and Zouros (1977) studied polymorphism of the X-linked phosphoglucomutase−1 gene in California populations of *Drosophila persimilis*. They found two alleles ($Pgm{-}1^A$ and $Pgm{-}1^B$) at respective frequencies 0.25 and 0.75. Assuming random-mating proportions, what are the expected genotype frequencies in males and females?

ANSWER In *D. persimilis* males, $Pgm{-}1^A = 0.25$; $Pgm{-}1^B = 0.75$. In females, $Pgm{-}1^A/Pgm{-}1^A = (0.25)^2 = 0.0625$; $Pgm{-}1^A/Pgm{-}1^B = 2(0.25)(0.75) = 0.3750$; and $Pgm{-}1^B/Pgm{-}1^B = (0.75)^2 = 0.5625$.

Before leaving the subject of X-linkage, it is necessary to point out that certain species—among them birds, moths, and butterflies—have the sex chromosome situation "backwards." In these species, females are XY and males XX. The consequences of random mating are the same, except that the sexes are reversed.

Linkage and Linkage Disequilibrium

With random mating, the alleles of any gene rapidly acquire random combinations into genotypes in the frequencies given by the Hardy–Weinberg proportions. To be specific, imagine a gene with two alleles, call them A_1 and A_2, and suppose that the gene is not X-linked; such a gene is an **autosomal gene**. If the frequencies of the A_1 and A_2 alleles are p_1 and p_2, respectively, with $p_1 + p_2 = 1$, then the Hardy–Weinberg principle tells us that genotypes A_1A_1, A_1A_2, and A_2A_2 occur in the proportions p_1^2, $2p_1p_2$, and p_2^2, respectively, provided that mating is random. Similarly, if we consider a different autosomal gene with alleles B_1 and B_2 at frequencies q_1 and q_2, respectively, where $q_1 + q_2 = 1$, then the Hardy–Weinberg principle again tells us that the genotype frequencies of B_1B_1, B_1B_2, and B_2B_2 occur in the proportions q_1^2, $2q_1q_2$, and q_2^2, respectively, provided that mating is random. Thus, the A_1 allele is in random association with the A_2 allele, and the B_1 allele is in random association with the B_2 allele. Strange as it may seem, the alleles of the A gene may nevertheless fail to be in random association in gametes with the alleles of the B gene. The precise meaning of "random association in gametes" is illustrated in Figure 16. Here the combinations of A's and B's in the squares refer to gametes, not to genotypes as in earlier figures. When the alleles of the genes are in random association, the frequency of a gamete carrying any particular combination of alleles equals the product of the frequencies of those alleles. Genes that are in random association are said to be in a state of **linkage equilibrium**, and genes not in random association are said to be in **linkage disequilibrium**.

With random mating and the other simplifying assumptions listed earlier (e.g., large population with no mutation, migration, or selection), linkage equilibrium between genes is eventually attained. The attainment of linkage equilibrium is gradual, however, and the rate of approach can be very slow. This slow approach is in contrast to the attainment of the Hardy–Weinberg proportions with alleles of a single gene, which typically requires just one or a small number of generations.

The rate of approach to linkage equilibrium depends on the relative proportions of the various types of gametes that can be formed by an individual heterozygous for two genes. To be specific, consider an individual of genotype A_1A_2 B_1B_2. This individual is heterozygous for two genes, but the gen-

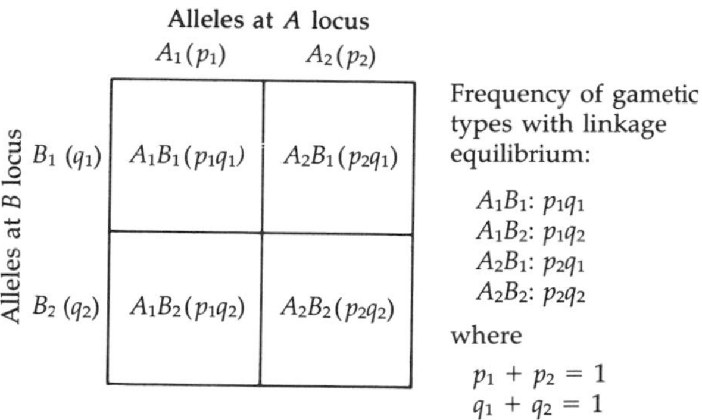

Figure 16. Random association between two alleles of each of two genes, showing expected gametic frequencies when the alleles are in linkage equilibrium.

otypic symbolism is ambiguous because it fails to distinguish between two important cases. In the first case, the individual could have been formed by the union of an A_1 B_1 gamete with an A_2 B_2 gamete; this case is symbolized by writing the genotype of the individual as A_1 B_1/A_2 B_2. In the second case, the individual could have been formed by the union of an A_1 B_2 gamete with an A_2 B_1 gamete, which is symbolized by writing the genotype of the individual as A_1 B_2/A_2 B_1. For the moment, suppose that the true genotype of the individual is A_1 B_1/A_2 B_2.

An individual of genotype A_1 B_1/A_2 B_2 produces four types of gametes: (1) A_1 B_1; (2) A_2 B_2; (3) A_1 B_2; and (4) A_2 B_1. Gametic types 1 and 2 are known as **nonrecombinant gametes** because their alleles are associated in the same manner as in the previous generation (i.e., A_1 with B_1, and A_2 with B_2). Gametic types 3 and 4 are known as **recombinant** gametes because their alleles are associated differently than in the previous generation (i.e., A_1 with B_2, and A_2 with B_1).

Because of Mendelian segregation during the formation of gametes, the frequency of gametic type 1 equals that of type 2, and the frequency of gametic type 3 equals that of type 4. That is, the two nonrecombinant gametes occur with equal frequency, and the two recombinant gametes occur with equal frequency. However, the overall frequency of recombinant gametes (i.e., type 3 + type 4) does not necessarily equal the overall frequency of nonrecombinant gametes (i.e., type 1 + type 2) except in special cases. The term **recombination fraction**, usually symbolized r, refers to the proportion of recombinant gametes produced by an individual. Suppose, for example, that an individual of genotype A_1 B_1/A_2 B_2 produces gametes A_1 B_1, A_2 B_2, A_1 B_2, and A_2 B_1 in the proportions 0.38, 0.38, 0.12, and 0.12, respectively. Then the recombination fraction between the genes is $r = 0.12 + 0.12 = 0.24$.

The recombination fraction between two genes depends on the position of the genes in the chromosomes. If they are in different chromosomes, the recombination fraction is $r = 0.5$; that is, the four possible gametic types are produced in equal frequency. If the two genes are in the same chromosome, the recombination fraction depends on their distance apart; this is because each chromosome aligns side-by-side with its partner chromosome during the formation of gametes and can undergo a sort of breakage and reunion resulting in an exchange of parts between the partner chromosomes. The closer two genes are, the less likely that such a breakage and reunion occurs between them; the farther apart they are, the more likely such an event occurs. The smallest possible recombination fraction is $r = 0$, which would imply that the two genes are so close together that a break never occurs between them. The largest possible recombination fraction is $r = 0.5$, which occurs when genes are very far apart in the same chromosome or, as noted above, when they are in different chromosomes altogether. Genes for which the recombination fraction is less than 0.5 must necessarily be in the same chromosome, and such genes are said to be **linked**.

To sum up, if the recombination fraction between the A and B genes is denoted r, then an individual of genotype $A_1\ B_1/A_2\ B_2$ produces the following types of gametes:

$A_1\ B_1$ with frequency $(1 - r)/2$
$A_2\ B_2$ with frequency $(1 - r)/2$
$A_1\ B_2$ with frequency $r/2$
$A_2\ B_1$ with frequency $r/2$

The situation in $A_1\ B_2/A_2\ B_1$ individuals is much the same, but there is one important difference: In this case, the $A_1\ B_1$ and $A_2\ B_2$ gametes are the *recombinant types*, and the $A_1\ B_2$ and $A_2\ B_1$ gametes are the *nonrecombinant types*. Thus, an individual of genotype $A_1\ B_2/A_2\ B_1$ produces the following types of gametes:

$A_1\ B_1$ with frequency $r/2$
$A_2\ B_2$ with frequency $r/2$
$A_1\ B_2$ with frequency $(1 - r)/2$
$A_2\ B_1$ with frequency $(1 - r)/2$

PROBLEM 16 The human MN gene and the Ss gene discussed in Problem 7 are known to be close together in the same chromosome. The recombination fraction between the genes is certainly less than $r = 0.02$, but the true value is unknown. Assume for the sake of argument that the true value of r equals 0.01. What types and frequencies of gametes would be produced in an *M S/N s* individual? In an *M s/N S* individual?

ANSWER The *M S/N s* individual produces gametic types *M S*, *N s*, *M s*, and *N S* in proportions (1 − 0.01)/2 = 0.495, (1 − 0.01)/2 = 0.495, 0.01/2 = 0.005, and 0.01/2 = 0.005, respectively. The *M s*/*N S* individual produces exactly the same gametic types, but their frequencies are 0.005, 0.005, 0.495, and 0.495, respectively.

The recombination fraction between genes is important in population genetics because the rate of approach to linkage equilibrium depends on the recombination fraction. To be more precise about the effect that recombination has on the approach to linkage equilibrium, suppose that the actual frequencies of $A_1\, B_1$, $A_1\, B_2$, $A_2\, B_1$, and $A_2\, B_2$ gametes are P_{11}, P_{12}, P_{21}, and P_{22}, respectively, where $P_{11} + P_{12} + P_{21} + P_{22} = 1$. In these symbols, linkage equilibrium is defined as the state in which $P_{11} = p_1q_1$, $P_{12} = p_1q_2$, $P_{21} = p_2q_1$, and $P_{22} = p_2q_2$ (see Figure 16).

Suppose for the moment that the genes are not in linkage equilibrium. We want to find out the gametic frequencies in the next generation. Consider first the $A_1\, B_1$ gamete. In any one generation, there are only two things that could have happened to a chromosome carrying A_1 and B_1: it could have undergone recombination between the genes (an event that occurs with probability r, where r again denotes the recombination fraction), or the chromosome might have failed to undergo recombination between the genes (an event with probability $1 - r$). Among those chromosomes that did not undergo recombination, the frequency of $A_1\, B_1$ is the same as it was in the previous generation; among the chromosomes that did undergo recombination, the frequency of $A_1\, B_1$ chromosomes is simply the frequency of $A_1\, B_?/A_?\, B_1$ genotypes in the previous generation (the question mark subscript denotes any allele of the gene), and this is p_1q_1 because mating is random. Putting this together, the frequency of $A_1\, B_1$ in any generation, call it P_{11}', is related to the frequency P_{11} in the previous generation by the equation

$$P_{11}' = (1 - r)P_{11} \text{ [for the nonrecombinants]} + rp_1q_1 \text{ [for the recombinants]}$$

Subtraction of p_1q_1 from both sides leads to

$$P_{11}' - p_1q_1 = (1 - r)(P_{11} - p_1q_1) \quad \textbf{(1.11)}$$

Equation 1.11 becomes simplified somewhat by defining D as the difference $P_{11} - p_1q_1$. Then D_n is the value of D in the nth generation, and Equation 1.11 implies that $D_n = (1 - r)D_{n-1}$. The solution of this equation is found by successive substitution as

$$D_n = (1 - r)D_{n-1} = (1 - r)^2D_{n-2} = \ldots = (1 - r)^nD_0 \quad \textbf{(1.12)}$$

where D_0 is the value of D in the founding population. Because $1 - r < 1$, then $(1 - r)^n$ goes to zero as n gets large, but how rapidly $(1 - r)^n$ goes to zero depends on r; the closer r is to zero, the slower the rate. (Recall here that $r = 0.5$ corresponds either to genes far apart in the same chromosome or to genes in different chromosomes.) In any case, since $(1 - r)^n$ goes to zero, D goes to zero, and therefore P_{11} goes to p_1q_1. Analogous arguments hold for gametes containing $A_1\ B_2$, $A_2\ B_1$, or $A_2\ B_2$, so P_{12}, P_{21}, and P_{22} go to p_1q_2, p_2q_1, and p_2q_2, respectively. Thus, linkage equilibrium is attained at a rate determined by the value of r.

The value of D which holds for $P_{11} - p_1q_1$ also holds for the other possible gametes, as follows

$$\begin{aligned}
P_{11} &= p_1q_1 + D \\
P_{12} &= p_1q_2 - D \\
P_{21} &= p_2q_1 - D \\
P_{22} &= p_2q_2 + D
\end{aligned}$$

The quantity D is often called the **linkage disequilibrium parameter**. In terms of the gametic frequencies, D can be shown to satisfy

$$D = P_{11}P_{22} - P_{12}P_{21} \tag{1.13}$$

With random mating, the value of D changes according to Equation 1.12, and $D = 0$ corresponds to linkage equilibrium. For any prescribed allele frequencies p_1, p_2, q_1, q_2, the smallest (D_{min}) and largest (D_{max}) possible values of D are as follows:

$$\begin{aligned}
D_{min} &= -p_1q_1 \text{ or } -p_2q_2 \text{ (whichever is larger)} \\
D_{max} &= p_1q_2 \text{ or } p_2q_1 \text{ (whichever is smaller)}
\end{aligned} \tag{1.14}$$

In studies of linkage disequilibrium, estimation of the gametic frequencies P_{11}, P_{12}, P_{21} and P_{22} usually requires complex statistical procedures rather than straightforward chromosome-counting methods because there are ten genotypes ($A_1\ B_1/A_2\ B_2$ and $A_1\ B_2/A_2\ B_1$ must be distinguished), but usually no more than nine phenotypes.

An example of linkage disequilibrium is found in the genes controlling the MN and Ss blood groups in humans. In an earlier section, we cited data from 1000 Britishers with respect to the MN blood groups and showed that the genotypes *MM*, *MN*, and *NN* are in Hardy–Weinberg proportions. In Problem 7, data from the same 1000 people were analyzed with respect to the Ss blood groups, and genotypes *SS*, *Ss*, and *ss* were found to satisfy the Hardy–Weinberg proportions. In order to discuss linkage disequilibrium between the genes, it will be convenient to use the symbols p_1 and p_2 for the allele frequencies of *M* and *N*, respectively, and the symbols q_1 and q_2 for the allele frequencies of *S* and *s*, respectively. Recall from the earlier discussions that $\langle p_1 \rangle = 0.5425$, $\langle p_2 \rangle = 0.4575$, $\langle q_1 \rangle = 0.3080$, and $\langle q_2 \rangle = 0.6920$. Were the loci in linkage equilibrium, the gametic frequencies would be p_1q_1 for *M S*,

p_1q_2 for $M\ s$, p_2q_1 for $N\ S$, and p_2q_2 for $N\ s$. Therefore, among the 1000 people (representing 2000 chromosomes), the expected numbers are as shown in the third column below (the second column gives the observed numbers):

$M\ S$	474	$(0.5425)(0.3080)(2000) = 334.2$
$M\ s$	611	$(0.5425)(0.6920)(2000) = 750.8$
$N\ S$	142	$(0.4575)(0.3080)(2000) = 281.8$
$N\ s$	773	$(0.4575)(0.6920)(2000) = 633.2$

The χ^2 for goodness of fit is 184.7 with one degree of freedom [i.e., 4 (to start with) − 1 − 1 (for estimating p_1 from the data) − 1 (for estimating q_1 from the data) = 1]. The associated probability is so small as to be off the chart in Figure 12, so it is very much less than 0.0001. This means that chance alone would produce a fit as poor or poorer substantially less than one time in 10,000, so the hypothesis that the loci are in linkage equilibrium can confidently be rejected.

In order to quantify the amount of linkage disequilibrium, we must estimate the gametic frequencies P_{11}, P_{12}, P_{21}, and P_{22}.

$M\ S$	$\langle P_{11}\rangle = 474/2000 = 0.2370$
$M\ s$	$\langle P_{12}\rangle = 611/2000 = 0.3055$
$N\ S$	$\langle P_{21}\rangle = 142/2000 = 0.0710$
$N\ s$	$\langle P_{22}\rangle = 773/2000 = 0.3865$

Thus, D can be estimated as $\langle D\rangle = P_{11}P_{22} - P_{12}P_{21} = (0.2370)(0.3865) - (0.3055)(0.0710) = 0.07$. From Equation 1.14, D_{max} is given by p_1q_2 or p_2q_1, whichever is smaller; in this case, $p_1q_2 = 0.38$ and $p_2q_1 = 0.14$, so $D_{max} = 0.14$. Therefore, $\langle D\rangle/D_{max} = 0.07/0.14 = 50$ percent, so we conclude that there is linkage disequilibrium between the MN and Ss genes, and the amount of disequilibrium is about 50 percent of its theoretical maximum.

In most natural populations of sexual organisms that regularly avoid extreme inbreeding, such as self-fertilization, values of D are typically zero or close to zero (indicating linkage equilibrium) unless the genes are very closely linked. This overall conclusion is exemplified in the following problems.

PROBLEM 17

As mentioned for the data of Mukai, Watanabe, and Yamaguchi (1974) in Problem 3, the order of genes in the third chromosome is *E6*–*EC*–*Odh*. The recombination fraction between *E6* and *EC* is 0.122, that between *EC* and *Odh* is 0.002 (i.e., *E6* and *EC* are rather loosely linked, *EC* and *Odh* are tightly linked). Carry out an analysis as for the *MN* and *Ss* genes to determine whether there is linkage disequilibrium between *E6* and *EC*. If there is linkage disequilibrium, how much is there in terms of *D*, relative to the theoretical maximum (or minimum) value?

ANSWER Observed numbers of the four chromosomal types $E6^F EC^F$, $E6^F EC^S$,

$E6^SEC^F$, and $E6^SEC^S$ are (from Problem 3) 159, 16, 277, and 37, respectively. Allele frequencies of $E6^F$, $E6^S$, EC^F, and EC^S are 0.3579, 0.6421, 0.8916, and 0.1084, respectively. Expected numbers of the four chromosomal types (assuming linkage equilibrium) are 156.0, 19.0, 280.0, and 34.0, respectively. The χ^2 with one degree of freedom is 0.828, for which the associated probability is about 0.4. Thus, there is no reason to reject the hypothesis that *E6* and *EC* are in linkage equilibrium.

PROBLEM 18 Carry out an analysis of linkage disequilibrium for the genes *EC* and *Odh*, using the data in Problem 3. (Note: A shortcut to obtaining the χ^2 value is first to calculate

$$\rho = \langle D \rangle / (\langle p_1 \rangle \langle p_2 \rangle \langle q_1 \rangle \langle q_2 \rangle)^{1/2}$$

because $\chi^2 = \rho^2 N$, where N is the total number of chromosomes examined; ρ is an estimate of the correlation between alleles present in the same chromosome.)

ANSWER Observed numbers of the chromosomal types $EC^F\ Odh^F$, $EC^F\ Odh^S$, $EC^S\ Odh^F$, and $EC^S\ Odh^S$ are (from Problem 3) 416, 20, 44, and 9, respectively. Allele frequencies of EC^F, EC^S, Odh^F, and Odh^S are 0.8916, 0.1084, 0.9407, and 0.0593, respectively, and $\langle D \rangle = [(416)(9) - (20)(44)]/489^2 = 0.0120$. Thus, $\rho = \langle D \rangle / [(0.8916)(0.1084)(0.9407)(0.0593)]^{1/2} = 0.1631$, and $\chi^2 = (0.1631)^2(489) = 13.0$, for which the associated probability is 0.0004. Thus, there is significant linkage disequilibrium between the genes. The value of D_{max} is the smaller of 0.053 and 0.102, so $D_{max} = 0.053$. The amount of linkage disequilibrium is $\langle D \rangle / D_{max} = 0.012/0.053 = 22.6$ percent of its theoretical maximum. The χ^2 can also be calculated from the expected numbers of the four gametic types, which are, in the same order as above, 410.1, 25.9, 49.9, and 3.1.

PROBLEM 19 Use the formula for χ^2 in Problem 18 to evaluate the statistical significance of the linkage disequilibrium between alleles of the alcohol dehydrogenase gene in *Drosophila melanogaster* and the presence or absence of an *Eco*RI restriction site located 3500 nucleotides downstream. In the population referred to in Problem 4, the following data were obtained: Adh^F *Eco*RI$^+$ 22, Adh^F *Eco*RI$^-$ 3, Adh^S *Eco*RI$^+$ 4, Adh^S *Eco*RI$^-$ 5 (Cross and Birley 1986).

ANSWER $\langle D \rangle = 0.085$ and $\chi^2 = \rho^2 N = (0.453)^2(34) = 7.0$ with one degree of freedom, for which the associated probability value is approximately 0.01. The linkage disequilibrium is statistically significant and is 49 percent of its maximum possible value.

Linkage disequilibrium such as seen in the preceding examples can be caused by linkage disequilibrium in the founding population that has not yet had time to dissipate due to the small value of r. Another possible cause is admixture of populations with differing gametic frequencies. A third possibility is the occurrence of a sufficient intensity of natural selection in favor of certain genotypes to overcome the natural tendency for D to go to zero.

Two cases in which linkage disequilibrium typically occurs in natural populations should be mentioned here. One case concerns plants that ordinarily undergo self-fertilization, and an example will be given later in connection with the discussion of inbreeding. The other case involves certain inversions that are polymorphic in populations of *Drosophila pseudoobscura*, *D. subobscura*, and their relatives. A chromosome with an **inversion**, as the name implies, has a certain segment of its genetic loci in reverse of the normal order. Because of this inverted segment, the process of breakage and reunion of chromosomes during gamete formation cannot occur in the normal manner, with the result that the genes in the inverted segment rarely undergo recombination. Because inversions prevent recombination, each inversion represents a sort of "supergene," and natural selection accumulates beneficially interacting alleles within each inversion. The beneficially interacting alleles are said to show **genetic coadaptation**.

ASSORTATIVE MATING

Although random mating is the most important mating system in many natural populations, certain departures from random mating can also be important. Major types of departure are listed and defined in Table 4. When choice of mates is based on phenotypes, mating is said to be *assortative*. In **positive assortative mating** (often called simply **assortative mating**), individuals tend to choose mates that are phenotypically like themselves. In **negative assortative mating** (also called **disassortative mating**), individuals tend to choose mates that are phenotypically unlike themselves. Of course, even with random mating, some mating pairs will be phenotypically similar or dissimilar, so assortative mating refers only to those situations in which mating partners are phenotypically more similar or dissimilar than would be expected by chance in a random-mating population.

Table 4. Characteristics of several mating systems.

Mating system	Defining feature
Random mating	Choice of mates independent of genotype and phenotype
Positive assortative mating	Mates phenotypically more similar than would be expected by chance
Negative assortative mating (disassortative)	Mates phenotypically more dissimilar than would be expected by chance
Inbreeding	Mating between relatives

An example of negative assortative mating is found in the polymorphism known as **heterostyly**, which occurs in most species of primroses (*Primula*) and their relatives. In most primrose populations, there are approximately equal proportions of two types of flowers, one known as *pin*, which has a tall style and short stamens, and the other known as *thrum*, which has a short style and tall stamens. (In botanical terminology the style carries the stigma, which is the organ that receives pollen, whereas the stamens carry the anthers, which produce pollen.) In heterostyly, insect pollinators that work high on the flowers will pick up mostly *thrum* pollen and deposit it on *pin* stigmas, whereas pollinators that work low in the flowers will pick up mostly *pin* pollen and deposit it on *thrum* stigmas. Negative assortative mating therefore occurs because *pins* mate preferentially with *thrums*.

Pollination biology of flowering plants also provides examples of positive assortative mating. For example, when the length of time during which any individual plant flowers is short relative to the total duration of the flowering season, then plants that flower early in the season are preferentially pollinated by other early flowering plants, and those that flower late are preferentially pollinated by other late flowering ones. Thus, there is positive assortative mating for flowering time.

In humans, positive assortative mating occurs for height, IQ score, and certain other traits, although the amount of assortative mating varies in different populations and does not occur in some. As might be expected, positive assortative mating occurs for certain socioeconomic variables. This effect is seen in one study in the United States, in which the highest correlation found between married couples was in the number of rooms in their parents' homes. Negative assortative mating is apparently quite rare, so we shall ignore it, but one obvious example is sex, in which mating pairs are always phenotypically different.

In certain species of *Drosophila*, a curious type of nonrandom mating is found in a phenomenon called *minority male mating advantage*, in which females mate preferentially with males with rare phenotypes. For example, in one study involving experimental populations of *D. pseudoobscura* containing flies homozygous for either the recessive orange-eye-color mutation or the recessive purple-eye-color mutation, Ehrman (1970) found that when 20 percent of the males were orange, the orange-eyed males participated in 30 percent of the observed matings; conversely, when 20 percent of the males were purple, the purple-eyed males participated in 40 percent of the observed matings.

The consequences of positive assortative mating are complex and are dependent on the number of genes that influence the trait in question, on the number of different possible alleles of the genes, on the number of different phenotypes, on the sex performing the mate selection, and on the criteria for mate selection. Rarely are traits with assortative mating determined by the alleles of a single gene, however; most such traits are polygenic, so reasonably realistic models of assortative mating tend to be rather complex. Here we should note one obvious, qualitative consequence of positive assortative mating: since like phenotypes tend to mate, assortative mating generally increases the frequency of homozygous genotypes in the population at the expense of heterozygous genotypes, and thus the phenotypic variance in the population increases. (Negative assortative mating generally has the opposite effect.)

INBREEDING

Mating between relatives is **inbreeding**. Like positive assortative mating, inbreeding increases the homozygosity in a population. Unlike assortative mating, which affects only those genes on which mate selection is based (and genes in linkage disequilibrium with them), inbreeding affects all genes. In humans, the closest degree of inbreeding that commonly occurs in most societies is first-cousin mating, but many plants regularly undergo self-fertilization.

Genotype Frequencies with Inbreeding

The principal effect of inbreeding in a population is to increase the frequency of homozygous genotypes at the expense of the frequency of heterozygous genotypes. This effect can most easily be seen in the case of repeated self-fertilization. Consider a self-fertilizing population of plants that consists of (1/4)*AA*, (1/2)*Aa*, and (1/4)*aa* genotypes. These genotype frequencies are in Hardy–Weinberg proportions. Because each plant undergoes self-fertilization, the *AA* and *aa* genotypes produce only *AA* and *aa* offspring, respectively, and the *Aa* genotypes produce (1/4)*AA*, (1/2)*Aa*, and (1/4)*aa* offspring. After one generation of self-fertilization, therefore, the genotype frequencies of *AA*, *Aa*, and *aa* are [(1/4)(1) + (1/2)(1/4)]*AA*, [(1/2)(1/2)]*Aa*, and [(1/4)(1) + (1/2)(1/4)]*aa*, respectively, or (3/8)*AA*, (2/8)*Aa*, and (3/8)*aa*. These genotype frequencies are no longer in the Hardy-Weinberg proportions because there is a deficiency of heterozygous genotypes and an excess of homozygous genotypes.

After a second generation of self-fertilization, the genotype frequencies are (7/16)*AA*, (2/16)*Aa*, and (7/16)*aa*, which represents an even greater deficiency of heterozygotes. Note, however, that the allele frequency of *A* (call it p) remains constant. In the initial population $p = (1/4) + (1/2)(1/2) = 1/2$; after one generation of selfing, $p = (3/8) + (1/2)(2/8) = 1/2$; and after two generations of selfing, $p = (7/16) + (1/2)(2/16) = 1/2$. This self-fertilization example illustrates the general principle that inbreeding, by itself, does not change allele frequency. (The constancy of allele frequency with inbreeding occurs only when all genotypes have an equal likelihood of survival and reproduction, which is to say that no natural selection occurs. If selection does occur, then allele frequencies can change with inbreeding, or even with random mating.)

The effects of inbreeding are easily made quantitative in terms of the reduction in heterozygosity that occurs. That is, we can measure the amount of inbreeding by comparing the actual proportion of heterozygous genotypes in the population with the proportion of heterozygous genotypes that would occur with random mating. To be precise, consider a locus with two alleles, *A* and *a*, at respective frequencies p and q (with $p + q = 1$). Suppose the actual frequency of heterozygous genotypes in the population is denoted H. Were the population undergoing random mating for the gene, the frequency of heterozygous genotypes would be $2pq$ because of the Hardy–Weinberg principle, but for the sake of generality we will denote the random-mating heterozygosity by the symbol H_0. The effects of inbreeding can be defined in terms of the quantity $(H_0 - H)/H_0$, which is usually denoted in population genetics by the symbol F and called the **inbreeding coefficient**. Thus

$$F = (H_0 - H)/H_0 \qquad \textbf{(1.15)}$$

In biological terms, F measures the fractional reduction in heterozygosity, relative to a random-mating population with the same allele frequencies. Because $H_0 = 2pq$, the actual frequency of heterozygous genotypes in the inbred population can be written in terms of F as $H = H_0 - H_0F = H_0(1 - F) = 2pq(1 - F)$.

The frequency of *AA* homozygous genotypes in the inbred population can also be expressed in terms of F. Suppose that the proportion of *AA* genotypes is actually P. Because the allele frequency of *A* is p, we must have, by Equation 1.5, $P + H/2 = p$. But $H = 2pq(1 - F)$, thus $P = p - 2pq(1 - F)/2$.

PROBLEM 20 Use the equation $P = p - 2pq(1 - F)/2$ and the fact that $p + q = 1$ to show that $P = p^2(1 - F) + pF$.

ANSWER $P = p - pq(1 - F) = p - p(1 - p)(1 - F) = p - (p - p^2)(1 - F) = p - p + p^2 + pF - p^2F = p^2(1 - F) + pF$.

Problem 20 implies that the frequency of *AA* genotypes in an inbred population equals $p^2(1 - F) + pF$. In a similar manner, it can be shown that the frequency of *aa* genotypes will be $q^2(1 - F) + qF$.

In summary, in a population with inbreeding coefficient F, the genotype frequencies occur in the proportions

$$\begin{array}{ll} AA & p^2(1-F) + pF = p^2 + pqF \\ Aa & 2pq(1-F) = 2pq - 2pqF \\ aa & q^2(1-F) + qF = q^2 + pqF \end{array} \qquad \textbf{(1.16)}$$

The expressions at the far right in Equation 1.16 facilitate comparison with the genotype frequencies in the Hardy–Weinberg principle. Notice that, when $F = 0$, which represents no inbreeding (i.e., random mating), the genotype frequencies are in the familiar Hardy–Weinberg proportions p^2, $2pq$, and q^2. When $F = 1$, which represents complete inbreeding, the population consists entirely of AA and aa homozygotes in the frequencies p and q, respectively.

If the population contains multiple alleles $A_1, A_2, \ldots, A_n$ at respective frequencies $p_1, p_2, \ldots, p_n$ (with $p_1 + p_2 + \ldots + p_n = 1$), then in a population with inbreeding coefficient F, the frequencies of A_iA_i homozygotes and A_iA_j heterozygotes are as follows:

$$\begin{array}{l} p_i^2(1-F) + p_iF \text{ for } A_iA_i \text{ homozygotes} \\ 2p_ip_j(1-F) \text{ for } A_iA_j \text{ heterozygotes} \end{array} \qquad \textbf{(1.17)}$$

We are now in a position to use Equations 1.16 and 1.17 in actual situations.

PROBLEM 21

For a population of plants that are self-compatible (i.e., able to undergo self-fertilization), if each plant undergoes self-fertilization a fraction s of the time and otherwise mates randomly, then it can be shown that F very quickly attains the value $F = s/(2 - s)$. (For derivation and extensions, see Crow and Kimura, 1970, and Hedrick and Cockerham, 1986). *Phlox cuspidata*, mentioned in Problem 13, is self-compatible, and for this species the amount of self-fertilization is estimated at approximately $\langle s \rangle = 0.78$ (Levin 1978); thus we can estimate the inbreeding coefficient as $\langle F \rangle = 0.78/(2 - 0.78) = 0.64$. In a Texas population of *P. cuspidata*, Levin (1978) found two alleles of the phosphoglucomutase-2 gene, designated $Pgm{-}2^a$ and $Pgm{-}2^b$. In a sample of 35 individuals, Levin found 15 $Pgm{-}2^a/Pgm{-}2^a$, 6 $Pgm{-}2^a/Pgm{-}2^b$, and 14 $Pgm{-}2^b/Pgm{-}2^b$ individuals. Are these numbers consistent with the estimate $F = 0.64$? (Note: The χ^2 here has one degree of freedom because only the allele frequency is estimated from the data; if F also were estimated from the data, rather than being calculated independently from the degree of self-fertilization, then there would be zero degrees of freedom and no goodness-of-fit test would be possible.)

ANSWER Allele frequencies of $Pgm{-}2^a$ and $Pgm{-}2^b$ are estimated as $(30 + 6)/70 = 0.514$ and $1 - 0.514 = 0.486$, respectively. The hypothesis is that $F = 0.64$, so $1 - F = 0.36$. Expected numbers of the three genotypes, given in the same order as above, are

$$[(0.514)^2(0.36) + (0.514)(0.64)](35) = 14.8,$$
$$[2(0.514)(0.486)(0.36)](35) = 6.3, \text{ and}$$
$$[(0.486)^2(0.36) + (0.486)(0.64)](35) = 13.9.$$

The χ^2 is 0.02 with one degree of freedom, and the associated probability is about 0.96. The fit to the inbreeding model is excellent.

PROBLEM 22 Assuming $F = 0.64$ in Texas populations of *Phlox cuspidata*, calculate the expected genotype frequencies for the *Adh* gene using the data in Problem 13.

ANSWER *Adh*−1/*Adh*−1 = 0.0748, *Adh*−1/*Adh*−2 = 0.0665, *Adh2*/*Adh2* = 0.7916, *Adh*−1/*Adh*−3 = 0.0008, *Adh*−2/*Adh*−3 = 0.0060, *Adh*−3/*Adh*−3 = 0.0064, *Adh*−1/*Adh*−4 = 0.0032, *Adh*−2/*Adh*−4 = 0.0242, *Adh*−3/*Adh*−4 = 0.0003, *Adh*−4/*Adh*−4 = 0.0262.

The Inbreeding Coefficient

The inbreeding coefficient *F* has an alternative interpretation in terms of probability, and this interpretation is extremely useful for calculation of *F* from pedigrees. To express the inbreeding coefficient in terms of probability, imagine the two alleles present at homologous loci in some single inbred individual. These two alleles could be **identical by descent**, that is, they could both have been derived by replication of a single allele in some ancestral population. If the two alleles of the gene in question are identical by descent, then the alleles are said to be **autozygous**. On the other hand, the alleles may not be replicas of a single ancestral allele, in which case the alleles are not identical by descent, and they are said to be **allozygous**. The interpretation of *F* in terms of probability is this: *The inbreeding coefficient* F *is the probability that the two alleles of a gene in an individual are identical by descent (autozygous).*

Seen in terms of the probability of autozygosity, the inbreeding coefficient is clearly a relative concept; it measures the amount of autozygosity relative to that in some ancestral population. That is, we arbitrarily assume that all alleles in the ancestral population are not identical by descent, and the inbreeding coefficient of an individual in the present population is then the probability that the two alleles of a gene in the individual arose by replication of a single allele more recently than the time at which the ancestral population existed. The ancestral population need not be remote in time from the present one. Indeed, the ancestral population, presumed to be noninbred ($F = 0$), typically refers to the population existing just a few generations previous to the present one, and *F* in the present population then measures inbreeding that has occurred in the span of these few generations. Because the span of time involved is usually short, the possibility of mutation can safely be ignored. Autozygous individuals must therefore be homozygous for some allele at the locus under consideration. On the other hand, allozygous individuals can be either homozygous or heterozygous.

Figure 17 illustrates how the concepts of autozygosity and allozygosity are related to those of homozygosity and heterozygosity. The essential point is that two alleles can be chemically identical (in terms of having the same sequence of nucleotides along the DNA) without being identical by descent; the concept of identity by descent pertains to the ancestral origin of an allele and not to its chemical makeup. Although, as shown in Figure 17, two distinct alleles that are chemically identical (two A_1s or two A_2s, for example) may come together in an individual and thereby make the individual homozygous, the alleles in the ancestral population are, by definition, not identical by descent, so the individual is allozygous. Similarly, although a heterozygous individual must be allozygous (ignoring mutation), a homozygous individual may be either autozygous or allozygous (see Figure 17).

While the inbreeding coefficient F is the probability that the two alleles of a gene in an individual are identical by descent, the idea can readily be extended to alleles that are not necessarily in the same individual. That is to say, one could define another sort of "inbreeding coefficient" as the proba-

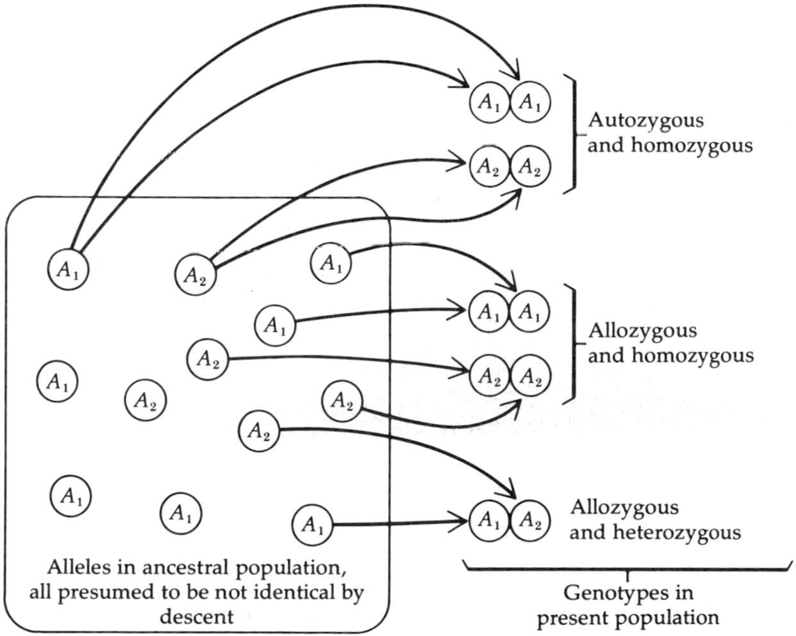

Figure 17. In an autozygous individual, homologous alleles are derived from replication of a single DNA sequence in an ancestor, and they are therefore identical by descent. In an allozygous individual, homologous alleles are not identical by descent. As shown here, allozygous individuals may be heterozygous or homozygous, but autozygous individuals must be homozygous (except in the unlikely event that one allele undergoes a mutation).

bility that any two randomly chosen alleles in a population (not necessarily in the same individual) are identical by descent. This rather general comment is intended as a preliminary introduction to Chapter 2, where the inbreeding coefficient in this extended context is used in order to measure genetic relationships within and between populations. For the moment, we need the concept only in its simplest form: The inbreeding coefficient F is the probability that the two alleles of a gene in an individual are identical by descent.

To show that the inbreeding coefficient defined in terms of probability of autozygosity is equivalent to that defined in Equation 1.15 in terms of heterozygosity, we need only consider the implications of the probability definition for an entire population. For this purpose, imagine a population in which individuals have average inbreeding coefficient F. Focus on one individual, and consider the alleles of any gene in the individual. Either of two things must be true: the alleles must be either allozygous (probability $1 - F$) or autozygous (probability F). If the alleles are allozygous, then the probability that the individual has any particular genotype is simply the probability of that genotype in a random mating population, because, by chance, the inbreeding has not affected this particular gene. On the other hand, if the alleles are autozygous, then the individual must be homozygous, and the probability that the individual is homozygous for any particular allele is simply the frequency of that allele in the population as a whole. (Since the alleles in question are autozygous, knowing which allele is present in one chromosome immediately tells you that an identical allele is in the homologous chromosome.) These considerations hold regardless of the number of alleles, but to simplify matters, suppose there are only two alleles A and a at frequencies p and q (with $p + q = 1$). The probability that an individual has genotype AA is therefore $p^2(1 - F)$ [for cases in which the alleles are allozygous] + pF [for cases in which the alleles are autozygous]. Similarly, the probability that the individual has genotype aa is $q^2(1 - F) + qF$. Heterozygous Aa genotypes then occur in the frequency $2pq(1 - F)$, since alleles that are heterozygous must be allozygous. The genotype frequencies with inbreeding are summarized in Table 5. Note that the genotype frequencies are exactly the same as those given in Equation 1.16, which shows that the autozygosity definition of F and the heterozygosity definition of F, though superficially quite different, are actually equivalent.

Effects of Inbreeding

In outcrossing species—those that regularly avoid inbreeding—close inbreeding is generally harmful. The effects are seen most dramatically when inbreeding is complete or nearly complete. Although nearly complete autozy-

Table 5. Genotype frequencies with inbreeding.

	FREQUENCY IN POPULATION				
Genotype	With inbreeding coefficient F			With $F = 0$ (random mating)	With $F = 1$ (complete inbreeding)
AA	$p^2(1 - F)$	+	pF	p^2	p
Aa	$2pq(1 - F)$			$2pq$	0
aa	$q^2(1 - F)$	+	qF	q^2	q
	Allozygous genes		Autozygous genes		

gosity can be approached in most species by many generations of brother-sister mating, autozygosity of individual chromosomes can easily be accomplished in *Drosophila* by the sort of mating scheme shown in Figure 18. *Cy* (*Curly* wings) and *Pm* (*Plum*-colored eyes) are dominant mutations on chromosome 2 and are present on certain laboratory second chromosomes (denoted by gray and black) that carry several long inversions to prevent recombination. Single wild males are mated individually to *Cy*/*Pm* females, and a single *Curly*-winged son, carrying one wildtype second chromosome (hatched), is selected from each mating (panel a). The *Curly*-winged male is backcrossed to *Cy*/*Pm* females, and *Curly*-winged heterozygous sons and daughters are isolated (panel b). These flies are mated with each other (panel c1) and the offspring counted. Because *Cy*/*Cy* homozygotes are lethal, one-third of the surviving offspring are expected to be homozygous for the isolated wildtype chromosome, unless, of course, the wildtype chromosome carries recessive mutations that reduce viability. At the same time as the matings in panel (c1) are carried out, heterozygous males from one strain are mated with heterozygous females carrying a different wildtype chromosome (stippling in panel c2). In this case, one-third of the offspring are expected to be heterozygous—that is, to carry two different wildtype chromosomes. For the matings in both panel (c1) and panel (c2), an estimate $\langle v \rangle$ of the viability (ability to survive) of the +/+ genotype, relative to that of the *Cy*/+ genotype, is given by

$$\langle v \rangle = \frac{2(+/+)}{(Cy/+) + (1/n)} \tag{1.18}$$

where (+/+) and (*Cy*/+) represent the observed proportions of wildtype (*straight* wings) and *Curly* offspring, respectively, and n is the total number of offspring observed (Haldane 1956). The term $1/n$ in the denominator makes

(a) Mate and select single *Curly*-winged son.

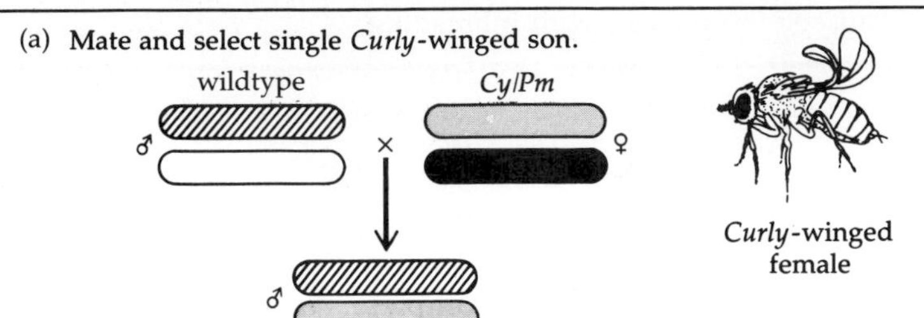

(b) Backcross the son from (a) and select *Curly* sons and daughters, which will be heterozygous.

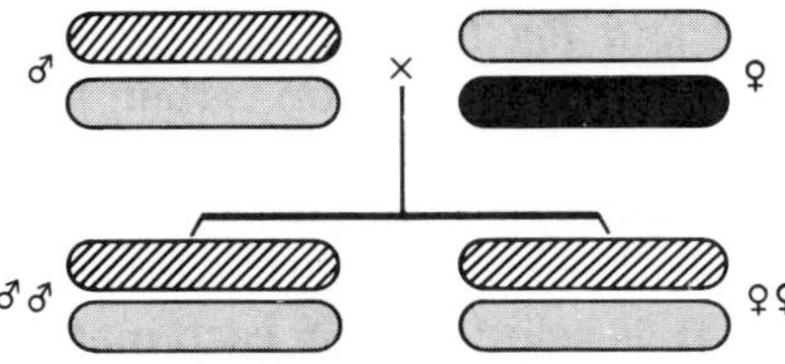

(c1) Mate heterozygotes from same strain and count proportion of *straight*-winged offspring.

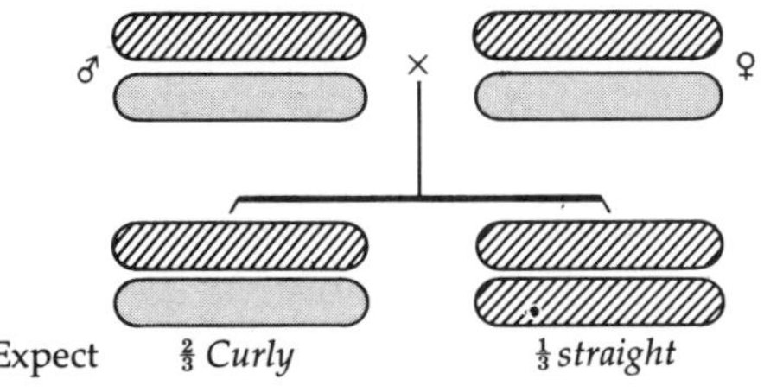

(c2) Mate heterozygotes from different strains and count proportion of *straight*-winged offspring.

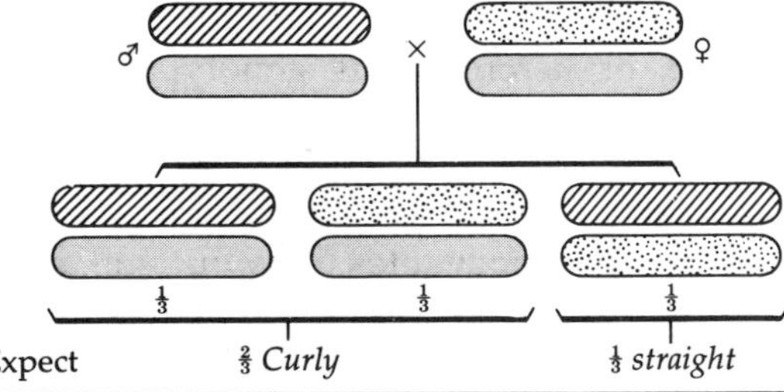

the estimate of v almost unbiased. When the sample size n is large, $\langle v \rangle$ is essentially equal to two times the number of wildtype offspring divided by the number of *Curly* offspring.

Results of an experiment using the procedure in Figure 18 are shown in Figure 19. It is evident that the homozygous genotypes (gray outline) are relatively poor in viability. In fact, about 37 percent of the homozygotes are

◄ **Figure 18. Mating scheme to extract single chromosomes (in this case, the second chromosome) from populations of *Drosophila melanogaster*. *Cy* (*Curly* wings) and *Pm* (*Plum* eyes) are dominant mutations contained in certain special laboratory chromosomes that have multiple inversions to prevent recombination. From each mating of the type in panel (a), a single son (containing one of his father's second chromosomes, indicated by the hatching) is selected. This son is backcrossed (panel b) in order to produce many replicas of that second chromosome. Brother-sister matings as in panel (c1) are expected to produce 1/4 *Cy*/*Cy*, 1/2 *Cy*/+, and 1/4 +/+ zygotes (the + denotes the wildtype second chromosome), but since *Cy*/*Cy* zygotes do not survive, the expected offspring are 2/3 *Cy*/+ (*Curly* wings) and 1/3 +/+ (normal, *straight* wings). Matings as in panel (c2), between a male containing one wildtype second chromosome (hatched) and a female carrying a different one (stippled), are also expected to produce 2/3 *Curly* wings and 1/3 *straight* wings. However, in panel (c1), the *straight*-wing flies are homozygous for a single wildtype second chromosome, whereas in panel (c2) the *straight*-wing flies are heterozygous for two different wildtype second chromosomes.**

lethal. Moreover, among the homozygotes that have viabilities within the normal range of heterozygotes (black outline), virtually all can be shown to have reduced fertility (Sved 1975, Simmons and Crow 1977). Inbreeding so close as to make entire chromosomes homozygous is rare in outcrossing species, except in the kind of experiment in Figure 19, but the effects are clearly very harmful and provide a new dimension of genetic diversity. In

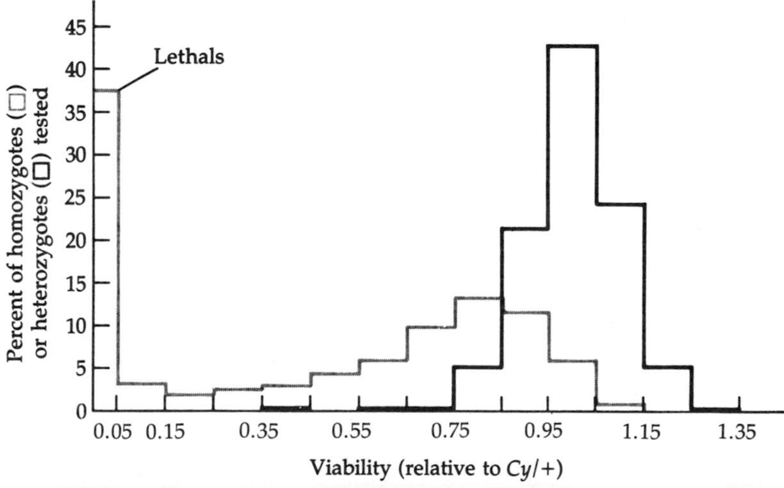

Figure 19. Viability distributions of wildtype homozygotes (gray outline) and wildtype heterozygotes (black outline) of second chromosomes extracted from *Drosophila melanogaster* according to the mating scheme in Figure 18. The histograms depict results of testing 691 homozygous combinations and 688 heterozygous combinations. Note that in this sample, nearly 37 percent of the wildtype chromosomes are lethal when homozygous, and many more have viabilities substantially below normal. (Data from Mukai and Yamaguchi 1974.)

the case of allozymes, genetic diversity is due to common alleles that do not perceptibly impair viability or fertility when homozygous. In the case of inbreeding, the effects are mainly due to rare alleles that are severely detrimental when homozygous. (That the alleles are rare is shown by the small proportion of lethal or near-lethal heterozygotes.) Figure 19 shows that natural populations of *Drosophila* contain considerable hidden genetic variation in the form of rare deleterious recessive alleles.

Detrimental effects of inbreeding (called **inbreeding depression**) are found in virtually all outcrossing species, and the more intense the inbreeding, the more harmful the effects (Charlesworth and Charlesworth in press). Inbreeding in humans is generally harmful also, but the effect is difficult to measure because the degree of inbreeding is less than what is possible to achieve in experimental organisms, and the effects may also vary from population to population. Nevertheless, children of first-cousin matings are on the average less capable than noninbred children in any number of ways (e.g., higher rate of mortality, lower IQ scores), though it should be emphasized that many such children are within a normal range of abilities. As in most organisms, inbreeding depression is largely due to the increased homozygosity of rare recessive alleles, and inbreeding effects in humans are also seen most dramatically in the enhanced frequency of genetic abnormalities due to harmful recessive alleles among the children of first cousin matings. The enhanced frequency of such conditions results from the genotype frequencies given in Table 5; if a denotes a rare deleterious recessive allele, then among the children of first cousin matings, the frequency of aa is $q^2(1 - 1/16) + q(1/16)$, because for these children $F = 1/16$, as will be shown in the next section. On the other hand, with random mating, the frequency of recessive homozygotes is q^2. Thus the risk of an affected offspring from a first cousin mating relative to that from a mating of nonrelatives is given by

$$[q^2(1 - 1/16) + q(1/16)]/q^2 = 0.9375 + 0.0625/q \qquad \textbf{(1.19)}$$

For example, when $q = 0.01$, the increased risk is approximately 7. That is, a first-cousin mating has seven times the chance of producing a homozygous recessive child as compared to a mating between nonrelatives when the frequency of the harmful recessive allele is 0.01. This is clearly a dramatic inbreeding effect, and the rarer the frequency of the deleterious recessive allele, the greater the effect.

PROBLEM 23 Calculate the risk of a homozygous recessive offspring from a mating of second cousins ($F = 1/64$) when the recessive allele frequency is $q = 0.01$, relative to the risk with random mating.

ANSWER $[q^2(1 - F) + qF]/q^2 = (1 - F) + F/q$. For $F = 1/64$, this becomes $0.9844 + 0.0156/q$, and the corresponding value for $q = 0.01$ is approximately 2.5.

Calculation of the Inbreeding Coefficient from Pedigrees

Computation of F from a pedigree is simplified by drawing the pedigree in the form shown in Figure 20a, where the lines represent gametes contributed by parents to their offspring. The same pedigree is shown in conventional form in Figure 20b. The individuals in gray in panel (b) are not represented in panel (a) because they have no ancestors in common and therefore do not contribute to the inbreeding of individual I. The inbreeding coefficient F_I of individual I is the probability that I is autozygous for alleles at the autosomal locus under consideration. The first step in calculating F_I is to locate all the common ancestors in the pedigree, because an allele could become autozygous in I only if it were inherited through both of I's parents from a common ancestor; in this case, there is only one common ancestor, namely, the individual labeled A. The next step in calculating F_I, which is carried out for each common ancestor, is to trace all the paths of gametes that lead from one of I's parents back to the common ancestor and then down again to the other parent of I. These paths are the paths along which an allele in a common ancestor could become autozygous in the individual I. In Figure 20a, there is only one such path: DB$\underline{\text{A}}$CE (the common ancestor is underlined for bookkeeping purposes, an especially useful procedure in complex pedigrees).

The third step in calculating F_I is to calculate the probability of autozygosity in I due to each of the paths in turn. For the path DB$\underline{\text{A}}$CE, the reasoning involved is illustrated in Figure 21. Here the black dots represent alleles transmitted along the gametic paths, and the number associated with each loop is the probability of identity by descent of the alleles indicated. For all

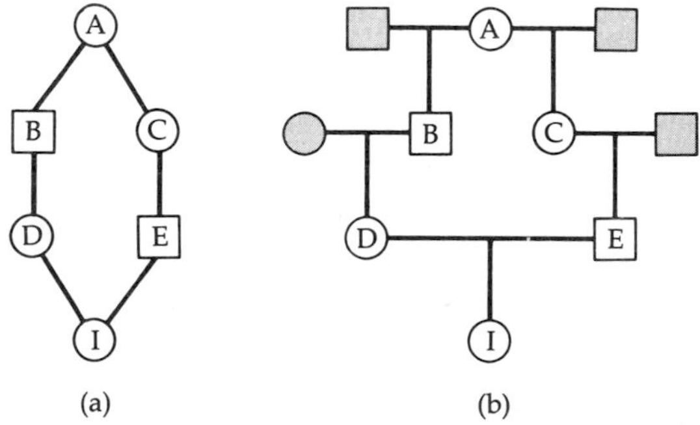

Figure 20. (a) Convenient way to represent pedigrees for calculation of the inbreeding coefficient. In this case, the pedigree shows a mating between half-first cousins. (b) Conventional representation of the same pedigree as in panel (a). Squares represent males, circles represent females, and the shaded individuals in (b) are not depicted in (a) because they do not contribute to the inbreeding of the individual designated I.

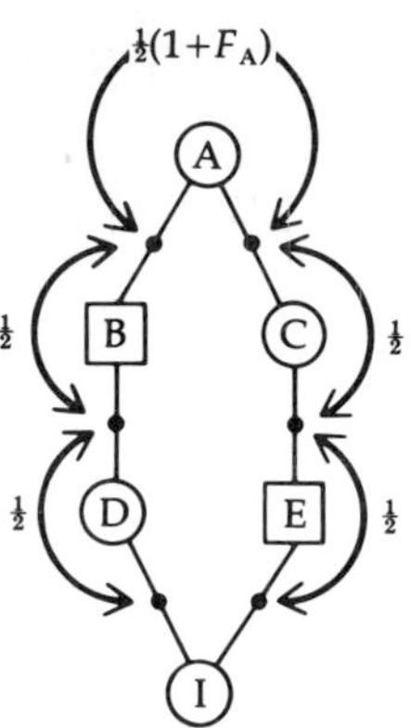

Figure 21. Loops for the pedigree in Figure 20(a), showing probabilities that designated alleles (solid dots) are identical by descent. Each loop is independent of the others, so their probabilities multiply. Thus, the inbreeding coefficient of individual I is $F_I = (1/2)^5(1 + F_A)$, where F_A represents the inbreeding coefficient of the common ancestor.

individuals except the common ancestor, the probability is one-half because, with Mendelian segregation, the probability that a particular allele present in a parent is transmitted to a specified offspring is one-half. To see why $(1/2)(1 + F_A)$ is the probability associated with the loop around the common ancestor, denote the alleles in the common ancestor as α_1 and α_2. (These symbols are used to avoid confusion with conventional allele symbols designating chemical types of alleles, such as *A* and *a* for dominant and recessive, respectively.) The pair of gametes contributed by individual A could contain $\alpha_1\alpha_1$, $\alpha_2\alpha_2$, $\alpha_1\alpha_2$, or $\alpha_2\alpha_1$, each with a probability of 1/4 because of Mendelian segregation. In the first two cases, the alleles are clearly identical by descent; in the second two cases, the alleles are identical by descent only if α_1 and α_2 are identical by descent, and α_1 and α_2 are identical by descent only if individual A is autozygous, which has probability F_A (the inbreeding coefficient of A). Altogether, the required probability for the loop around individual A is $1/4 + 1/4 + (1/4)F_A + (1/4)F_A = 1/2 + (1/2)F_A = (1/2)(1 + F_A)$. Now, since each of the loops in Figure 21 is independent of the others, the total probability of autozygosity in individual I due to this path is $1/2 \times 1/2 \times (1/2)(1 + F_A) \times 1/2 \times 1/2$, or $(1/2)^5(1 + F_A)$. Note that the exponent on the 1/2 is simply the number of individuals in the path. In general, if a path through a common ancestor A contains *i* individuals, the probability of autozygosity due to that path is

$$(1/2)^i(1 + F_A)$$

Thus, the inbreeding coefficient of individual I in Figure 20a is $(1/2)^5(1 + F_A)$, or, assuming $F_A = 0$, simply $(1/2)^5 = 1/32$.

In more complex pedigrees, there is more than a single path. The paths are mutually exclusive, because, if an individual is autozygous due to an allele inherited along one path, the individual cannot at the same time be autozygous due to an allele inherited along a different path. Thus the total inbreeding coefficient is the sum of the probability of autozygosity due to each separate path. The whole procedure for calculating F is summarized in an example involving a first-cousin mating in Figure 22. Here there are two common ancestors (A and B) and two paths (one each through A and B). The total inbreeding coefficient of I is the sum of the two separate contributions shown in Figure 22. If A and B are noninbred, then $F_A = F_B = 0$, and $F_I = (1/2)^5 + (1/2)^5 = 1/16$. This is the probability that I is autozygous at the specified locus. Alternatively, F_I can be interpreted as the average proportion of all genes in I in which the alleles are autozygous.

In general, for any autosomal gene, the formula for calculating the inbreeding coefficient F_I of an individual I is

$$F_I = \Sigma(1/2)^i(1 + F_A) \tag{1.20}$$

where the summation Σ is over all possible paths through all common ancestors, i is the number of individuals in each path, and A is the common ancestor in each path.

The procedure for calculating F in Equation 1.20 is valid only for autosomal (i.e., non-X-linked) genes. An analogous sort of procedure does exist for calculating F for X-linked genes (Hartl 1980). You might find it an interesting challenge to consider how the method should be revised to apply to X linkage.

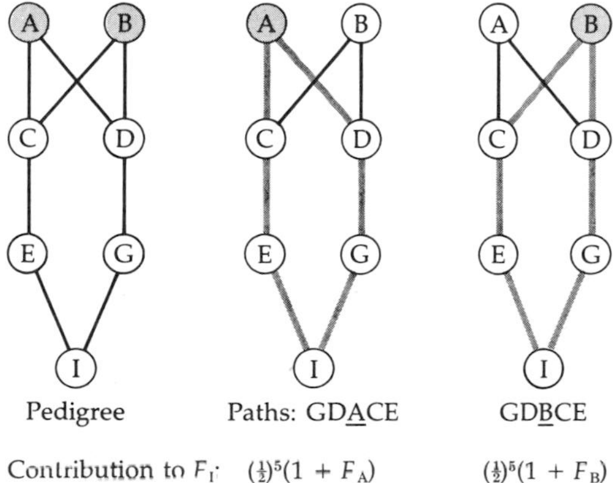

Figure 22. On the left is a pedigree of an individual I, the child of a first-cousin mating. On the right are the two paths through common ancestors (heavy lines) involved in calculating the inbreeding coefficient of I. Below each path is the contribution to F_I due to that path, calculated as in Figure 21. Each path is mutually exclusive of the others, so their probabilities add. Thus, the total inbreeding coefficient of I is the sum of the two separate contributions.

PROBLEM 24 The pedigree shown here depicts two generations of brother–sister mating. Calculate the inbreeding coefficient of individual I, assuming that none of the common ancestors is inbred. (Hint: There are four common ancestors and six paths altogether.)

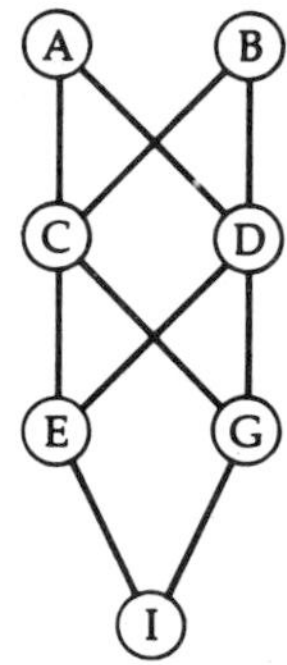

ANSWER Paths are GCE, GDE, GDACE, GCADE, GDBCE, and GCBDE. Thus, using Equation 1.20, $F_I = (1/2)^3(1 + F_C) + (1/2)^3(1 + F_D) + (1/2)^5(1 + F_A) + (1/2)^5(1 + F_A) + (1/2)^5(1 + F_B) + (1/2)^5(1 + F_B)$. When the common ancestors are assumed to be noninbred, $F_A = F_B = F_C = F_D = 0$, and then $F_I = 3/8$.

Regular Systems of Mating

In plant and animal breeding, it is often useful to know how rapidly the inbreeding coefficient increases when a strain is propagated by a regular system of mating, such as repeated self-fertilization, sib mating, or backcrossing to a standard strain. The reasoning involved in calculating the inbreeding coefficient for any generation is illustrated in Figure 23 for repeated self-fertilization. In this figure, the labels $t - 1$ and t refer to the individuals after $t - 1$ and t generations of self-fertilization. The loop around the individual in generation $t - 1$ designates the probability that the two indicated alleles are identical by descent. Here the formula in Equation 1.20 applies with only one path and only one individual in that path, so $F_t = (1/2)^1(1 + F_{t-1})$, where F_t is the inbreeding coefficient in generation t. This equation is easy to solve in terms of the quantity $1 - F_t$, which is often called the **panmictic index**. (**Panmixia** is a synonym for random mating.) Multiplying both sides of the equation for F_t by -1 and then adding $+1$ to each side leads to $1 - F_t = 1 - (1/2)(1 + F_{t-1}) = 1 - (1/2) - (1/2)F_{t-1} = (1/2)(1 - F_{t-1})$, or

$$1 - F_t = (1/2)^t(1 - F_0) \qquad \textbf{(1.21)}$$

where F_0 is the inbreeding coefficient in the initial generation. Self-fertilization therefore leads to an extremely rapid increase in the inbreeding coefficient. When $F_0 = 0$, then $F_1 = 1/2$, $F_2 = 3/4$, $F_3 = 7/8$, $F_4 = 15/16$, and so on.

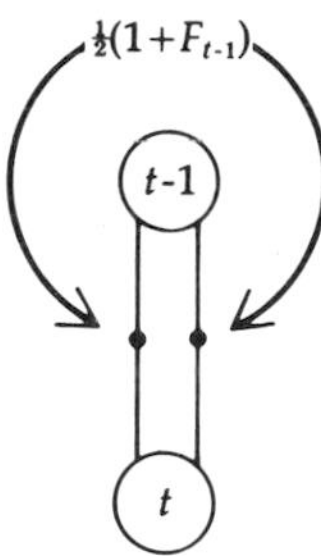

Figure 23. Increase of F resulting from continued self-fertilization. The individual in generation t is the offspring of self-fertilization of the individual in generation $t - 1$. The loop shows that $F_t = (1/2)(1 + F_{t-1})$.

How F increases when $F_0 = 0$ under self-fertilization and several other regular systems of mating is shown in Figure 24.

Many plants reproduce predominantly by self-fertilization, including crop plants such as soybeans, sorghum, barley and wheat. As expected of highly self-fertilizing species, each individual plant is highly homozygous for alleles such as those determining allozymes. Yet the proportion of polymorphic genes is comparable to that found in outcrossing species; self-fertilization simply reorganizes the genetic variation into homozygous genotypes. On the other hand, self-fertilizing species contain fewer deleterious recessives than do outcrossing species, presumably because the increased homozygosity permits harmful recessives to be eliminated from the population by natural selection. One other important point about naturally self-fertilizing species: the high homozygosity of all genes implies that recombination rarely results

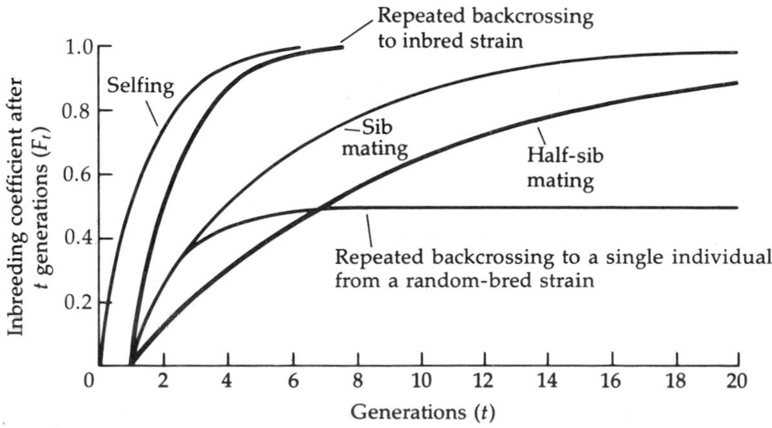

Figure 24. Theoretical increase in inbreeding coefficient F for regular systems of mating: selfing, sib mating, half-sib mating, and repeated backcrossing.

in new gametic types not already present in the parent. Predominance of selfing in a species has the effect of retarding the approach to linkage equilibrium, the reason being that the approach to linkage equilibrium is through recombination in double heterozygotes ($A\ B/a\ b$ and $A\ b/a\ B$ in the case of two alleles at each locus), and with extreme inbreeding, such double heterozygotes are rare. Indeed, the most extreme examples of linkage disequilibrium have been found in predominantly self-fertilizing species such as barley (*Hordeum vulgare*) and wild oats (*Avena barbata*).

Clegg et al. (1972) provide an example of extreme linkage disequilibrium between two unlinked ($r = 1/2$) esterase genes in *Hordeum vulgare*, a species that regularly undergoes more than 99 percent self-fertilization. The population had originated as a complex cross and had been maintained for 26 generations under normal agricultural conditions without conscious selection. At the esterase-B locus, they distinguished two alleles, B_1 and B_2; at the esterase-D locus, they distinguished two alleles, D_1 and D_2. They observed the following numbers of each of the gametic types, although for practical purposes these numbers also refer to homozygous genotypes because there is such close inbreeding.

$B_1\ D_1$	1501	(1642.6)
$B_1\ D_2$	754	(613.7)
$B_2\ D_1$	720	(577.1)
$B_2\ D_2$	74	(215.6)

(The numbers in parentheses are the expected numbers based on the assumption of linkage equilibrium, calculated exactly as for the MNSs example given earlier.) The χ^2 value in this case is 172.7 with one degree of freedom. The associated probability is much less than 0.0001, so there certainly is significant linkage disequilibrium. For the above data, the linkage disequilibrium parameter (Equation 1.12) is $D = -0.046$, which is about 66 percent of its theoretical minimum (Equation 1.14).

One of the dramatic successes of plant breeding has come from the crossing of inbred lines to produce high-yielding hybrid corn. Yield of a genetically heterogeneous, outcrossing variety of corn can be improved by selecting the plants with the highest yields in each generation to be the progenitors of the next generation; such artificial selection results in only gradual improvement, however (see Chapter 4). If a large number of self-fertilized lines are established from a heterogeneous population, each line declines in yield as inbreeding proceeds, due to the forced homozygosity of deleterious recessives. Many lines become so inferior that they have to be discontinued. Self-fertilized lines are not likely to become homozygous for exactly the same set of deleterious recessives, however, so when different lines are crossed to produce a hybrid, the hybrid becomes heterozygous for these genes. Alleles favoring high yield in corn are generally dominant, and there may also be genes in which the heterozygous genotypes have a more favorable effect on yield than do the homozygous genotypes; in any case, the hybrid has a much higher yield than either inbred parent. The phenomenon of enhanced hybrid performance is called **hybrid vigor** or **heterosis**. In practice, inbred lines are crossed in many combinations to identify those that produce the best hybrids. Yields of hybrid corn are typically 15 to 35 percent greater than yields of outcrossing varieties, and the successful introduction of hybrid corn has been remarkable. Virtually all corn acreage in the United States today is planted with hybrids, as compared to 0.4 percent of the acreage in 1933 (Sprague 1978).

SUMMARY

The focus of this chapter is on the detection of genetic variation and the organization of genetic variation in natural populations. Visible genetic variation affecting traits such as height or weight is widespread, but these traits are influenced by the alleles of many genes as well as by the effects of environment, so phenotypic differences for such traits can rarely be traced to the effects of particular genes. Phenotypic variation that can be traced to the effects of particular genes does occur, but it is relatively uncommon; only a few percent of the individuals in most natural populations exhibit such simply inherited phenotypic variation.

Although visible genetic variation is relatively rare, hidden genetic variation is extremely common, so much so that no two individuals in a population are likely to have the same genotype at all loci (unless they are identical twins). Hidden genetic variation can be detected by means of protein electrophoresis, and the various electrophoretic forms of a protein that can occur are known as allozymes. In most outcrossing populations, the proportion of polymorphic genes ranges between 15 and 40 percent, and a typical individual is heterozygous at between 4 and 14 percent of its genes.

Hidden genetic variation at the level of DNA sequences can be detected by direct DNA sequencing of cloned genes from different individuals. Variation in the presence or absence of cleavage sites for restriction enzymes, or in the occurrence of insertions or deletions of DNA sequences, can be determined by means of Southern transfer experiments, in which DNA restriction fragments separated by electrophoresis are transferred and attached to a sheet of membrane and mixed with radioactive probe DNA which combines with the sample sequences of interest.

Allozyme and DNA polymorphisms are useful as genetic markers in studies of natural populations, in inferring the ancestral history of populations or in assessing genetic relationships among subpopulations in a species, and in determining types of mating systems. They are also important in human population genetics as potential genetic markers linked in the chromosome with genes that cause inherited diseases.

The effects of some types of hidden genetic variation can be studied by inbreeding (mating between relatives), which increases the proportion of homozygous genotypes at the expense of heterozygous genotypes and thus permits rare recessive alleles to become homozygous. Much of the genetic variation revealed by inbreeding is harmful in causing death or severe physical or mental disability.

The association of genes into genotypes is determined largely by the mating system. For a population undergoing random mating for a gene with two alleles A and a at respective frequencies p and q ($p + q = 1$), the Hardy–Weinberg proportions p^2, $2pq$, and q^2 give the expected genotype frequencies of AA, Aa, and aa, respectively. When there is inbreeding, these proportions become modified to $p^2(1 - F) + pF$, $2pq(1 - F)$, and $q^2(1 - F) + qF$, respectively, where F is the inbreeding coefficient and represents the probability that the two homologous alleles in an individual are identical by descent (autozygous).

Nonrandom association between the alleles of different genes is measured by the linkage disequilibrium parameter D. Ordinarily, D is expected to go to zero (indicating random association, also called linkage equilibrium) at a rate determined by the recombination fraction between the genes. With unlinked genes, the amount of linkage disequilibrium decreases by one-half in each generation. Significant linkage disequilibrium is not usually found in natural populations unless the genes are tightly linked, unless they are within or near an inverted segment of chromosome, or unless they occur in a plant species that regularly undergoes self-fertilization.

The mating system determines the organization of genes into genotypes. By itself, the mating system does not change allele frequency. The next chapter discusses the major forces that do change allele frequency from one generation to the next.

FURTHER READINGS

Ayala, F. J. 1982. *Evolutionary and Population Genetics*. Benjamin-Cummings, Menlo Park, California.

Bodmer, W. F. and J. F. Kingman (eds.). 1984. *Mathematical Genetics*. The Royal Society Press, London.

Cavalli-Sforza, L. L. and W. F. Bodmer. 1971. *The Genetics of Human Populations*. W. H. Freeman, San Francisco.

Chakravarti, A. 1984. *Human Population Genetics*. Van Nostrand Reinhold, New York.

Cook, L. M. 1976. *Population Genetics*. Halsted Press, New York.

Crow, J. F. 1986. *Basic Concepts in Population, Quantitative and Evolutionary Genetics.* W. H. Freeman, San Francisco.

Crow, J. F. and M. Kimura. 1970. *An Introduction to Population Genetics Theory.* Harper & Row, New York.

Dobzhansky, Th. 1970. *Genetics of the Evolutionary Process.* Columbia University Press, New York.

Dobzhansky, Th. 1982. *Dobzhansky's Genetics of Natural Populations: I–XLIII.* R. C. Lewontin et al. (eds.). Columbia University Press, New York.

Elandt-Johnson, R. C. 1971. *Probability Models and Statistical Methods in Genetics.* John Wiley and Sons, New York.

Ennos, R. A. 1983. Maintenance of genetic variation in plant populations. *Evol. Biol.* 16: 129–156.

Ewens, W. J. 1979. *Mathematical Population Genetics.* Springer-Verlag, New York.

Feldman, M. and F. B. Christiansen. 1985. *Population Genetics.* Blackwell, Oxford.

Felsenstein, J. 1981. *Bibliography of Theoretical Population Genetics.* Van Nostrand Reinhold, New York.

Gale, J. S. 1980. *Population Genetics.* Halsted Press, New York.

Hartl, D. L. 1980. *Principles of Population Genetics.* Sinauer Associates, Sunderland, Massachusetts.

Hayes, H. K. 1963. *A Professor's Story of Hybrid Corn.* Burgess, Minneapolis.

Hedrick, P. W. 1983. *Genetics of Populations.* Jones & Bartlett, Portolla Valley, California.

Lewontin, R. C. 1974. *The Genetic Basis of Evolutionary Change.* Columbia University Press, New York.

Li, C. C. 1976. *First Course in Population Genetics.* Boxwood Press, Pacific Grove, California.

Maynard Smith, J. 1982. *Evolution Now: A Century After Darwin.* W. H. Freeman, San Francisco.

Mourant, A. E. 1983. *Blood Relations: Blood Groups and Anthropology.* Oxford University Press, Oxford.

Provine, W. B. 1971. *The Origins of Theoretical Population Genetics.* University of Chicago Press, Chicago.

Provine, W. B. 1986. *Sewall Wright and Evolutionary Biology.* University of Chicago Press, Chicago.

Smith, M. H. and J. Joule (eds.). 1980. *Mammalian Population Genetics.* University of Georgia Press, Athens, Georgia.

Spiess, E. B. 1977. *Genes in Populations.* John Wiley and Sons, New York.

Sprague, G. F. 1978. Introductory remarks to the session on the history of hybrid corn. *In* D. B. Walden (ed.), *Maize Breeding and Genetics,* John Wiley and Sons, New York, pp. 11–12.

Stebbins, G. L. 1976. Chromosome, DNA and plant evolution. *Evol. Biol.* 9: 1–34.

Vogel, F. and A. G. Motulsky. 1986. *Human Genetics,* 2nd ed. Springer-Verlag, New York.

Wallace, B. 1981. *Basic Population Genetics.* Columbia University Press, New York.

Wright, S. 1968. *Evolution and the Genetics of Populations,* Vol. 1: *Genetic and Biometric Foundations.* University of Chicago Press, Chicago.

Wright, S. 1969. *Evolution and the Genetics of Populations,* Vol. 2: *The Theory of Gene Frequencies.* University of Chicago Press, Chicago.

CAUSES OF EVOLUTION 2

Populations are extremely complex entities. The Hardy–Weinberg model discussed in Chapter 1, useful as it is as a first approximation, ignores most of the complexities of actual populations. Populations are not infinitely large, and their sizes are rarely constant, so fluctuations in allele frequency can occur by chance. Populations are also subject to the systematic evolutionary forces of migration, mutation, and natural selection, all of which cause nonrandom or directional changes in allele frequency. Changes in allele frequency are changes in the genetic makeup of a population. Since **evolution** may be defined as cumulative change in the genetic makeup of a population resulting in increased adaptation to the environment, the fundamental process in evolution is change in allele frequency. In this chapter we focus on the various forces that change allele frequency. These forces hold the key to understanding the origin and maintenance of genetic variation.

RANDOM GENETIC DRIFT

Random genetic drift refers to chance fluctuations in allele frequency which occur, particularly in small populations, as a result of random sampling among gametes. To be specific, a population of nine diploid organisms arises from a sample of just 18 gametes out of an essentially infinite pool of gametes. Because small samples are frequently not representative, an allele frequency in the sample may differ from that in the entire pool of gametes. In fact, if the number of gametes in a sample is represented as $2N$ (in the above example, $2N = 18$), the probability that the sample contains exactly 0, 1, 2, . . . , $2N$ alleles of type A is given by the successive terms in the expansions of $(pA + qa)^{2N}$, where p and q are the respective allele frequencies of A and a in the parental generation, and $p + q = 1$. The probability that the sample contains exactly i alleles of type A can be shown to equal

$$\frac{(2N)!}{i!(2N - i)!} p^i q^{2N-i} \qquad \textbf{(2.1)}$$

Equation 2.1 is often called the **binomial probability**, and the symbol $i!$ (**i factorial**) represents the product of all integers up to and including i; that is, $i! = i(i - 1)(i - 2) \ldots (3)(2)(1)$. Similarly, $(2N)! = (2N)(2N - 1)(2N - 2) \ldots (3)(2)(1)$. (For the sake of generality, the number 0! is defined to equal 1.)

To consider a concrete example, suppose a monoecious population of size $N = 9$ contains exactly 16 A alleles. Then $p = 16/(2N) = 16/18 = 8/9$, and $q = 1 - p = 1/9$. In the next generation, the probability that A becomes **fixed** (i.e., $p = 1$) is given by Equation 2.1 with $i = 18$, namely

$$\frac{18!}{18!0!} (8/9)^{18}(1/9)^0 = 0.12$$

Similarly, the probability that A is **lost** (i.e., that a becomes fixed) is given by Equation 2.1 with $i = 0$. This probability is very much smaller, namely

$$\frac{18!}{0!18!} (8/9)^0(1/9)^{18} = 6.7 \times 10^{-18}$$

Thus, the probability that the population remains **unfixed** or **segregating** (i.e., neither A nor a becomes fixed) in the next generation equals $1 - 0.12 - (6.7 \times 10^{-18}) = 0.88$.

PROBLEM 1 A college student admires *Phlox cuspidata* and always keeps two plants in her dormitory room, which are grown from two seeds chosen at random from the plants of the previous year. (Because she is careful to ensure that mating is random, including an equal likelihood of cross- and self-fertilization, her method of perpetuation is equivalent to choosing four gametes at random and uniting them to form the two plants of the next generation.) Suppose that in one year her plants contain two Adh^a and two Adh^b alleles (Adh = alcohol dehydrogenase gene). Use Equation 2.1 to calculate the various probabilities that the next year's population contains exactly 0, 1, 2, 3, or 4 Adh^a alleles.

ANSWER Use Equation 2.1 with $N = 2$ (so $2N = 4$) and p the allele frequency of Adh^a. In the initial population, $p = 2/4 = 1/2$, so $q = 1 - p = 1/2$. In Equation 2.1, i represents the number of A (i.e., Adh^a) alleles in the next generation. The probability that $i = 0$ is thus $(4!/0!4!)(1/2)^0(1/2)^4 = 1/16 = 0.0625$; the probability that $i = 1$ is $(4!/1!3!)(1/2)^1(1/2)^3 = 4/16 = 0.25$; the probability that $i = 2$ is $(4!/2!2!)(1/2)^2(1/2)^2 = 6/16 = 0.375$; similarly, the probability that $i = 3$ is 0.25, and the probability that $i = 4$ is 0.0625.

The hypothetical population of *Phlox cuspidata* with $N = 2$ in Problem 1 is very small indeed, yet it is instructive because it typifies many of the important features of random genetic drift. Suppose there were many such populations, each of size $N = 2$. The calculations in Problem 1 imply that, in only one generation of random genetic drift, 6.25 percent of the populations would lose the Adh^a allele, 6.25 percent would become fixed for the Adh^a allele, and the other 87.5 percent would remain segregating. However, among the segregating populations, there is a spread in allele frequency: some populations have $p = 1/4$, some have $p = 1/2$, and some have $p = 3/4$. Of course, random genetic drift continues generation after generation. Each of the unfixed populations produces a spread of allele frequencies in the following generation, including some cases in which the Adh^a allele is newly fixed or newly lost. The fixed populations, whether fixed for Adh^a or Adh^b, remain fixed except in the unlikely event of mutation. Thus, as time goes on, a smaller and smaller fraction of the populations continues to segregate both alleles.

PROBLEM 2

For the population in Problem 1 in which $p = 1/2$, what is the probability that the population is still segregating after two generations of random genetic drift? (Hint: Use the proportions calculated in Problem 1, along with Equation 2.1, to calculate the probability that A is fixed or lost in two generations; then determine the probability that the population is still segregating by subtraction.)

ANSWER The allele frequency of A after one generation of random genetic drift is either 0, 1/4, 1/2, 3/4, or 1 with respective probabilities 0.0625, 0.25, 0.375, 0.25 and 0.0625, as determined in Problem 1. When $p = 0$ or $p = 1$ the allele frequency remains 0 or 1, respectively. The probability of loss when $p = 1/4$, 1/2, or 3/4 is $(4!/0!4!) \times (1/4)^0(3/4)^4 = 0.31641$, $(4!/0!4!)(1/2)^0(1/2)^4 = 0.0625$, and $(4!/0!4!)(3/4)^0(1/4)^4 = 0.00391$, respectively. The overall probability of loss in two generations is therefore $(0.0625)(1) + (0.25)(0.31641) + (0.375)(0.0625) + (0.25)(0.00391) + (0.0625)(0) = 0.16602$. In a similar manner, it can be calculated that the probability of fixation in two generations is also 0.16602. Thus, the proportion of populations still expected to be segregating after two generations is $1 - 0.16602 - 0.16602 = 66.8$ percent. This compares with the 87.5 percent still segregating after one generation.

After random genetic drift has continued for a sufficient number of generations, most populations become fixed for one allele or the other. After only four generations of random genetic drift in the population with $N = 2$, for example, the proportion of unfixed populations is down to 38 percent. Moreover, the populations that are still segregating are nearly as likely to have any one allele frequency as any other; in the $N = 2$ example, the ratio of unfixed populations having $p = 1/4$ to those with $p = 1/2$ to those with $p = 3/4$ after four generations is 0.32:0.36:0.32. Fixation continues as time goes on. After 19 generations with $N = 2$, fully 99.6 percent of the populations have become fixed; among the remaining 0.4 percent of unfixed populations, however, the ratio of those having 1, 2, or 3 *A* alleles is still 0.32:0.36:0.32.

The progressive spreading out of allele frequencies due to random genetic drift and the accompanying accumulation of populations in which one or another allele has become fixed also occurs in larger populations, but in larger populations a correspondingly longer time is required for the effects to become pronounced. A computer-generated example based on random numbers is shown in Figure 1. Each small graph gives the number of *A* alleles in 19 successive generations of random genetic drift in a population of size $N = 9$ (so $2N = 18$). As is apparent from the figure, individual populations behave erratically. In seven populations (29 percent), the *A* allele became fixed; in six populations (25 percent), the *A* allele was lost. The remaining 11 populations (46 percent) were still segregating after 19 generations. (Compare this 46 percent with the corresponding value of 0.4 percent for populations with $N = 2$.) Among the segregating populations, however, the final allele frequency was as likely to be one value as any other. The principal conclusion from the example with $N = 2$ is that the overall pattern of allele frequency change due to random genetic drift is predictable—the

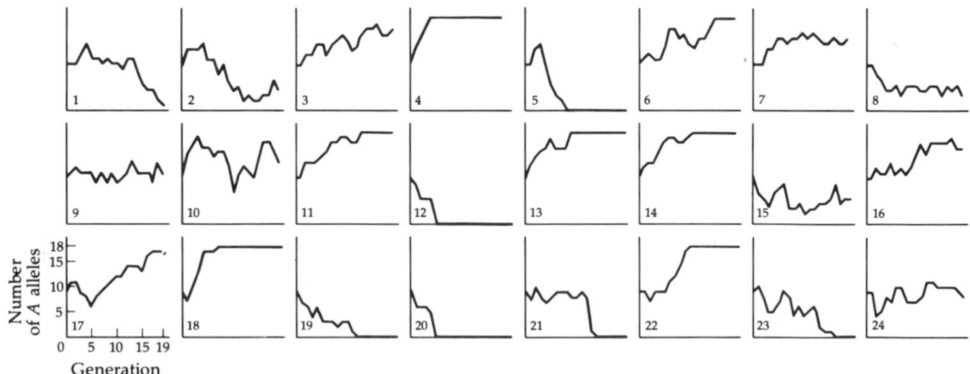

Figure 1. Change of allele frequency by random genetic drift over 19 generations in 24 hypothetical populations of size $N = 9$.

allele frequencies progressively spread out as the proportion of fixed populations steadily increases. The overall conclusion from Figure 1 is that allele frequencies in any single population behave so erratically that prediction of allele frequency change in any one population is virtually impossible.

In view of the unpredictability of changes in allele frequency in any single population, the effects of random genetic drift on allele frequency are often discussed in terms of the expected changes in allele frequency when averaged over a very large number of populations. A framework for thinking about the problem from this perspective is illustrated in the model in Figure 2. Here an infinitely large initial population is imagined to be split up into a large number of local populations or subpopulations, each of size N, which

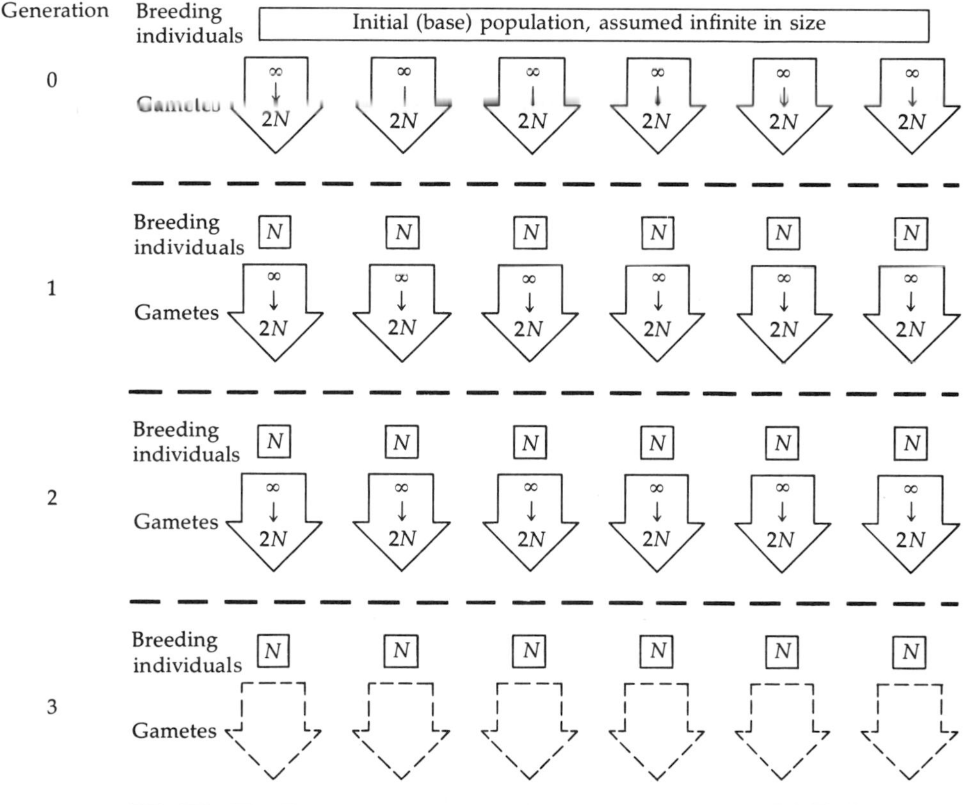

Figure 2. Model for analyzing effects of random genetic drift. Each of the subpopulations founded from the initial large population (shown by vertical columns of boxes and arrows) is assumed to be genetically isolated from the others. Each subpopulation produces an infinite number of gametes, of which $2N$ are chosen at random to form the next generation's breeding population. Random genetic drift results from sampling error in this process.

correspond to the populations in Figure 1. (As will be explained shortly, N is technically the **effective population number**.) The subpopulations are assumed to satisfy the Hardy–Weinberg assumptions of diploid organism, sexual reproduction, nonoverlapping generations, no mutation, and no selection. In addition, we assume that

1. Many independent subpopulations exist, each of constant size N.
2. Random mating occurs within each subpopulation.
3. No migration occurs among the subpopulations.
4. Each subpopulation has an equal number of males and females.
5. Every individual has an equal chance of contributing successful gametes to the next generation.
6. The number of successful gametes per individual conforms to a Poisson (random) distribution. [With this assumption, the probability that an individual contributes i successful gametes, when the mean number per individual is λ, is given by $(\lambda^i/i!)\exp(-\lambda)$.]

In short, the subpopulations in Figure 2 are theoretically uncomplicated or *ideal*, and reference will be made to such ideal populations throughout this chapter. The importance of the concept of ideal populations is that they provide a standard of comparison for other populations that violate the above assumptions.

Results of the kind of population structure in Figure 2 are shown for an actual case in Figure 3, which records the history of 19 generations of random genetic drift in 107 subpopulations of *Drosophila melanogaster* (Buri 1956). Each population was initiated with 16 bw^{75}/bw heterozygotes (bw = *brown* eyes) and maintained at a constant size of 16 by randomly choosing 8 males and 8 females as the breeding population in each generation. The horizontal axis in Figure 3 gives the number of populations having 0, 1, 2, . . . , 32 bw^{75} alleles. The overall pattern of change in allele frequency in Figure 3 is apparent: the initially humped distribution of allele frequency (in the background) gradually becomes flat (foreground) as populations fixed for bw^{75} or bw begin to pile up. By about 18 generations, half the populations are fixed for one allele or the other, and among the unfixed populations, the distribution of allele frequencies is essentially flat.

Consequences of Random Genetic Drift

The observed pattern of change in allele frequency shown in Figure 3 is very nearly that expected theoretically for an ideal population. Although the full-blown theory of random genetic drift is conceptually simple (it involves mathematical methods for the repeated application of the binomial formula to each subpopulation, as was illustrated for the case N = 2 in Problems 1

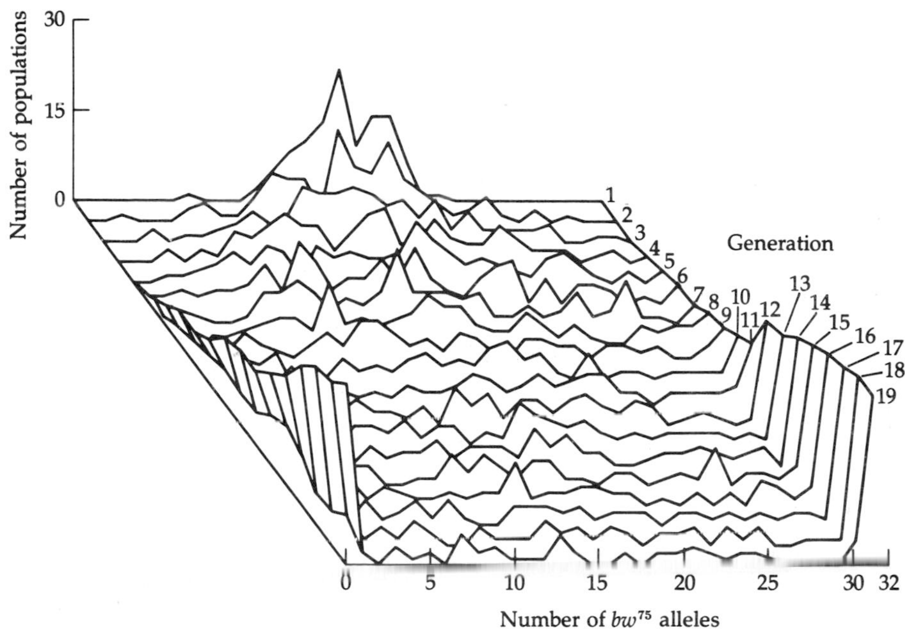

Figure 3. Random genetic drift in 107 actual populations of *Drosophila melanogaster*. Each of the initial 107 populations consisted of 16 bw^{75}/bw heterozygotes (N = 16; *bw* = *brown* eyes). From among the progeny in each generation, eight males and eight females were chosen at random to be the parents of the next generation. The horizontal axis of each curve gives the number of bw^{75} alleles in the population, and the vertical axis gives the corresponding number of populations. (Data from Buri 1956.)

and 2), the actual calculations are formidable (see, for example, Crow and Kimura, 1970, and Kimura and Ohta, 1971). Some results of the theory are summarized in the two families of curves in Figure 4, which represent the theoretical distributions of allele frequency among unfixed (segregating) populations after various times (t) measured in units of N generations. In Figure 4(a), all populations have an initial allele frequency of 0.5 (as in the actual populations in Figure 3); after about $t = 2N$ generations, the distribution of allele frequency is essentially flat, and by this time about half the populations are still unfixed. The distributions in Figure 4 refer only to those populations that are unfixed. As time goes on, more and more of the populations do become fixed, and thus the distributions progressively pile up at 0 and 1, as demonstrated in Figure 3. Indeed, in Figure 4, the area under each curve is equal to the proportion of unfixed populations, and this area becomes progressively smaller as time goes on. Figure 4(b) shows what happens when the initial allele frequency is 0.1; here the distributions are highly asymmetrical, and the distribution of allele frequency does not become flat until about

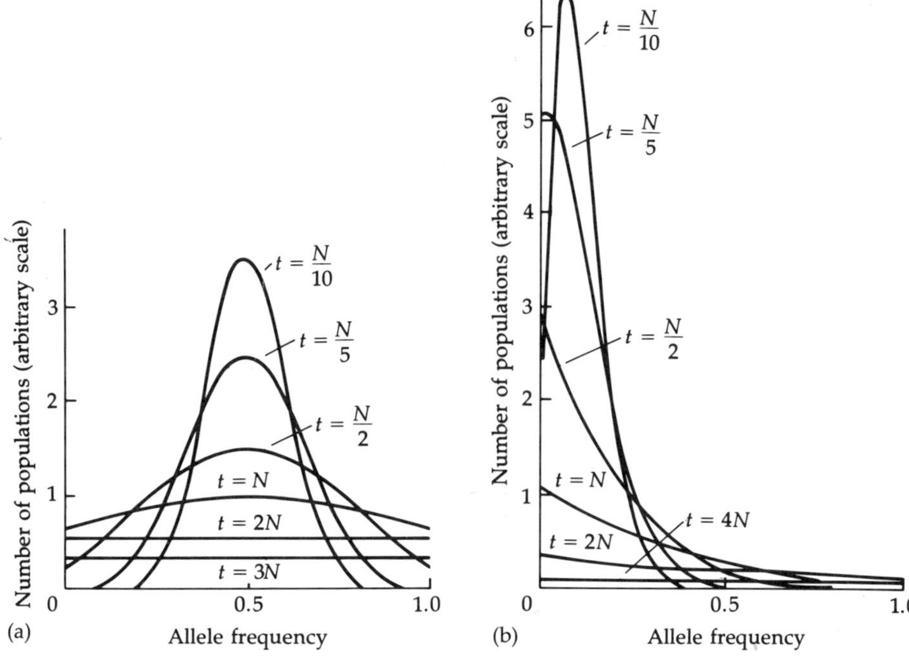

Figure 4. Theoretical results of random genetic drift. (a) Initial allele frequency = 0.5. (b) Initial allele frequency = 0.1. The curves have been scaled so that the area under each curve is equal to the proportion of populations in which fixation or loss has not yet occurred. The curves are therefore the distributions of allele frequencies among segregating populations. (From Kimura 1955.)

$t = 4N$ generations, by which time only about 10 percent of the populations remain unfixed. Once a flat distribution of allele frequency is reached, the distribution remains flat, but random drift continues until fixation or loss has occurred in all populations. The average time required for fixation or loss to occur depends on the initial allele frequency (p_0). When $p_0 = 0.5$, the average time that a population remains segregating is $2.8N$ generations, and when $p_0 = 0.1$, the average time is $1.3N$ generations. Thus, populations remain segregating longer when they are large or when the allele frequency is near 0.5.

Many of the important consequences of random genetic drift follow from the fact that the population structure in Figure 2 entails a peculiar sort of "inbreeding," but the nature of the inbreeding is very subtle because the population structure in Figure 2 can be considered on two distinct levels—the level of the individual subpopulation and the level of the total population (i.e., the aggregate of all subpopulations).

Consider first any one of the subpopulations in Figure 2. Within this subpopulation (call it subpopulation number i), mating is random because of assumption (2) in the earlier list. If the allele frequencies of A and a in the ith subpopulation are denoted p_i and q_i, then the genotype frequencies of AA, Aa, and aa are given by the familiar Hardy–Weinberg principle as p_i^2, $2p_iq_i$, and q_i^2, respectively. Furthermore, think about the situation in Figure 2 at a time so advanced that all subpopulations have become fixed for one allele or the other. When the ith subpopulation is fixed, either p_i equals 0 or p_i equals 1. The genotype frequencies of AA, Aa, and aa in the subpopulation are either 0, 0, and 1 (if $p_i = 0$), or 1, 0, and 0 (if $p_i = 1$). These genotype frequencies, though extreme, still satisfy the Hardy–Weinberg principle. Thus, within any one subpopulation in Figure 2, the frequency of heterozygotes equals the frequency expected with random mating.

The situation regarding the total population in Figure 2 is very different, however, as there is an overall deficiency of heterozygotes. The meaning of "total population" in this context can be made clear by considering a simple example. Suppose the subpopulations in question were colonies of mice in a large barn. (Were the mice real, we would have to worry about migration between subpopulations, but these are hypothetical mice which stay where they are put.) Suppose further that we are unaware of the existence of such subpopulations but instead think that the barn contains a single randomly mating population. To study the "total population" of the barn, we trap mice at random. If time is so advanced that a fraction p of the subpopulations are fixed for A and a fraction q are fixed for a (remember that $q = 1 - p$), then a fraction p of the time, we would trap an AA mouse and a fraction q of the time, an aa mouse. Since the overall allele frequency of A among the trapped animals is p, we would naively expect a fraction $2pq$ of the animals to be heterozygous. In fact, we would have caught no heterozygotes at all!

This rather paradoxical result—that there is a deficiency of heterozygotes in the total population even though random mating occurs within each subpopulation—is a consequence of the random genetic drift of allele frequencies among subpopulations due to their finite size. Were the subpopulations so large that random drift could be ignored, each subpopulation would have the same allele frequencies and Hardy–Weinberg genotype frequencies as any other. In such a case, to return to the mouse example, it would not affect the proportion of heterozygotes if mice were sampled from the total population or from any one of the subpopulations. For the model in Figure 2, however, random genetic drift is important. The total population has a deficiency of heterozygotes, much as if there were inbreeding. It is this inbreeding-like effect of population subdivision that we now set out to quantify.

Levels of Population Structure

Because population subdivision entails an inbreeding-like effect, it is convenient to measure the effect in terms of decrease in the proportion of heterozygous genotypes, much as the effects of nonrandom mating were measured in Chapter 1. A subdivided population has three distinct levels of complexity, however—individual organisms (*I*), subpopulations (*S*), and the total population (*T*). To avoid potential confusion of these various levels, it is useful to adopt the terminology below. We will also allow the possibility of nonrandom mating within subpopulations, so assumption (2) in the earlier list can be violated.

H_I = the heterozygosity of an *individual* in a subpopulation
H_S = the expected heterozygosity of an individual in an equivalent random mating *subpopulation* **(2.2)**
H_T = the expected heterozygosity of an individual in an equivalent random mating *total* population

As noted in Chapter 1, H_I can be interpreted as the average heterozygosity of all the genes in an individual or as the probability of heterozygosity of any one gene. H_S represents the level of heterozygosity that would be found in a subpopulation if the subpopulation were undergoing random mating; therefore, H_S always equals $2p_iq_i$ for a subpopulation with allele frequency p_i. The quantity H_T represents what the heterozygosity would be were all subpopulations pooled together and mated randomly; if the average allele frequency among subpopulations is denoted p_0, then $H_T = 2p_0q_0$.

The **inbreeding coefficient** defined in Chapter 1 measures the reduction in heterozygosity of an individual due to nonrandom mating within its subpopulation. To avoid confusion, we will call this inbreeding coefficient F_{IS} (rather than simply F as in Chapter 1). Thus,

$$F_{IS} = \frac{H_S - H_I}{H_S} \qquad \textbf{(2.3)}$$

The effects of population subdivision are measured by a quantity called the **fixation index** (symbolized F_{ST}), which is the reduction in heterozygosity of a subpopulation due to random genetic drift. Thus,

$$F_{ST} = \frac{H_T - H_S}{H_T} \qquad \textbf{(2.4)}$$

The overall inbreeding coefficient of an individual (denoted F_{IT}) includes a contribution due to actual nonrandom mating within subpopulations (F_{IS}) and another contribution due to the subdivision itself (F_{ST}). The quantity F_{IT}

measures the reduction in heterozygosity of an individual relative to the total population, so

$$F_{IT} = \frac{H_T - H_I}{H_T} \tag{2.5}$$

The **hierarchical *F*-statistics** defined in Equations 2.3, 2.4, and 2.5 are all types of "inbreeding coefficients," but they differ in the reference populations. F_{IS} is concerned with inbreeding in individuals (*I*), relative to the subpopulation (*S*) to which they belong; F_{ST} is concerned with inbreeding in subpopulations (*S*), relative to the total population (*T*) of which they are a part; and F_{IT} is concerned with inbreeding in individuals (*I*), relative to the total population (*T*). F_{IT} is the most inclusive measure of inbreeding in that it takes into account both the effects of nonrandom mating within subpopulations (F_{IS}) and the effects of population subdivision (F_{ST}). The mathematical relation between the three types of inbreeding coefficient is demonstrated in the following problem.

PROBLEM 3

Show that F_{IS}, F_{ST}, and F_{IT} are related by the equation $(1 - F_{IS})(1 - F_{ST}) = (1 - F_{IT})$.

ANSWER From Equation 2.3, $F_{IS} = 1 - (H_I/H_S)$, or $1 - F_{IS} = H_I/H_S$. Equation 2.4 implies that $F_{ST} = 1 - (H_S/H_T)$, or $1 - F_{ST} = H_S/H_T$. And Equation 2.5 implies that $F_{IT} = 1 - (H_I/H_T)$, or $1 - F_{IT} = H_I/H_T$. Now multiply the expressions for $1 - F_{IS}$ and $1 - F_{ST}$ together to obtain $(1 - F_{IS})(1 - F_{ST}) = (H_I/H_S)(H_S/H_T) = H_I/H_T = (1 - F_{IT})$.

Equations 2.3–2.5 can be applied to data of Levin (1978) on natural subpopulations of *Phlox cuspidata* in Texas. Altogether, 43 subpopulations were studied with respect to the phosphoglucomutase-2 (*Pgm*−2) gene, and two alleles ($Pgm{-}2^a$ and $Pgm{-}2^b$) were found. In 40 subpopulations, the $Pgm{-}2^b$ allele was fixed; in the remaining three subpopulations, the allele frequencies of $Pgm{-}2^b$ were estimated as 0.49, 0.83, and 0.91, respectively. In the three polymorphic populations, the observed frequencies of heterozygotes were 0.17, 0.06, and 0.06, respectively.

The individual heterozygosity (H_I) can be estimated from the observed data as the average heterozygosity of individuals within subpopulations; thus $\langle H_I \rangle = [(40)(0) + 0.17 + 0.06 + 0.06]/43 = 0.0067$, where the first term comes from the 40 subpopulations in which the observed heterozygosity was 0. The subpopulation heterozygosity (H_S) can be estimated from the estimated allele frequencies within subpopulations and the expected frequency of heterozygotes under the supposition of random mating, thus $\langle H_S \rangle = [(40)(0) + 2(0.49)(0.51) + 2(0.83)(0.17) + 2(0.91)(0.09)]/43 = 0.0220$, where again the first term comes from the 40 fixed subpopulations. Finally, the total heterozygosity (H_T) can be estimated from the average allele frequency in the combined subpopulations and the Hardy–Weinberg expectation of heterozygosity with random mating; the average allele frequency of the $Pgm-2^b$ allele is $[(40)(1) + 0.49 + 0.83 + 0.91]/43 = 0.9821$, so that $\langle H_T \rangle = 2(0.9821)(0.0179) = 0.0352$.

Now the various F-statistics can be estimated from Equations 2.3–2.5: $\langle F_{IS} \rangle = (\langle H_S \rangle - \langle H_I \rangle)/\langle H_S \rangle = (0.0220 - 0.0067)/(0.0220) = 0.70$; this is the estimated inbreeding coefficient due to nonrandom mating within subpopulations, and note that it is in reasonable agreement with the value calculated in Problem 21 in Chapter 1 based on the fraction of self-fertilization in this species. The fixation index F_{ST} is estimated as $\langle F_{ST} \rangle = (\langle H_T \rangle - \langle H_S \rangle)/\langle H_T \rangle = (0.0352 - 0.0220)/(0.0352) = 0.38$; this is the amount of inbreeding due solely to population subdivision. Finally, the total inbreeding coefficient is estimated as $\langle F_{IT} \rangle = (\langle H_T \rangle - \langle H_I \rangle)/\langle H_T \rangle = (0.0352 - 0.0067)/(0.0352) = 0.81$; this is the amount of inbreeding due to the combined effects of nonrandom mating within subpopulations and to random genetic drift among subpopulations. Note that the total inbreeding coefficient is quite large (as expected from the substantial fraction of self-fertilization), but that an appreciable fraction of the total inbreeding is attributable to subdivision and random genetic drift.

The method of estimating the F-statistics by replacing the parameters in Equations 2.3–2.5 with their observed or estimated values is not necessarily the best, particularly with small samples. Ideally, estimates of the F-statistics should correct for the effects of sampling a limited number of subpopulations, as well as for the effects of sampling a limited number of individuals per subpopulation. Methods for making these corrections have been developed but are complex and raise additional issues. For an excellent discussion, see Weir and Cockerham (1984). Important issues are also addressed in Wright (1978, p. 86ff.), Curie-Cohen (1982), Nei and Chesser (1983), and Nei (1986). We will use the uncorrected estimation procedure used above, which is adequate for purposes of illustration. When applied to the *Pgm* data from *Phlox cuspidata*, for example, the more sophisticated method of Weir and Cockerham gives $F_{IS} = 0.70$, $F_{ST} = 0.36$ and $F_{IT} = 0.81$.

PROBLEM 4 Levin (1978) also studied the glutamate oxalacetate transaminase-2 (*Got*−2) gene in the 43 subpopulations of *Phlox cuspidata* discussed earlier. Three alleles were found: $Got-2^a$, $Got-2^b$ and $Got-2^c$. A total of 39 subpopulations were monomorphic for $Got-2^b$. One subpopulation contained $Got-2^a$ and $Got-2^b$ with respective allele frequencies 0.37 and 0.63, and the observed frequency of heterozygotes was 0.17. Three subpopulations contained only $Got-2^b$ and $Got-2^c$; the allele frequencies of $Got-2^b$ in these subpopulations were 0.87, 0.91, and 0.82, respectively, and the observed frequencies of heterozygotes were 0.09, 0.06, and 0.09, respectively. Estimate the various heterozygosities and F-statistics for this gene.

ANSWER $\langle H_I \rangle = [(39)(0) + 0.17 + 0.09 + 0.06 + 0.09]/43 = 0.00953$; $\langle H_S \rangle = [(39)(0) + 2(0.37)(0.63) + 2(0.87)(0.13) + 2(0.91)(0.09) + 2(0.82)(0.18)]/43 = 0.02678$. The overall estimated frequency of *Got*-2^a is $[(39)(0) + 0.37 + 3(0)]/43 = 0.0086$, that of *Got*$-2^b$ is $[(39)(1) + 0.63 + 0.87 + 0.91 + 0.82]/43 = 0.9821$, so that of *Got*$-2^c$ is $1 - 0.0086 - 0.9821 = 0.0093$. From the Hardy–Weinberg principle for multiple alleles (Chapter 1), $\langle H_T \rangle = 2(0.0086)(0.9821) + 2(0.0086)(0.0093) + 2(0.9821)(0.0093) = 0.03532$. Thus $\langle F_{IS} \rangle = 0.64$, $\langle F_{ST} \rangle = 0.24$, and $\langle F_{IT} \rangle = 0.73$. As you may have noted, the *F*-statistics vary somewhat from gene to gene; the range of $\langle F_{IS} \rangle$ for five genes studied in *Phlox cuspidata* was 0.60 to 0.77 (average 0.67), and that of $\langle F_{ST} \rangle$ was 0.17 to 0.86 (average 0.41).

Except for plants that have a high frequency of self-fertilization or for certain insects that regularly undergo parent–offspring or brother–sister mating, values of F_{IS} in most natural populations are typically close to zero, which indicates random mating within subpopulations. (Within very small subpopulations, there may be mating between relatives—but as long as the mating between relatives occurs no more often than would be expected by chance in a subpopulation of that size, F_{IS} remains zero.) For example, Crow and Mange (1965) studied a South Dakota population of Hutterites, a group that is relatively isolated from surrounding communities because of their religious beliefs. In this population, F_{ST} was estimated as 0.04 (a relatively large value for a human population), yet F_{IS} was essentially zero. For many natural populations, therefore, particularly in animals or outcrossing plants, it is reasonable to assume $F_{IS} = 0$. This assumption is equivalent to the assumption of random mating within subpopulations (assumption 2 in the earlier list), which puts us back in the context of the model outlined in Figure 2. Furthermore, when $F_{IS} = 0$, then $F_{ST} = F_{IT}$, and there is no longer need for the subscripts. For the sake of typographical convenience, therefore, the symbol F will hereafter be used in place of F_{ST} unless otherwise specified. In some contexts, F_{ST} is denoted G_{ST} (Nei 1975, p. 151; Crow and Aoki 1984).

Increase in Fixation Index from Random Genetic Drift

As was true of the inbreeding coefficient F_{IS} in Chapter 1, the fixation index F_{ST} (or, as we are now calling it, simply F) can be interpreted in terms of identity by descent. Indeed, F is the probability that two alleles chosen at random from within the same subpopulation are identical by descent. As random genetic drift continues generation after generation, the value of F will change, so it will be necessary to use the symbol F_t to represent the average value of the fixation index F_{ST} among subpopulations in generation t. The value of F_t can be calculated from the reasoning illustrated in Figure 5. This figure shows the $2N$ alleles in a breeding population of generation

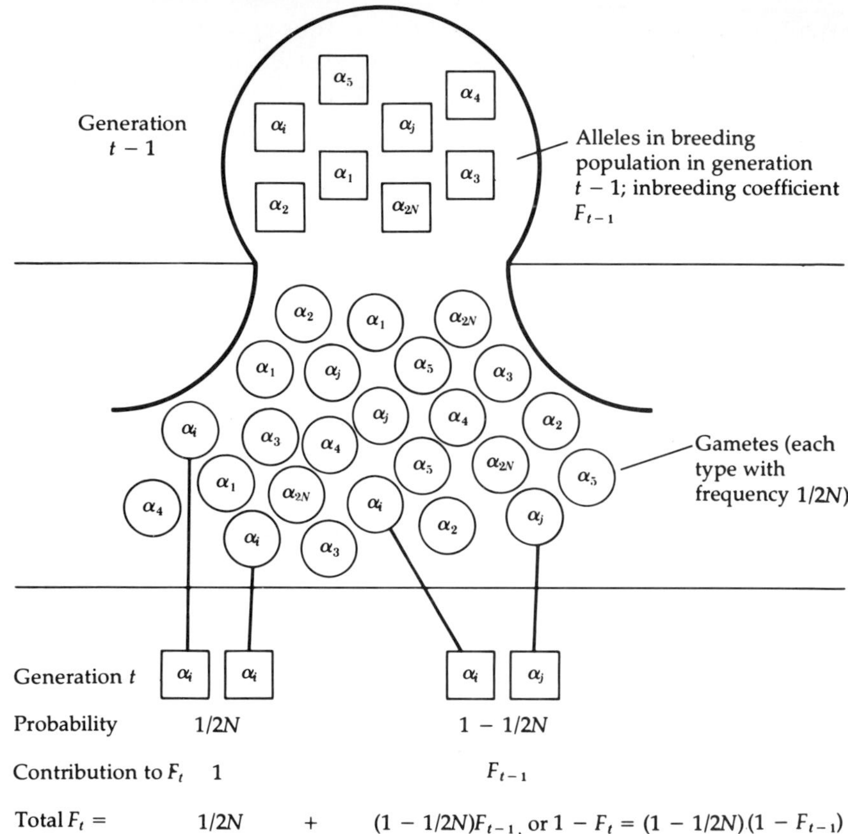

Figure 5. Random sampling in each of the subpopulations in Figure 2 implies the relation $F_t = (1/2N) + [1 - (1/2N)]F_{t-1}$.

$t - 1$ (labeled $\alpha_1, \alpha_2, \alpha_3, \ldots, \alpha_i, \ldots, \alpha_j, \ldots, \alpha_{2N}$ to avoid confusion with the conventional allele symbols A and a), the gametes derived from generation $t - 1$ (each gametic type represents a fraction $1/2N$ of the entire pool of gametes), and two randomly chosen pairs of alleles in the breeding population of generation t. The probability that the second allele chosen is of the same type as the first is $1/2N$, because this is the frequency of each allelic type in the gametic pool; the probability that the second allele is of a different type from the first is accordingly $1 - 1/2N$. In the first case ($\alpha_i\alpha_i$), the probability of identity by descent is 1; in the second case ($\alpha_i\alpha_j$), it is F_{t-1}. Altogether $F_t = 1/2N + (1 - 1/2N)F_{t-1}$. Multiplying both sides by -1 and adding $+1$ leads to

$$1 - F_t = 1 - 1/2N - (1 - 1/2N)F_{t-1}$$
$$= (1 - 1/2N)(1 - F_{t-1})$$

the solution of which is

$$1 - F_t = (1 - 1/2N)^t(1 - F_0) \tag{2.6}$$

In the case when $F_0 = 0$, Equation 2.6 simplifies to

$$F_t = 1 - \left(1 - \frac{1}{2N}\right)^t \tag{2.7}$$

Strictly speaking, Equations 2.6 and 2.7 apply only to populations (many plants, for example) that can undergo self-fertilization, because the theoretical considerations in Figure 5 make no stipulation that half of the successful gametes in any generation must come from males and the other half from females. The correction for organisms such as those with separate sexes that cannot undergo self-fertilization is minor, however: if there are equal numbers of males and females, simply replace N in Equations 2.6 and 2.7 with $N + 1/2$ (see Crow and Kimura 1970). Since $1/2(N + 1/2)$ is very nearly equal to $1/2N$ for realistic values of N, Equations 2.6 and 2.7 as they stand (without the correction) may be taken as a satisfactory approximation for organisms with separate sexes.

As an example of the use of Equation 2.6, consider the following question: What is the inbreeding coefficient of a mouse in a colony that was established from a much larger colony 30 generations ago and since that time has maintained a constant size of 20 individuals each generation? For the sake of simplicity, assume that the colony is ideal in the sense discussed earlier, particularly with regard to random mating within the colony. Because random mating within the colony is assumed, the total inbreeding coefficient F_{IT} of an animal is equal to the fixation index F_{ST} of the colony, and $F_{ST} = F_t$ in Equation 2.6. Thus, F_{ST} after 30 generations is given by Equation 2.6 with $t = 30$, which is $1 - F_{30} = (1 - 1/2N)^{30}(1 - F_0)$, where we may take $N = 20$ (as noted earlier, use of $N = 20.5$ would be a bit more accurate), and $F_0 = 0$ (justified because the foundation population was said to be large). Substituting, $1 - F_{30} = (1 - 1/40)^{30} = 0.468$, or $F_t = 0.532$. Note that there is substantial inbreeding even though mating is random and genotypic frequencies within the colony are given by the Hardy–Weinberg principle. The inbreeding does not result from actual nonrandom mating but from the fact that the population is small in size.

PROBLEM 5

From four large chicken barns near Ramona, California, Selander and Yang (1969) trapped wild mice (*Mus musculus*) and carried out an electrophoretic study of a large number of genes, including those for hexose-6-phosphate dehydrogenase, NADP-isocitrate dehydrogenase, and hemoglobin. Estimated values of F_{ST} for these genes were 0.10, 0.16, and 0.11, respectively, averaging $F_{ST} = 0.12$. Assuming that the mouse populations in each barn have about the same size (N), how long would it take for random genetic drift to result in a value of $F_{ST} = 0.12$ in the ideal (and undoubtedly oversimplified) case when migration between barns does not occur, assuming that $N = 20$? How long would it take when $N = 100$?

ANSWER In the case $N = 20$, we have $1 - F_t = (1 - 1/40)^t$ with $F_t = 0.12$, or $t = \ln(0.88)/\ln(0.975) = 5$ generations. For $N = 100$, the answer is $t = 26$ generations. As will be discussed later, migration has the effect of retarding the increase in F_t, so the times would be correspondingly longer with migration taken into account. On the other hand, founder effects (defined later) are probably significant in barn populations of mice, and these have the effect of accelerating genetic divergence due to the recurrent "bottlenecks" of population size.

Several important consequences of the population structure in Figure 2 are summarized in Table 1. First, although each subpopulation is finite in number, we can imagine so many of them that the total population number is effectively infinite. For an infinite population that obeys the assumptions of an ideal population discussed earlier, the allele frequencies remain constant. That is, even though the allele frequencies within the subpopulations may change willy-nilly due to random genetic drift, the overall average allele frequency of A among subpopulations always remains p_0, where p_0 represents the allele frequency of A in the base population.

Secondly, when mating is random within subpopulations, F_t is the probability of autozygosity of a gene in an individual in generation t. For an individual chosen at random from the total population in generation t, therefore, the probability of allozygosity is $1 - F_t$. Allozygous alleles are those that have escaped the effects of inbreeding, and in this subset of cases the genotypes occur in Hardy–Weinberg proportions. Autozygous alleles are identical by descent because of the inbreeding, and in this subset of cases the genotype frequencies are the same as with complete inbreeding. Therefore, in a total population split into subpopulations as in Figure 2, in which p_0 is the overall average allele frequency of A, the genotype frequencies, averaged across all subpopulations, are given by

$$\begin{aligned} AA&: \; p_0^2(1 - F_t) + p_0F_t \\ Aa&: \; 2p_0q_0(1 - F_t) \\ aa&: \; q_0^2(1 - F_t) + q_0F_t \end{aligned} \qquad \textbf{(2.8)}$$

Equation 2.8 implies that the genotype frequencies in the total population are given by the usual formula for inbreeding (see Table 5 in Chapter 1). However, *within any one subpopulation, the genotype frequencies still obey the Hardy–Weinberg principle because of random mating within the subpopulation.* When F_t is substituted from Equation 2.7 into the expression for the frequency of heterozygous genotypes in Equation 2.8, we see that the average hetero-

Table 1. Consequences of population subdivision and random genetic drift on genotype frequency.

		GENERATION		
		0	t	∞
Inbreeding coefficient (F_t) (Average over all populations)		0	$1 - (1 - 1/2N)^t$	1
Genotype frequency (Average over all populations)	*AA*:	p_0^2	$p_0^2(1 - F_t) + p_0F_t$	p_0
	Aa:	$2p_0q_0$	$2p_0q_0(1 - F_t)$	0
	aa:	q_0^2	$q_0^2(1 - F_t) + q_0F_t$	q_0
Allele frequency (Average over all populations)	*A*:	p_0	p_0	p_0
	a:	q_0	q_0	q_0

zygosity among subpopulations at time *t* decreases according to the equation

$$2p_0q_0(1 - F_t) = 2p_0q_0(1 - 1/2N)^t \quad \textbf{(2.9)}$$

Third, and finally, since F_t eventually goes to 1, all subpopulations eventually become fixed for one allele or the other. Because the average allele frequency of *A* remains p_0 even when all subpopulations have become fixed, the proportion of subpopulations that eventually become fixed for *A* is p_0 (and the proportion that eventually become fixed for *a* is q_0). Stated another way, the probability of ultimate fixation of a designated allele in any ideal subpopulation is equal to the frequency of the allele in the initial population.

Effective Population Number

No real population can be expected to obey all of the assumptions pertaining to ideal populations. In any actual case, there must be corrections for age-related differences in reproductive rate, for unequal numbers of males and females, and for unequal family size (see Crow and Kimura, 1970, for a good review). To take a specific example, consider the experiment depicted in Figure 3, for which $p_0 = q_0 = 1/2$. The total heterozygosity is given by $H_T = 2(1/2)(1/2) = 0.5$, and this value remains constant throughout the experiment, as expected. Furthermore, the theory predicts that the average subpopulation heterozygosity (H_S) is given by Equation 2.9 as follows:

$$H_S = 2p_0q_0(1 - F_t) = 2p_0q_0(1 - 1/2N)^t = (0.5)(1 - 1/2N)^t$$

The actual size (N_a) of the subpopulations in Figure 3 is $N_a = 16$, and the predicted decrease in H_S with this population size is shown in Figure 6, along

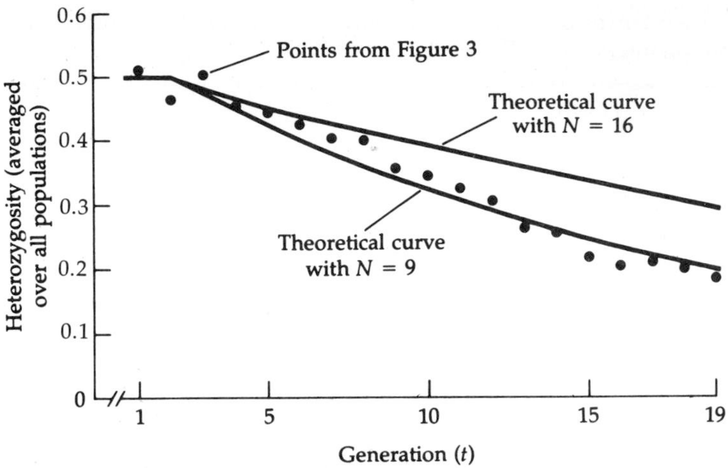

Figure 6. Theoretical curves for average heterozygosity with $N = 9$ or $N = 16$, along with actual values from the experiment in Figure 3. (Data from Buri 1956.)

with the observed values. It is evident that the fit is not very impressive, especially for the later generations. However, the pattern of decrease in H_S does follow the theoretical prediction if we assume that $N = 9$ (see Figure 6). That is, the observed decrease in average heterozygosity in the actual populations matches what would be expected in ideal populations of size $N = 9$. The number $N = 9$ is called the *inbreeding effective number* (or simply the *effective number*) of the populations in question.

To say the same thing somewhat more formally, the **inbreeding effective number** N_e of an actual (non-ideal) population is the population number of an ideal population which undergoes the same rate of increase in F (decrease in H_S) as the actual population. The use of Equation 2.9 to estimate the effective population number is illustrated in the following problem.

PROBLEM 6 The neotropical giant toad *Bufo marinus* became established in Hawaii and Australia through the release of a large number of individuals in the 1930s. Allele frequencies of ten polymorphic enzyme genes were studied in individuals from ten local populations in Hawaii and five local populations in Australia (Easteal 1985). Estimated values of F_{ST} were 0.056 in Hawaiian populations, 0.063 in Australian populations, and 0.059 combined. The species has one generation per year, and there have been approximately 45 years since the introductions. Use these data to estimate the effective population number in the Hawaiian, Australian, and combined populations.

ANSWER Equation 2.9 requires a little manipulation in order to obtain $1/2N = 1 - \exp[(1/t)\ln(1 - F_{ST})] = \omega$ (say), and then $N = (2\omega)^{-1}$. Substituting the estimates of $\langle F_{ST} \rangle$ and $\langle t \rangle = 45$ generations yields $\omega = 0.001280$ for Hawaii, 0.001445 for Australia and

0.001350 combined, for $\langle N \rangle$ = 391, 346 and 370, respectively. Easteal has estimated the 95 percent confidence intervals as (119, 812), (104, 719), and (112, 770), respectively. Estimates of population number based on dispersal rates and population densities have given estimates ranging from 4000 to 20,000, which emphasizes the principle that the long-term average effective population number can be very different than the actual population number.

In the consideration of random genetic drift, the effective population number (N_e) is usually of greater interest than the actual population number (N_a). There are three important instances in which the two quantities are related in a reasonably simple manner. The first case involves populations that fluctuate in size from generation to generation, the second case concerns populations with unequal numbers of males and females, and the third case pertains to populations that are spread out uniformly in two dimensions rather than being clustered into discrete colonies. These cases are considered in the following paragraphs.

Variable population number. Imagine a population that is ideal in all respects except that its number changes from generation to generation. Suppose, to be precise, that its effective number in generation 1 is N_1, in generation 2 is N_2, and in generation t is N_t. In this situation, the overall increase in F is given by $1 - F_t = (1 - 1/2N_e)^t$, where

$$1/N_e = (1/t)(1/N_1 + 1/N_2 + \ldots + 1/N_t) \qquad \textbf{(2.10)}$$

Equation 2.10 says that the average effective population number is calculated as the reciprocal of the average of the reciprocals, which is a special sort of average called the **harmonic mean**. The harmonic mean tends to be dominated by the smallest terms. Suppose, for example, that $N_1 = 1000$, $N_2 = 10$, and $N_3 = 1000$ in a population that underwent a severe temporary reduction in population number (a **bottleneck**) in generation 2. Then, using Equation 2.10, $1/N_e = (1/3)(1/1000 + 1/10 + 1/1000) = 0.034$, or $N_e = 1/0.034 = 29.4$. The average effective number over the three-generation period is only 29.4, whereas the average actual number is $(1/3)(1000 + 10 + 1000) = 670$. A severe population bottleneck often occurs in nature when a small group of emigrants from an established subpopulation founds a new subpopulation. The random genetic drift accompanying such a founder event is known as a **founder effect**.

PROBLEM 7

A certain highly isolated colony of the moth *Panaxia dominula* near Oxford, England, has been intensively studied by Ford and collaborators over the period 1928 to 1968 (see Ford and Sheppard 1969). This species has one generation per year, and estimates of population size were carried out yearly beginning in 1941. For the years 1950 to 1961, inclusive, estimates of population size were as follows:

1950	4100	1956	11,000
1951	2250	1957	16,000
1952	6000	1958	15,000
1953	8000	1959	7000
1954	11,000	1960	2500
1955	2000	1961	1400

Assuming that the actual size of the population in any year equals the effective size in that year, use Equation 2.10 to estimate the average effective number over the entire 12-year period.

ANSWER $\langle 1/N_e \rangle = (1/12)[(1/4100) + (1/2250) + \ldots + (1/2500) + (1/1400)] = 0.00025$, so $\langle N_e \rangle = 1/0.00025 = 4000$. Wright (1978) has argued that this is still a substantial overestimate of the true effective population number, being too large by a factor of about 10, because the effective number in any generation is only about one-tenth of the actual number owing to large differences in reproductive success among individuals.

Unequal sex ratio. A second important case in which the effective number of a nonideal population can readily be calculated concerns sexual populations in which the number of males and females is unequal. This inequality creates a peculiar sort of "bottleneck": because half of the alleles in any generation must come from each sex, any departure of the sex ratio from equality enhances the opportunity for random genetic drift. This situation is important in wildlife management, where for many game animals (pheasants and deer come immediately to mind) the legal bag limit for males is much larger than for females. Although some management considerations are served by such hunting regulations (for example, the species involved are usually polygamous, so a single male can fertilize many females and overall actual population size can be maintained), the resulting inequality in sex ratio reduces the effective population number. Specifically, if a sexual population consists of N_m males and N_f females, the actual size is $N_a = N_m + N_f$. However, the effective population number is given by

$$N_e = \frac{4N_mN_f}{N_m + N_f} \qquad \textbf{(2.11)}$$

To take a realistic example, if hunting is permitted to a level at which the number of surviving males is one-tenth the number of females, then $N_m = 0.1N_f$ and $N_a = N_m + N_f = 1.1N_f$. From Equation 2.11, the effective number is $N_e = 4N_mN_f/(N_m + N_f) = (0.4/1.1)N_f = 0.33N_a$. In this case, the effective population number is a mere one-third of the actual number.

A dairy farmer has a herd consisting of 200 cows and 2 bulls. What is the effective size of the population?

PROBLEM 8

ANSWER $N_e = 4(2)(200)/(2 + 200) = 8$. Random genetic drift will have a marked effect on the allele frequencies in the population as time goes on, and with such a small effective population number may even result in an increase in the frequency of harmful alleles.

Uniform population dispersion. The remaining case of effective-size calculations to consider pertains to populations spread out uniformly in two dimensions, such as on a prairie, rather than being clumped into discrete colonies. In this case the effective size depends on two quantities: (1) the number of breeding individuals per unit area, usually denoted by the symbol δ, and (2) the amount of dispersion between an individual's own birthplace and that of its offspring. The latter quantity, usually denoted by the symbol σ^2, is called the *one-way variance of distance between birth and breeding sites.* If dispersion follows a normal, bell-shaped curve in both dimensions (like an inverted salad bowl), then σ^2 has the following biological interpretation: 39 percent of all individuals have their offspring within a circle of radius σ centered at their own birthplace; 87 percent have their offspring within a circle of radius 2σ and 99 percent have their offspring within a circle of radius 3σ. In terms of δ and σ, the effective size of the population (often called **neighborhood size** in this context) is given by

$$N_e = \pi\delta\sigma^2 \quad \textbf{(2.12)}$$

where $\pi = 3.14159$.

Equation 2.12 can be applied to data on the abundant prairie deer mouse *Peromyscus maniculatus*. In a large area in southern Michigan, Dice and Howard (1951) estimated the density of breeding individuals to be between $\langle\delta\rangle = 5$/ha and $\langle\delta\rangle = 7$/ha (ha = hectare = 10,000 m^2—about 2.5 acres). By following the movement of marked animals from birth to breeding sites, they estimated $\langle\sigma^2\rangle = 1.3$ ha (excluding woodlands, which this organism avoids). Applying Equation 2.12, the neighborhood size is estimated to be between $\langle N_e\rangle = 4(3.14159)(5)(1.3) = 82$ and $\langle N_e\rangle = 4(3.14159)(7)(1.3) = 114$, or approximately 80 to 110. The neighborhood size is surprisingly small for such an abundant animal.

Phlox pilosa and *Liatris cylindracea* are insect-pollinated, perennial plants that are very abundant in undisturbed prairies in Illinois. In plots with densities of 9 plants/m^2 (*P. pilosa*) and 5 plants/m^2 (*L. cylindracea*), the one-way variances for dispersal of gametes, averaged over both pollen and seed, were estimated as 3.9 m^2 and 2.6 m^2,

PROBLEM 9

respectively (Levin and Kerster 1968, Schaal and Levin 1978). In the Texas bluebonnet (*Lupinus texensis*), estimates of δ and σ^2 were $\langle\delta\rangle = 15/\text{m}^2$ and $\langle\sigma^2\rangle = 0.5\ \text{m}^2$ (Schaal 1980). Use Equation 2.12 to estimate the neighborhood size for each species.

ANSWER $\langle N_e\rangle = 4\pi\langle\delta\rangle\langle\sigma^2\rangle = 4(3.14159)(9)(3.9) = 441$ individuals for *P. pilosa*, $\langle N_e\rangle = 163$ individuals for *L. cylindracea*, and $\langle N_e\rangle = 94$ individuals for *L. texensis*. The estimated effective sizes are all rather small considering the abundance of the organisms, and they give ample scope for the accumulation of local genetic differences.

Having discussed the relationship between effective population number and actual population number in the preceding special cases, we now drop the subscript e from N_e with the understanding that, hereafter, N always means N_e unless otherwise specified. Bear in mind, however, that the effective number of a population is usually smaller than the actual number, and sometimes much smaller.

A rare situation in which N_e can be greater than N occurs when the distribution of offspring among individuals is more even (smaller variance) than would be expected from the Poisson distribution. This possibility is sometimes of use to animal breeders, who, by deliberately evening out the offspring distribution among animals, can help maximize the effective size of the population.

Genetic Divergence among Subpopulations

The fixation index F_{ST} defined in Equation 2.4 serves as a convenient and widely used measure of genetic differences among populations. In an ideal population with no mutation, migration, or selection, the value of F_{ST} can easily be interpreted in terms of random genetic drift. Interpretations are not so easy for natural populations, because observed values of F_{ST} are influenced not only by random drift but also by mutation and especially by migration and natural selection. Difficulties in interpretation do not, however, invalidate the usefulness of F_{ST} as an index of genetic differentiation.

Although F_{ST} has a theoretical minimum of 0 (indicating no genetic divergence) and a theoretical maximum of 1 (indicating fixation for alternative alleles in the subpopulations), the observed maximum is usually much less than 1. Wright (1978) suggests the following qualitative guidelines for the interpretation of F_{ST}:

1. The range 0 to 0.05 may be considered as indicating *little* genetic differentiation.
2. The range 0.05 to 0.15 indicates *moderate* genetic differentiation.
3. The range 0.15 to 0.25 indicates *great* genetic differentiation.
4. Values of F_{ST} above 0.25 indicate *very great* genetic differentiation.

However, to quote Wright (1978), who developed the concept of F_{ST}, "Differentiation is by no means negligible if F_{ST} is as small as 0.05 or even less."

Amounts of genetic divergence among human subpopulations and among subpopulations of several other species are presented in Table 2. The values of F_{ST} imply that genetic divergence among human subpopulations is quite small. Of the total genetic variation found in three major races (Caucasoid, Negroid, and Mongoloid), only 0.07 (7 percent) is ascribable to genetic differences among races. That is, 93 percent of the total genetic variation is found within races. Again, of the total genetic variation found in the native Yanomama Indians of Venezuela and Brazil, only 0.077 (7.7 percent) is due to differences in allele frequency among villages, which implies that 92.3 percent of the total genetic variation is found within any single village. Values of F_{ST} for other organisms are quite variable, presumably because F_{ST} is influenced by the effective size of the subpopulations, by the amount and pattern of migration between subpopulations, and by other factors, including natural selection.

Table 2 provokes a brief discussion of the sensitive term *race* because the term is prone to misunderstanding or misuse. In population genetics, a **race** is a group of individuals in a species who are genetically more similar to each other than they are to the members of other such groups. Populations

Table 2. Total heterozygosity (H_T), average heterozygosity (H_S) among subpopulations, and fixation index (F_{ST}) for various organisms.

Organism	Number of populations	Number of loci	H_T	H_S	F_{ST}[a]
Human (major races)	3	35	0.130	0.121	0.069
Human, Yanomama Indian villages	37	15	0.039	0.036	0.077
House mouse (*Mus musculus*)	4	40	0.097	0.086	0.113
Jumping rodent (*Dipodomys ordii*)	9	18	0.037	0.012	0.676
Drosophila equinoxialis	5	27	0.201	0.179	0.109
Horseshoe crab (*Limulus*)	4	25	0.066	0.061	0.076
Lycopod plant (*Lycopodium lucidulum*)	4	13	0.071	0.051	0.282

(Data from Nei 1975.)

[a] F_{ST} is calculated as $(H_T - H_S)/H_T$.

PROBLEM 10 The data below are the allele frequencies of several genes in three human subpopulations: Blacks from West Africa; Blacks from Claxton, Georgia; and Whites from Claxton, Georgia (data from Adams and Ward 1973). Each gene has two predominant alleles and may, for purposes of this problem, be considered to have only two alleles. The genes control the MN blood group (alleles M and N), the Ss blood group (alleles S and s), the Duffy blood group (alleles Fy^a and Fy^b), the Kidd blood group (alleles Jk^a and Jk^b), the Kell blood group (alleles Js^a and Js^b), the enzyme glucose-6-phosphate dehydrogenase (alleles $G6PD^-$ and $G6PD^+$), and beta-hemoglobin (alleles β^S and β^+).

	Blacks (West Africa)	**Blacks (Claxton)**	**Whites (Claxton)**
M	0.474	0.484	0.507
S	0.172	0.157	0.279
Fy^a	0	0.045	0.422
Jk^a	0.693	0.743	0.536
Js^a	0.117	0.123	0.002
$G6PD^-$	0.176	0.118	0
β^S	0.090	0.043	0

For each gene, use Equation 2.4 to estimate F_{ST} for the comparison West African Blacks versus Claxton Blacks and for the comparison West African Blacks versus Claxton Whites. Classify the F_{ST} for each gene and comparison according to the qualitative guidelines enumerated earlier.

ANSWER Let West African Blacks, Claxton Blacks and Claxton Whites be populations A, B and C, respectively. For the MN blood group in the A–B comparison, $\langle H_S \rangle$ for A is 2(0.474)(0.526) = 0.49865, $\langle H_S \rangle$ for B is 2(0.484)(0.516) = 0.49949, average $\langle H_S \rangle$ = (0.49865 + 0.49949)/2 = 0.49907; average estimated allele frequency of M is (0.474 + 0.484)/2 = 0.479, so $\langle H_T \rangle$ = 2(0.479)(0.521) = 0.49912. Replacing the parameters provided in Equation 2.4 by their estimated values results in $\langle F_{ST} \rangle$ = (0.49912 − 0.49907)/(0.49912) = 0.0001. For the MN blood group in the A–C comparison, $\langle H_S \rangle$ for C is 2(0.507)(0.493) = 0.4999, average $\langle H_S \rangle$ = (0.49865 + 0.4999)/2 = 0.49928; the average estimated allele frequency of M is (0.474 + 0.507)/2 = 0.4905, so $\langle H_T \rangle$ = 2(0.4905)(0.5095) = 0.49982. Thus $\langle F_{ST} \rangle$ = (0.49982 − 0.49928)/(0.49982) = 0.00108. Calculations for the other genes are carried out similarly. Overall $\langle F_{ST} \rangle$ estimates and their qualitative interpretations are as follows:

	A versus B	**A versus C**
MN	0.0001 (little)	0.0011 (little)
Ss	0.0004 (little)	0.0164 (little)
Duffy	0.0230 (little)	0.2674 (very great)
Kidd	0.0031 (little)	0.0260 (little)
Kell	0.0001 (little)	0.0591 (moderate)
G6PD	0.0067 (little)	0.0965 (moderate)
β	0.0089 (little)	0.0471 (little)
Average	0.0060 (little)	0.0734 (moderate)

that have undergone some degree of genetic divergence as measured by, for example, F_{ST}, therefore qualify as races. Using this definition, the human population contains many races. Each Yanomama village represents, in a certain sense, a separate "race," and the Yanomama as a whole also form a distinct "race." Such fine distinctions are rarely useful, however. It is usually more convenient to group populations into larger units that still qualify as races in the definition given. These larger units often coincide with races based on physical characteristics such as skin color, hair color, hair texture, facial features, and body conformation, as defined by anthropologists (modern anthropologists also take cultural and linguistic similarities into account).

Here it must be pointed out that the data in Table 2, which indicate much more genetic variation within than among human races, may be misleading. The conclusion was based primarily on genes determining allozymes, and it certainly is not true for genes influencing skin color, hair color, hair texture, and other traits that most people think of in connection with the word "race." However, skin color and other prominent racial characteristics are used to delineate races precisely because racial differences for these traits are rather large, so the genes involved cannot be representative of the entire genome. On the other hand, allozyme loci may not be very representative of the genome either. See Nei and Roychoudhury (1982) for a review of the genetic relationship and evolution of human races.

MUTATION

Mutation is the ultimate source of genetic variation. This is an important statement, but hardly a profound one. New genetic variation is created by changes in the genetic material, and heritable changes in the genetic material are, by definition, **mutations**. The word *mutation* is used here in its widest sense to mean all genetic changes, including visible chromosome abnormalities, such as inversions and translocations, and polyploidy. **Polyploidy** refers to organisms or species having multiple sets of chromosomes. Stebbins (1976) has shown that polyploidy is of immense importance in the origin of new species of higher plants. However, the vast bulk of genetic variation within virtually all natural populations is presently thought to consist of changes in DNA which are invisible in the microscope. At the molecular level, one class of mutations results from the replacement of one pair of nucleotides with another, which includes the mutations responsible for allozymes. Another class results from the deletion or duplication of nucleotide sequences. Still another class results from the transposition of DNA sequences from one location in the genome to another. Transposition mutations often result from the movement of particular DNA sequences called **transposable elements**, most of which contain the genes that promote their own transposition. Mutations resulting from transposable elements constitute an important source of genetic variation, and the implications of transposable elements in population genetics are discussed in Chapter 3.

Because spontaneous mutation rates are typically rather small (on the order of 10^{-4} to 10^{-6} mutations per gene per generation), the tendency for allele frequencies to change as a result of recurrent mutation (**mutation pressure**) is very small over the course of a few generations. On the other hand, the cumulative effects of mutation over long periods of time can become appreciable. A useful model for thinking about the problem is the Hardy–Weinberg model of Chapter 1, but with mutation allowed. For the moment we focus on mutations that are **selectively neutral**—mutations that have so little effect on the ability of the organism to survive and reproduce that natural selection does not appreciably influence their frequency.

Changes in Allele Frequency with Mutation Pressure

Consider a gene with two alleles A and a, and suppose that the mutation rate per generation from A to a is μ and that the rate from a to A is ν. Let p_t and q_t (with $q_t = 1 - p_t$) denote the allele frequencies of A and a in generation t. An A allele in generation t can originate in only two ways: it could have been an A allele in generation $t - 1$ that escaped mutation to a (probability $1 - \mu$), or it could have been an a allele that mutated to A (probability ν). Symbolically,

$$p_t = p_{t-1}(1 - \mu) + (1 - p_{t-1})\nu$$

The solution of this equation gives p_t in terms of p_0 (the initial frequency of A), and it is

$$p_t = \frac{\nu}{\mu + \nu} + \left(p_0 - \frac{\nu}{\mu + \nu}\right)(1 - \mu - \nu)^t \qquad \textbf{(2.13)}$$

To understand the biological meaning of Equation 2.13, consider first the case when t is small, for example less than 100 generations. Because μ and ν are ordinarily very small, the term $(1 - \mu - \nu)^t$ is approximately equal to $1 - t(\mu + \nu)$. Suppose also that $p_0 = 0$ (i.e., that the initial population is fixed for a). With these provisions (small t, $p_0 = 0$), Equation 2.13 implies that $p_t = t\nu$. That is, the frequency of the A allele increases linearly with time, and the slope of the line equals ν. Because ν is small, however, the linear increase in p_t is difficult to detect experimentally except in very large populations. Large population sizes can be attained in bacterial **chemostats**, which are devices for maintaining populations of bacteria in a continuous state of growth and cell division (Figure 7). The linear increase in p_t that occurs in chemostats from mutation pressure is shown in Figure 8. Note the

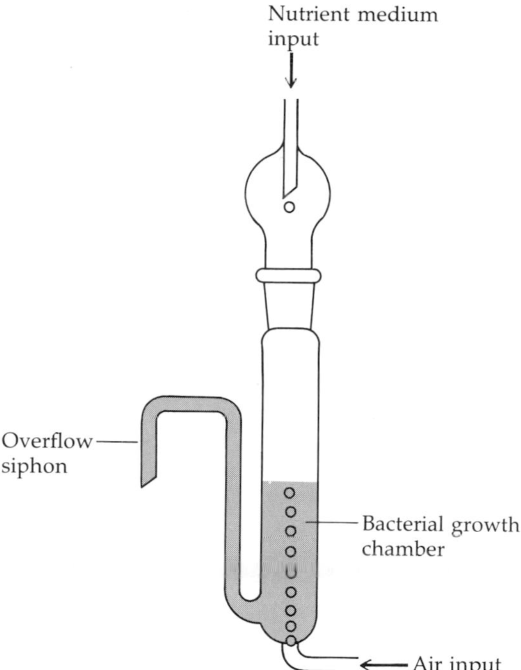

Figure 7. Diagram of a bacterial chemostat. Nutrient medium drips in at the top, but a constant volume is maintained by means of an overflow siphon. The air coming in at the bottom provides oxygen. At the steady state, the rate of nutrient inflow equals the rate of outflow. Cells within the chemostat are in a continuous state of division, but the population does not increase in size because, in any interval of time, the number of new cells produced by division is balanced by the number washing out through the siphon.

PROBLEM 11

A genetic factor has been described in *Drosophila mauritiana* which results in the spontaneous deletion of the transposable genetic element *mariner* at a frequency of approximately one percent per generation for each copy (Bryan et al., 1987). In a population containing a locus at which a *mariner* insertion is fixed (homozygous), how many generations would be required for the frequency of individuals which are homozygous for a deletion of the element to exceed 5 percent? Assume that the population is large, that mating is random, that the excision factor is fixed, and that deletion of the element does not affect survival or reproduction.

ANSWER: Let p_t be the frequency of chromosomes in which the *mariner* element remains undeleted in generation t, and $\mu = 0.01$ be the probability of deletion of the element per generation. With $\nu = 0$ and $p_0 = 1$, Equation 2.13 implies that $p_t = (1 - \mu)^t$. The frequency of deletion homozygotes is greater than 5 percent when $(1 - p_t)^2 >$ 0.05, or $p_t < 1 - (0.05)^{1/2} = 0.776$. Thus, t should be greater than $\ln(0.776)/\ln(0.99) =$ 25.2 generations.

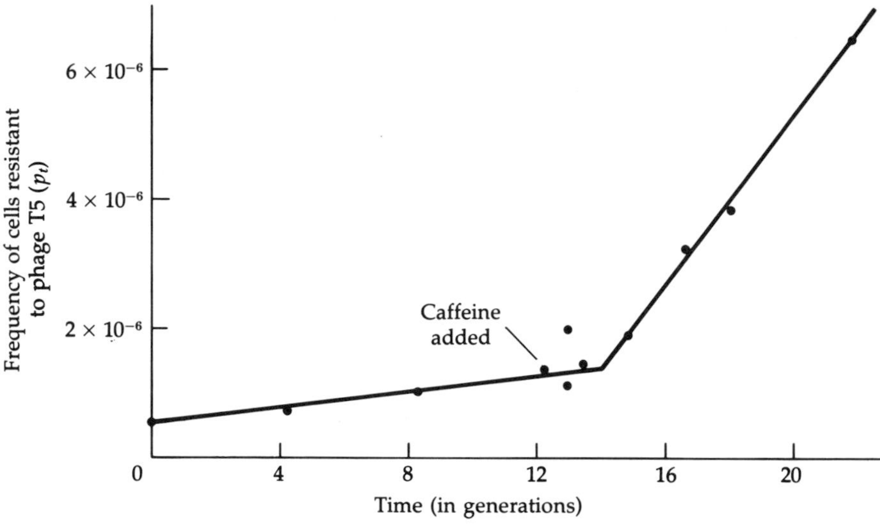

Figure 8. Measurement of mutation rates in bacterial chemostats. This example concerns the rate of mutation of a gene for resistance to bacteriophage T5. The mutation rate is estimated as the slope of the straight-line segments. Prior to addition of caffeine, the slope was $\langle \nu \rangle = 1.3 \times 10^{-8}$ per hour. After addition of caffeine at a concentration of 150 mg/l, the slope increased to $\langle \nu \rangle = 12 \times 10^{-8}$ per hour. Mutation rates per hour in bacterial chemostats are constant for generation times between 2 and 12 hours. The present experiment had a generation time of 5.5 hours. (From Novick 1955.)

abrupt increase in mutation rate (indicated by the increase in slope) shortly after the addition of caffeine, a bacterial mutagen.

To see what happens to allele frequency in the long run, consider Equation 2.13 in the case when t is very large, for example 10^5 or 10^6 generations. Even though $1 - \mu - \nu$ is ordinarily close to 1, the value of t eventually becomes so large that $(1 - \mu - \nu)^t$ becomes approximately 0. Thus, the whole right-hand term in Equation 2.13 drops out, and p_t attains a value that remains the same generation after generation. Such a value of p is called an **equilibrium** value, which we will denote by $\hat{p}$. In case of mutation, the equilibrium is given by

$$\hat{p} = \frac{\nu}{\mu + \nu} \quad \textbf{(2.14)}$$

The manner in which p_t converges to its equilibrium value is shown in Figure 9 for the case $\mu = 10^{-4}$ and $\nu = 10^{-5}$. Note that, whatever the initial frequency of A, the allele frequency of A eventually goes to $\hat{p}$, which in the example in the figure equals $0.00001/(0.0001 + 0.00001) = 1/11 = 0.091$. Figure 9 indicates that mutation pressure is usually very weak in changing allele frequency, inasmuch as the population requires thousands or tens of thousands of generations to reach equilibrium.

PROBLEM 12

Stocker (1949) studied a case in the bacterium *Salmonella typhimurium* in which the mutation rates were sufficiently large that Equation 2.13 can be tested. The gene in question controls a protein component of the cellular flagella, and there are two alleles, which we can call *A* (for the "specific-phase" flagellar property) and *a* (for the "group-phase" flagellar property). The mutation rate from *A* to *a* was estimated as $\langle\mu\rangle = 8.6 \times 10^{-4}$ per generation, and that of *a* to *A* as $\langle\nu\rangle = 4.7 \times 10^{-3}$ per generation. These mutation rates are orders of magnitude larger than typically observed for other genes. The reason is that the change from *A* to *a* and back again does not involve mutation in the conventional sense, but results from intrachromosomal recombination (Simon et al., 1980). Formally, however, we can treat the system as one with reversible mutation. In cultures initially established with the frequency of *A* at $p_0 = 0$, Stocker found that the frequency increased to $p = 0.16$ after 30 generations and to $p = 0.85$ after 700 generations. In cultures initiated with $p_0 = 1$, the frequency decreased to 0.88 after 388 generations and to 0.86 after 700 generations. How do these values agree with those calculated from Equation 2.13 using the estimated mutation rates? What is the predicted equilibrium frequency of *A*?

ANSWER Note that $\langle\nu\rangle/(\langle\mu\rangle + \langle\nu\rangle) = 0.845$. This is the predicted equilibrium frequency (Equation 2.14). Also, $1 - \langle\mu\rangle - \langle\nu\rangle = 0.99444$, and this quantity determines the rate of approach to equilibrium. For the cultures with $p_0 = 0$, predicted values are $p_{30} = 0.845 - (0.845)(0.99444)^{30} = 0.13$ and $p_{700} = 0.845 - (0.845)(0.99444)^{700} = 0.83$. For the cultures with $p_0 = 1$, predicted values are $p_{388} = 0.845 + (0.155)(0.99444)^{388} = 0.86$ and $p_{700} = 0.845 + (0.155)(0.99444)^{700} = 0.85$. The predicted values are in good agreement with the observations.

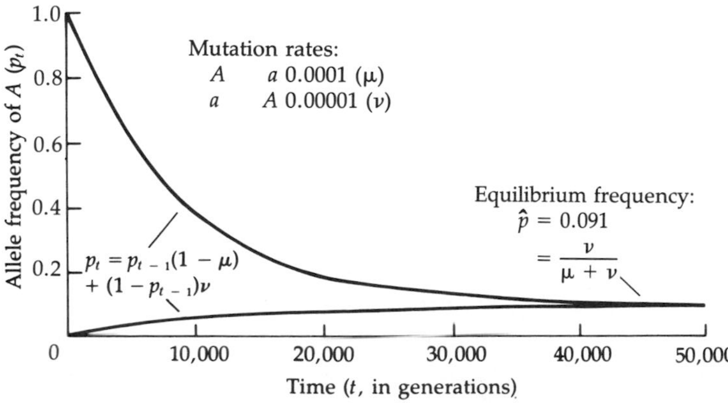

Figure 9. Theoretical change in allele frequency under pressure of reversible mutation. Note that attainment of near-equilibrium values requires tens of thousands of generations for realistic mutation rates.

Number of Alleles Maintained in Populations

Recall from Chapter 1 that genes controlling allozymes often have more than two alleles. It is therefore of some importance to determine how many alleles can be maintained by mutation pressure. If allozyme genes tend to have more alleles than would be expected from mutation pressure alone, then other forces that operate in nature must tend to accumulate alleles. On the other hand, if there are fewer alleles than expected, then other forces must tend to eliminate alleles. There is a technical problem in calculating the number of alleles that can be maintained by mutation. If the procedure for distinguishing among alleles has low resolving power, few alleles will be detected no matter how many may actually be present in the population. Electrophoresis augmented with other procedures may have satisfactory resolving power for some types of studies, but many alleles may still remain undetected. Nevertheless, in principle, we can imagine for the sake of argument that all alleles are distinguishable, as is the case when their DNA nucleotide sequences are determined. Since an average protein contains roughly 300 amino acids, the average length of the amino acid coding part of an average gene is roughly 900 nucleotides. The number of possible alleles is therefore very large—4^{900} to be exact, which equals about 10^{542}. Hence, we can suppose that every time a mutation occurs, it creates a new allele that does not already exist in the population. This is called the **infinite-alleles model** of mutation. The infinite-alleles model is but one way to specify the characteristics of new mutations. Although it represents a somewhat simplified view of mutation, it nevertheless provides a useful standard of comparison for other models or for observed allele frequencies.

In the infinite-alleles model, we must also assume that the population in question is finite, because in an infinite population new alleles would continue to accumulate forever. Consequently, the model of interest is one in which both random genetic drift and mutation occur. In order to evaluate whether actual populations contain more or fewer alleles than would be expected from mutation pressure alone in the infinite-alleles model, we must determine how many alleles can be maintained in the infinite-alleles model. The question is, for what number of alleles is the creation of new alleles by mutation balanced exactly by the loss of old alleles due to random genetic drift? For the infinite-alleles model, the answer is surprisingly simple. It can be calculated by expressing the proportion of homozygous genotypes in two ways: in terms of the allele frequencies, and in terms of the fixation index (F_t in Equation 2.7). Setting these two proportions equal to each other yields the desired result. Accordingly, let the effective population number be N and the mutation rate be μ. Because of the infinite-alleles assumption, each allele in the population arises only once, and homozygotes for any allele are therefore autozygous. For n alleles with frequencies $p_1, p_2, p_3, \ldots p_n$, the homozygosity in any subpopulation is

$$\Sigma p_i^2 = p_1^2 + p_2^2 + \ldots + p_n^2 \qquad \textbf{(2.15)}$$

because mating is random within each subpopulation. While Σp_i^2 equals the homozygosity in terms of allele frequency, the autozygosity in the infinite-alleles model also equals the fixation index F. The calculation of F_t in Figure 5 is still correct for this model providing neither allele (α_i or α_j) undergoes mutation in the passage of one generation. Therefore, $F_t = [1/2N + (1 - 1/2N)F_{t-1}](1 - \mu)^2$. The equilibrium value of F_t, call it $\hat{F}$, is found by solving $\hat{F} = [1/2N + (1 - 1/2N)\hat{F}](1 - \mu)^2$, from which

$$\hat{F} = \frac{1}{4N\mu + 1} \qquad \textbf{(2.16)}$$

to an excellent approximation. Therefore, the number of selectively neutral alleles increases under mutation pressure until $\hat{F}$ satisfies Equation 2.16.

We thus have two measures of homozygosity: Σp_i^2 in Equation 2.15 and $\hat{F}$ in Equation 2.16. Since they measure the same thing, they must equal each other, so $\Sigma p_i^2 = \hat{F} = 1/(4N\mu + 1)$. (Alternative approaches leading to essentially the same result are discussed in Sved and Latter, 1977.) One complication is that any number of distributions of allele frequency can result in the same homozygosity. For example, an equilibrium population with four alleles at frequencies $p_1 = 0.7$, $p_2 = 0.1$, $p_3 = 0.1$, and $p_4 = 0.1$ has a homozygosity of $\hat{F} = p_1^2 + p_2^2 + p_3^2 + p_4^2 = (0.7)^2 + (0.1)^2 + (0.1)^2 + (0.1)^2 = 0.52$; likewise a population with two alleles at frequencies $p_1 = 0.6$, $p_2 = 0.4$ has a homozyogosity of 0.52, because $p_1^2 + p_2^2 = (0.6)^2 + (0.4)^2 = 0.52$. The problem that many distributions of allele frequency can result in the same homozygosity can be sidestepped by assuming that all alleles are equally frequent. If the population contains n equally frequent alleles, then $p_1 = p_2 = p_3 = \ldots = p_n = 1/n$, and the homozygosity is given in Equation 2.15 as $n(1/n)^2 = 1/n$. At equilibrium, therefore, $1/n = \hat{F} = 1/(4N\mu + 1)$, or $n = 4N\mu + 1$. The number n of equally frequent alleles is called the **effective number of alleles**, often symbolized as n_e. Diverse distributions of allele frequency can therefore be compared in terms of their effective number of alleles. The four-allele population and the two-allele population above that have identical homozygosities of 0.52 have the same effective number of alleles, namely $n_e = 1/0.52 = 1.92$. Biologically speaking, n_e is the number of equally frequent alleles in an ideal population that would be required to produce the same homozygosity as in an actual population. At equilibrium, $n_e = 4N\mu + 1$.

PROBLEM 13 In a large electrophoretic study of Caribbean populations of *Drosophila willistoni*, Ayala and Tracy (1974) found five alleles of the leucine amino peptidase-5 gene (*Lap*−5), with allele frequencies estimated as 0.494, 0.429, 0.057, 0.014, and 0.006 in the Santo Domingo population, and 0.801, 0.177, 0.014, 0.004, and 0.004 in the Yunque population. The Yunque population contained eight alleles of the xanthine dehydrogenase gene (*Xdh*), with frequencies estimated as 0.446, 0.406, 0.092, 0.034, 0.014, 0.004, 0.002, and 0.002. The Yunque population had estimated allele frequencies of 0.574, 0.309, 0.114, and 0.003 of alleles of the adenylate kinase-1 gene (*Adk*−1). Estimate the effective number of alleles in each of these cases.

ANSWER For *Lap*−5 in Santo Domingo, $\langle n_e \rangle = 2.32$; for *Lap*−5 in Yunque, $\langle n_e \rangle = 1.49$; for *Xdh*, $\langle n_e \rangle = 2.68$; for *Adk*−1, $\langle n_e \rangle = 2.28$. Note that the effective number of alleles is determined more by the uniformity of allele frequencies than by the actual number of alleles.

PROBLEM 14 Show that an infinite population having a gene with an infinite number of alleles at frequencies 1/2, 1/4, 1/8, 1/16, 1/32, . . . has an effective number of alleles equal to 3. (Hint: The infinite sum $x + x^2 + x^3 + x^4 + \ldots$ equals $x/(1 - x)$ for any value of x less than 1.)

ANSWER Homozygosity equals $(1/2)^2 + (1/4)^2 + (1/8)^2 + \ldots$, which is $1/4 + 1/16 + 1/64 + \ldots$, or $x + x^2 + x^3 + \ldots$ with $x = 1/4$. Thus, homozygosity equals $x/(1 - x) = (1/4)/(3/4) = 1/3$. Effective number of alleles is the reciprocal of the homozygosity, so $n_e = 3$.

The Neutrality Hypothesis

The hypothesis that observed allozyme polymorphisms result from selectively neutral alleles maintained in a balance between mutation and random genetic drift is known as the **neutrality hypothesis** or the hypothesis of **selective neutrality** (Kimura 1968; King and Jukes 1969). In essence, the neutrality hypothesis states that many mutations have so little effect on the organism that their influence on survival and reproduction is negligible. The frequencies of neutral alleles are not determined by the forces of natural selection, but are instead determined by other forces such as migration (discussed later) and random genetic drift. Because the neutrality hypothesis is of fundamental importance in population genetics and evolution, it has been (and still is) a subject of considerable discussion. If the hypothesis is true, or approximately true, then observed polymorphisms—of allozyme genes, for example—may have no particular significance in the adaptation of organisms to their present environments. From this perspective, selectively neutral polymorphisms are mere evolutionary "noise," and regardless of how much their study may reveal about population structure and random genetic drift, they tell us little or nothing about adaptive genetic changes in evolution. If, on the other hand, the neutrality hypothesis is false, the next important task would be to study in detail the actual adaptive significance of genetic polymorphisms.

To assess the plausibility of the neutrality hypothesis, many aspects of the hypothesis must be compared with the situation in actual populations. One aspect of the hypothesis developed in the preceding section concerns the homozygosity to be expected with the infinite-alleles model. Using observed homozygosities, we can estimate the effective number of alleles $\langle n_e \rangle$, and from the expression $n_e = 4N\mu + 1$, estimate the corresponding values of $\langle N\mu \rangle$. If the resulting values are grossly unreasonable, we can safely reject the infinite-alleles version of the neutrality hypothesis (or at least argue that actual populations cannot be in equilibrium).

Recall from Chapter 1 that observed values of heterozygosity of allozyme genes range from 0.04 to 0.14 in most organisms (see Figure 6 in Chapter 1). Observed homozygosities therefore range from $1 - 0.04 = 0.96$ to $1 - 0.14 = 0.86$, which correspond to estimated $\langle n_e \rangle$ in the range $1/0.96 = 1.04$ to $1/0.86 = 1.16$. Estimates of $\langle N\mu \rangle$, calculated as $(\langle n_e \rangle - 1)/4$, therefore range from 0.01 to 0.04. That the maximum estimated value of $\langle N\mu \rangle$ differs from the minimum by a factor of only about four is surprising inasmuch as the population number in different species ranges over a factor of 10^4 or more. The apparently too-uniform distribution of homozygosities of allozyme genes among diverse organisms has been interpreted as meaning that the neutrality hypothesis is wrong. This conclusion implies that natural selection is involved somehow in the maintenance of genetic polymorphisms. On the other hand, rejection of the neutrality hypothesis on grounds of the estimated range of $\langle n_e \rangle$ is probably premature because routine electrophoresis does not distinguish all alleles (see Chapter 1 for discussion). Beyond that, estimates of effective population number $\langle N \rangle$ in natural populations are generally imprecise because the studies are very difficult, and estimates of $\langle \mu \rangle$ (which in this case is the mutation rate to *neutral* alleles) are even more uncertain.

Figure 10(a) shows a second type of test of the adequacy of the neutrality hypothesis in explaining observed levels of genetic variation of allozyme genes. The figure shows the observed distribution of heterozygosity of 74 genes in Caucasians (shaded) along with the computer-generated theoretical distribution expected with the infinite-alleles model (solid lines). The observed average heterozygosity is 0.099, and the theoretical heterozygosity is 0.091. The correspondence between the histograms is fairly good, but the observed distribution seems to include too many genes with heterozygosities in the range of 0.35 to 0.55 (for a possible explanation see Fuerst et al. 1977).

A third type of test of the neutrality hypothesis is shown in Figure 10(b), which presents data on the mean and variance of heterozygosity in 77 ver-

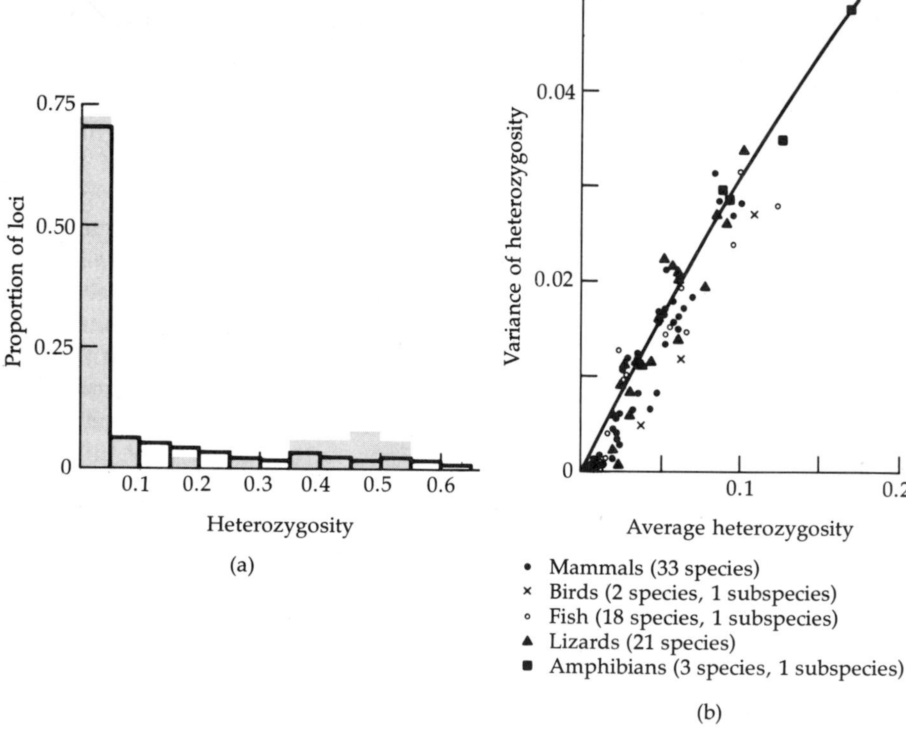

Figure 10. (a) Observed distribution of heterozygosity among genes in Caucasians (shaded) along with theoretical distribution for selectively neutral genes (solid lines). (b) Mean and variance of heterozygosity among genes in vertebrates. The solid line in (b) is the theoretical curve derived by Nei, Fuerst, and Chakraborty for the infinite-alleles model when the mutation rate to neutral alleles varies among genes in such a manner that the variance in mutation rate equals the square of the mean mutation rate. (After Nei et al. 1976.)

tebrate species. The curve is the theoretical expectation from the infinite-alleles model when the rate of selectively neutral mutation varies among genes (Nei et al. 1976). At first glance, the fit in Figure 10b is impressive. On the other hand, the observed points are sufficiently scattered that any number of other curves might fit at least as well. Evidently, statistical comparisons of this sort are too lacking in power to distinguish between the hypotheses.

A brief background on the phrase *lacking in power* may be in order. The neutral theory is useful in being a sort of starting point, or null hypothesis, which provides predictions about the relations among observed quantities that can be confirmed or rejected. Statistical tests of the neutral theory are similar to other types of statistical tests in that two distinct types of possible errors must be balanced. If the tests are too demanding, for example in failing to allow for the effects of random sampling error, then data may often result in rejection of the hypothesis, even when it is true. This is called *Type I error*. On the other hand, if the statistical test allows too much latitude in the data, then data will seldom result in rejection of the hypothesis, even when it is false. This is called *Type II error*. The trade-off between Type I error and Type II error is that the probability of Type I error cannot be decreased without increasing the probability of Type II error, and vice versa. By convention, statisticians usually adopt a five percent criterion for rejection of the null hypothesis. This is the familiar five-percent level of statistical significance, and it means that there is a five percent chance of rejecting a true hypothesis (Type I error). With this convention, the probability of a Type II error (failing to reject a false hypothesis) falls where it may, and tests which relatively high probabilities of Type II error are said to be *lacking in power*.

Figure 10 is a preliminary introduction to Chapter 3. It is relevant here because of the use of the effective number of alleles and average heterozygosity. Although the comparisons in the figure are lacking in power and hence are inclusive in their support of the neutrality hypothesis, many other observations and types of data have been brought to bear in assessing the hypothesis. These data often rely on comparisons of nucleotide sequences of DNA in different genes or in different species. These types of comparisons and the conclusions from them are discussed further in Chapter 3.

MIGRATION

In a subdivided population like the one in Figure 2, random genetic drift results in genetic divergence among subpopulations. **Migration**, which refers to the movement of individuals among subpopulations, is a sort of genetic glue that holds subpopulations together and sets a limit to how much genetic divergence can occur. To understand the homogenizing effects of migration, it is useful to study migration in several simple models of population structure.

The Island Model of Migration

In the **island model** of migration, a large population is split into many subpopulations dispersed geographically like islands in an archipelago. Examples of island population structure might include fish in freshwater lakes or slugs in dispersed garden plots. Each subpopulation is assumed to be so large that random genetic drift can be neglected. Consider two alleles A and a with average allele frequencies among the subpopulations equal to $\overline{p}$ and $\overline{q}$, respectively. Migration is assumed to occur in such a way that, in terms of allele frequency, the migrants are representative of the subpopulations. That is, among the migrants, the allele frequencies of A and a equal $\overline{p}$ and $\overline{q}$, respectively. The amount of migration is measured by a number m that equals the probability that a randomly chosen gene in any subpopulation comes from a migrant. Focus attention on a randomly chosen gene in any subpopulation in generation t. This allele could have come from the same subpopulation in generation $t - 1$ (with probability $1 - m$), in which case it is an A allele with probability p_{t-1}, where p_{t-1} represents the allele frequency of A in the subpopulation in question in generation $t - 1$. Alternatively, the allele could have come from a migrant in generation $t - 1$ (with probability m), in which case it is an A allele with probability $\overline{p}$. (Since random genetic drift, mutation, and natural selection are being ignored, $\overline{p}$ stays the same in all generations.) Altogether,

$$p_t = p_{t-1}(1 - m) + \overline{p}m \qquad \textbf{(2.17)}$$

Equation 2.17 is similar to Equation 2.13 for mutation, and its solution in terms of p_0 is

$$p_t = \overline{p} + (p_0 - \overline{p})(1 - m)^t \qquad \textbf{(2.18)}$$

where p_0 is the initial frequency of A in the subpopulation in question. As an example of the use of Equation 2.18, suppose there are only two populations with initial allele frequencies of A of 0.2 and 0.8, respectively, with $m = 0.10$ (i.e., 10 percent of the individuals in either subpopulation in any generation are migrants in which the allele frequency of A is $\overline{p} = (0.2 + 0.8)/2 = 0.5$). What is the allele frequency of A in the two populations after 10 generations? For the population with initial allele frequency 0.2, we substitute $p_0 = 0.2$, $\overline{p} = 0.5$, and $m = 0.10$ into Equation 2.18 to obtain $p_{10} = 0.5 + (0.2 - 0.5)(1 - 0.10)^{10}$, or $p_{10} = 0.395$; for the other population, we can substitute $p_0 = 0.8$, $\overline{p} = 0.5$, and $m = 0.10$, so that p_{10} $0.5 + (0.8 - 0.5)(1 - 0.10)^{10} = 0.605$.

Another example using Equation 2.18 is shown in Figure 11, where there are five subpopulations (initial frequencies 1, 0.75, 0.50, 0.25, and 0), again with $m = 0.10$. Note how rapidly the allele frequencies converge to the same value (in this case, 0.5). Although Equation 2.18 for migration is mathematically similar to Equation 2.13 for mutation, the biological implications are quite different. Because rates of migration are typically much greater than rates of mutation, changes in allele frequency generally occur much faster with migration.

One-Way Migration

When migration occurs predominantly from one population into another without an equal amount of migration in the reverse direction, then $\overline{p}$ in

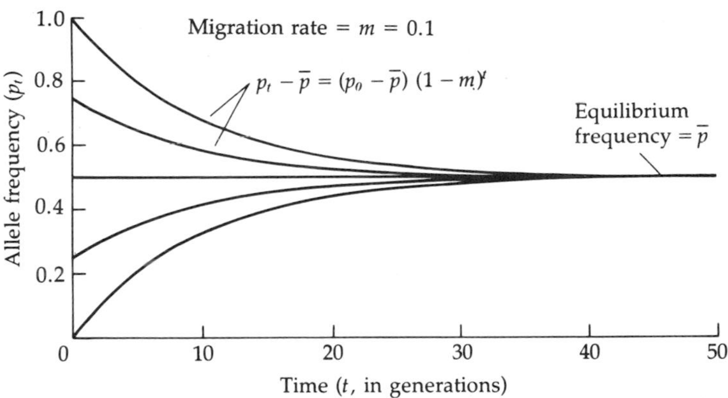

Figure 11. Change of allele frequency with time in five subpopulations exchanging migrants at the rate $m = 0.1$ per generation. Note the rapid convergence to a common equilibrium frequency.

Equation 2.18 should be interpreted as the allele frequency in the population that supplies the migrants. An example is found in racial admixture in the United States. In this case, genetic migration occurs mainly from Whites to Blacks because individuals of mixed racial ancestry are regarded as Blacks. The amount of genetic migration can be estimated using the allele-frequency data in Problem 10 and Equation 2.18. The case of the MN blood groups serves as an example. In West Africa (which may, for the purpose of this problem, be regarded as the ancestral Black population), $\langle p_0 \rangle = 0.474$ for the allele frequency of M. In present-day Claxton Blacks, $\langle p_t \rangle = 0.484$. The Claxton White population may reasonably be regarded as representative of the source of the migrants, and for Claxton Whites $\langle \bar{p} \rangle = 0.507$. Blacks were brought into the United States on a large scale from West Africa about 300 years ago, so $\langle t \rangle$ is about 10 generations. Thus, substituting these estimates into Equation 2.18, we obtain $0.484 = 0.507 + (0.474 - 0.507)(1 - \langle m \rangle)^{10}$, from which $1 - \langle m \rangle = [(0.484 - 0.507)/(0.474 - 0.507)]^{1/10} = 0.965$, or $\langle m \rangle = 0.035$ per generation. This estimate can be interpreted as implying that, in the genetic history of the population of Claxton Blacks, about 3.5 percent of the alleles of the MN gene in any generation were newly introduced by genetic migration from Whites.

PROBLEM 15

Estimate the amount of migration from Whites to Blacks using each of the other genes discussed in Problem 10.

ANSWER For Ss, $\langle m \rangle = -0.013$ per generation; for Duffy, $\langle m \rangle = 0.011$; for Kidd, $\langle m \rangle = -0.028$; for Kell, $\langle m \rangle = -0.005$; for G6PD, $\langle m \rangle = 0.039$; for hemoglobin β, $\langle m \rangle = 0.071$.

Problem 15 illustrates some of the difficulties in estimating racial admixture from allele frequencies. The positive values of $\langle m \rangle$ vary widely, and the negative values are not consistent with the proposed model of migration. Cavalli-Sforza and Bodmer (1971) remark that "The weakness of the analysis is mostly due to the uncertainty of the origin of black Americans . . . and the variability of gene frequencies in the probable area of the slave markets in West Africa. In addition, it is unavoidable that gene frequencies have changed somewhat from their original values, due to drift or, in some cases, selection. The opportunities for admixture, and the time available for it, must also have varied widely." The most reliable gene among those in Problem 15 is probably that for the Duffy blood groups, because the Fy^a allele is virtually nonexistent in all of West Africa. For this gene, the estimate of $\langle m \rangle$ is about one percent per generation, which is consistent with the average value for a large number of other genes (Cavalli-Sforza and Bodmer 1971).

Isolate Breaking and Wahlund's Principle

Isolate breaking refers to the fusion of formerly isolated subpopulations by migration. Fusion of populations reduces the frequency of homozygous genotypes, a phenomenon called **Wahlund's principle**. To reinforce this idea, imagine two isolated subpopulations with allele frequencies of a of q_1 and q_2. The average frequency of homozygous aa genotypes in the subpopulations is, from Table 1, $\bar{q}^2(1 - F) + \bar{q}F$, where $\bar{q}$ represents the average allele frequency of a over all subpopulations (denoted q_0 in Table 1) and F represents the fixation index F_{ST}. Were the subpopulations to fuse into one, the allele frequency of a in the fused population would be $\bar{q}$. After one generation of random mating, the frequency of homozygous aa genotypes is $\bar{q}^2$. Therefore, the average frequency of aa genotypes before population fusion is always greater than the frequency after fusion in the amount

$$\begin{aligned} \Delta R &= \bar{q}^2(1 - F) + \bar{q}F - \bar{q}^2 \\ &= \bar{q}^2F + \bar{q}F \\ &= \bar{q}(1 - \bar{q})F \end{aligned} \qquad \textbf{(2.19)}$$

In human populations, the main effect of isolate breaking (fusion of populations) is to decrease the overall frequency of children born with genetic defects resulting from homozygous recessive genes, particularly harmful recessive genes with frequencies that are, for whatever reasons, relatively high in one of the populations. Examples of harmful recessive genes at high frequency in certain human populations include the alleles for α_1-antitrypsin deficiency ($\langle q \rangle = 0.024$) and cystic fibrosis ($\langle q \rangle = 0.022$) in Caucasians, sickle-cell anemia in many Negro populations ($\langle q \rangle = 0.05$ in American Blacks, up to $\langle q \rangle = 0.1$ in some African populations), albinism in the Hopi and some other Southwest American Indians ($\langle q \rangle = 0.07$), and Tay-Sachs disease among Ashkenazi Jews ($\langle q \rangle = 0.013$).

To illustrate the effect of isolate breaking, imagine a population of gray squirrels that, because of a founder effect, has a high frequency of albinism, equal to 16 percent. (Albinism is inherited absence of pigment resulting from a homozygous recessive gene.) In a nearby forest is another population in which the albino mutation is absent, so the allele frequency in this population is 0. Overall, the average frequency of albinos in the two populations is $(0.16 + 0)/2 = 8$ percent. Were the two populations to fuse and undergo random mating, the allele frequency of the albino mutation in the fused population would be $[(0.16)^{1/2} + (0)^{1/2}]/2 = 0.2$, and the frequency of the homozygous genotype would equal $(0.2)^2 = 4$ percent. That is, the frequency of albinos in the fused population is smaller than the average frequency among the individual isolates.

PROBLEM 16

Tay-Sachs disease is an autosomal-recessive degenerative disorder of the brain that usually leads to death in infancy or early childhood. Among Ashkenazi Jews the incidence of the condition is about 1 in 6000 births, but among non-Jews the incidence is about 1 in 500,000 births (Myrianthopoulos and Aronson 1966). What incidence of the disease would be expected among the offspring of matings between Ashkenazi Jews and non-Jews? If these offspring were to mate randomly among themselves, what incidence of the disease would be expected among their offspring?

ANSWER If the allele frequencies of the Tay-Sachs mutation among Ashkenazi Jews and non-Jews are denoted q_1 and q_2, respectively, then, in the initial populations, $\langle q_1 \rangle = (1/6000)^{1/2} = 0.0129$ and $\langle q_2 \rangle = (1/500{,}000)^{1/2} = 0.0014$. In offspring of matings between Jews and non-Jews, the expected frequency of homozygous recessives is $\langle q_1 \rangle \langle q_2 \rangle = (0.0129)(0.0014) = 1.806 \times 10^{-5}$, or about 1 in 55,000 births. Within the hybrid population, the allele frequency of the harmful recessive is $\overline{q} = (1/2)[\langle q_1 \rangle(1 - \langle q_2 \rangle) + \langle q_2 \rangle(1 - \langle q_1 \rangle)] + \langle q_1 \rangle \langle q_2 \rangle = 0.0072$; the first term in this expression is for the heterozygous genotypes and the second term for the homozygous recessives. (Theoretically, the estimate should be adjusted to take into account the nonreproduction of homozygous recessives, but the correction is very small and can be neglected.) With random mating within the hybrid population, the frequency of homozygous recessives is $\overline{q}^2 = (0.0072)^2 = 5.184 \times 10^{-5}$, or about 1 in 19,000 births. Subsequent generations have approximately the same gene and genotype frequencies as the second generation.

As indicated in Equation 2.19, the reduction in frequency of recessive homozygotes due to population fusion can be expressed in terms of the fixation index as $\overline{pq}F$. The reduction in frequency can also be expressed in terms of the variance in allele frequency among subpopulations. The **variance** measures the degree of dispersion of a set of numbers from the mean, or how closely the individual numbers cluster around the mean. Among any set of numbers, the variance can be calculated as the mean of the squares (often called the **mean square**) minus the square of the mean. To take an extreme example, imagine two subpopulations with allele frequencies 1.0 and 0, respectively. The mean square is $(1/2)[(1)^2 + (0)^2] = 0.5$ and the mean is $(1/2)[1 + 0] = 0.5$, so the variance σ^2 equals 0.5 [the mean square] $- (0.5)^2$ [the squared mean] $= 0.25$. The allele frequencies in the two populations in this example are as different as they can possibly be, so the maximum possible variance in allele frequency for two populations is 0.25. As another example, suppose the two allele frequencies are 0.75 and 0.25; then $\sigma^2 = (1/2)[(0.75)^2 + (0.25)^2] - [(1/2)(0.75 + 0.25)]^2 = 0.0625$. Note in this example that the deviations of the allele frequencies from the mean are half as large as in the previous example—0.25 instead of 0.50—but the variance is only one-quarter as large. This disproportionate reduction in the variance occurs because the variance depends on the square of the deviation of each allele frequency from the mean. As a final example, suppose the allele frequencies in the two populations are both 0.5. Then $\sigma^2 = (1/2)[(0.5)^2 + (0.5)^2] - [(1/2)(0.5 + 0.5)]^2 = 0$, which is simply the mathematical manner of stating that the populations have the same allele frequency (i.e., there is no variation).

We are now in a position to express Wahlund's principle in terms of the variance in allele frequency. Consider the two populations in Figure 12 before they become fused. The average allele frequency of a, denoted $\bar{q}$, equals $(1/2)(q_1 + q_2)$. The mean square, denoted $\overline{q^2}$, is the average of the squares, which equals $(1/2)(q_1^2 + q_2^2)$. The variance in allele frequency σ^2 is numerically equal to the mean square minus the squared mean. In symbols, $\sigma^2 = \overline{q^2} - \bar{q}^2$. As shown in Figure 12, the average frequency of homozygous aa genotypes prior to population fusion is greater than the frequency after fusion by the amount σ^2. Because the decrease in homozygosity with population fusion is the same whether stated in terms of F or σ^2, it follows that $\bar{p}\bar{q}F = \sigma^2$, or

$$F_{ST} = \frac{\sigma^2}{\bar{p}\bar{q}} \qquad \textbf{(2.20)}$$

Equation 2.20 provides a convenient method of estimating F from allele-frequency data. For example, in the three pairs of hypothetical populations discussed earlier, $\bar{p} = \bar{q} = 0.5$ in all cases. For the populations with allele frequencies of 1 and 0, $F = \sigma^2/\bar{p}\bar{q} = 0.25/(0.5)(0.5) = 1$, which is simply a statement in terms of the fixation index F_{ST} that the populations are as different as they can be. Similarly, for the populations with allele frequencies of 0.75 and 0.25, $F = 0.0625/(0.5)(0.5) = 0.25$; and for the populations with allele frequencies of 0.5 and 0.5, $F = 0/(0.5)(0.5) = 0$.

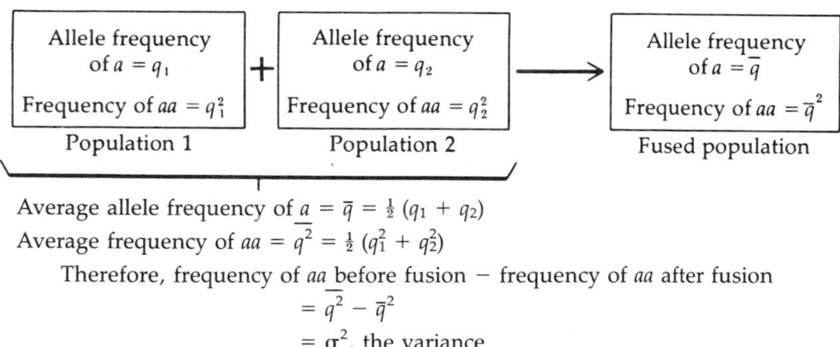

Figure 12. Illustration of Wahlund's principle (isolate breaking). The frequency of homozygous recessives upon population fusion and random mating is less than the average frequency before fusion by an amount equal to the variance in allele frequency. The variance is a measure of the dispersion or spread among a group of numbers, and it is numerically equal to the mean of the squares of the numbers minus the square of the mean of the numbers.

PROBLEM 17

For the MN and Duffy blood groups in Problem 10, verify that $F_{ST} = \sigma^2/\bar{p}\bar{q}$ in the comparisons "West African Blacks versus Claxton Blacks" and "West African Blacks versus Claxton Whites."

ANSWER West African Blacks versus Claxton Blacks: $\langle p_1 \rangle$ = estimated frequency M allele in West African Blacks = 0.474, $\langle p_2 \rangle$ = estimated frequency of M allele in Claxton Blacks = 0.484; $\langle \sigma^2 \rangle = [(p_1^2 + p_2^2)/2] - [(p_1 + p_2)/2]^2 = 2.5 \times 10^{-5}$; $\langle \bar{p} \rangle = (p_1 + p_2)/2 = 0.479$, so $\langle \bar{q} \rangle = 0.521$. $\langle F_{ST} \rangle = \langle \sigma^2 \rangle / \langle \bar{p} \rangle \langle \bar{q} \rangle = 0.0001$, which agrees with the answer given in Problem 10. For the Duffy blood group, $\langle p_1 \rangle = 0$, $\langle p_2 \rangle = 0.045$, $\langle \sigma^2 \rangle = 5.063 \times 10^{-4}$, $\langle \bar{p} \rangle = 0.0225$, $\langle \bar{q} \rangle = 0.9775$, $\langle F_{ST} \rangle = 0.0230$. West African Blacks versus Claxton Whites: for MN, $\langle p_1 \rangle = 0.474$, $\langle p_2 \rangle = 0.507$, $\langle \sigma^2 \rangle = 2.722 \times 10^{-4}$, $\langle \bar{p} \rangle = 0.4905$, $\langle \bar{q} \rangle = 0.5095$, $\langle F_{ST} \rangle = 0.0011$. For Duffy, $\langle p_1 \rangle = 0$, $\langle p_2 \rangle = 0.422$, $\langle \sigma^2 \rangle = 0.04452$, $\langle \bar{p} \rangle = 0.211$, $\langle \bar{q} \rangle = 0.789$, $\langle F_{ST} \rangle = 0.2674$. As in Problem 10, parameters have been estimated by equating them with observed values, which ignores certain small corrections for sampling variation.

How Migration Limits Genetic Divergence

It is remarkable how little migration is required to prevent significant genetic divergence among subpopulations resulting from random genetic drift. The effect can be seen quantitatively by considering the model of random drift depicted in Figure 2, but permitting migration at a rate m according to the island model. The expression for F_t derived in Figure 5 is still valid, provided that neither allele α_i or allele α_j has been replaced by a migrant allele.

Therefore $F_t = [1/2N + (1 - 1/2N)F_{t-1}](1 - m)^2$. At equilibrium, $F_t = F_{t-1} = \hat{F}$, so $\hat{F} = [1/2N + (1 - 1/2N)\hat{F}](1 - m)^2$, which for reasonably small values of m is approximately

$$\hat{F} = \frac{1}{4Nm + 1} \qquad \textbf{(2.21)}$$

Equation 2.21 for migration has the same form as Equation 2.16 for mutation, except that Nm replaces $N\mu$. This emphasizes the theoretical similarity between the effects of mutation and migration. However, it must be emphasized that rates of migration are typically much greater than rates of mutation, so the practical implications of Equations 2.21 and 2.16 for homozygosity are very different.

Since the actual number of migrants per generation equals Nm, Equation 2.21 implies that $\hat{F}$ decreases as the number of migrants increases. Indeed, the decrease in $\hat{F}$ with increasing Nm is extremely rapid. When $Nm = 0$, $\hat{F} = 1$. However, when

$Nm = 0.25$ (one migrant every fourth generation), $\hat{F} = 0.50$
$Nm = 0.5$ (one migrant every second generation), $\hat{F} = 0.33$
$Nm = 1$ (one migrant every generation), $\hat{F} = 0.20$
$Nm = 2$ (two migrants every generation), $\hat{F} = 0.11$

The implication is that migration is a potent force acting against genetic divergence resulting from genetic drift among subpopulations. For example, the mouse barns in Problem 5 would need only 1.8 migrant mice per barn per generation in order to maintain a value of $\hat{F} = 0.12$. [That is, $4Nm + 1 = (0.12)^{-1}$, so $Nm = 1.8$.]

Estimates of Migration

One method of estimating genetic migration in natural populations relies on the finding that, in theoretical models, the logarithm of Nm decreases approximately as a linear function of the average frequency of private alleles in samples from the subpopulations (Slatkin 1985a). *Private alleles* are those found in only one subpopulation. Data on the average frequency of private alleles in samples has been compiled and analyzed by Slatkin (1985a), and the resulting estimates of Nm and equilibrium values of F_{ST} are summarized in Table 3. There is obviously considerable variation in Nm among organisms. However, many of the values of Nm are approximately two or smaller, which gives considerable scope for genetic divergence resulting from random genetic drift.

A second kind of approach to estimating Nm in natural populations is illustrated in Figure 13, which gives the distribution of estimated values of

Table 3. Estimates of Nm and $\hat{F}_{ST}$.

Species	Type of organism	Estimated Nm	Estimated $\hat{F}_{ST}$
Stephanomeria exigua	Annual plant	1.4	0.152
Mytilus edulis	Mollusc	42.0	0.006
Drosophila willistoni	Insect	9.9	0.025
Drosophila pseudoobscura	Insect	1.0	0.200
Chanos chanos	Fish	4.2	0.056
Hyla regilla	Frog	1.4	0.152
Plethodon ouachitae	Salamander	2.1	0.106
Plethodon cinereus	Salamander	0.22	0.532
Plethodon dorsalis	Salamander	0.10	0.714
Batrachoseps pacifica ssp. 1	Salamander	0.64	0.281
Batrachoseps pacifica ssp. 2	Salamander	0.20	0.556
Batrachoseps campi	Salamander	0.16	0.610
Lacerta melisellensis	Lizard	1.9	0.116
Peromyscus californicus	Mouse	2.2	0.102
Peromyscus polionotus	Mouse	0.31	0.446
Thomomys bottae	Gopher	0.86	0.225

(Data from Slatkin 1985a.)

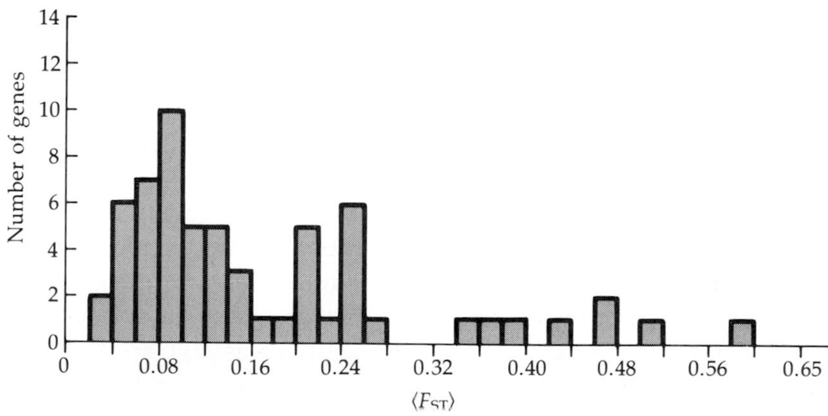

Figure 13. Distribution of estimated values of F_{ST} for 61 genes among natural populations of *Drosophila melanogaster*. While the average value of $\langle F_{ST} \rangle$ suggests migration at a level of Nm between 1 and 2, about one-third of the genes demonstrate $\langle F_{ST} \rangle$ values greater than 0.20. (From Singh and Rhomberg, personal communication.)

$\langle F_{ST} \rangle$ among 61 genes in natural populations of *Drosophila melanogaster* (Singh and Rhomberg, personal communication). The average of the estimated values is $F_{ST} = 0.16$, which, assuming equilibrium, estimates $(4Nm + 1)^{-1}$ (Equation 2.21). Therefore, $\langle Nm \rangle = [(1/0.16) - 1]/4 = 1.3$. This estimate is within the range for other *Drosophila* species in Table 3. However, there are many genes in Figure 13 which have $\langle F_{ST} \rangle$ values greater than 0.30.

Patterns of Migration

Migration in actual populations is more complex than is assumed in the island model of migration. In nature, migrants come primarily from nearby populations. To the extent that nearby populations have similar allele frequencies, the effects of migration are smaller, and sometimes much smaller, than predicted by the island model. Populations in nature may be strung out along one dimension, such as a river bank. Populations may also be distributed regularly in two dimensions, or there may be one large population with an internal genetic structure caused by the tendency for mating to occur between individuals born in the same region. Analysis of the effects of migration in such complex population structures is usually difficult (Slatkin 1985b). Among humans, migration rates depend on age, sex, marital status, socioeconomic status, population density, and many other factors. Migration rates can change rapidly, moreover, so a full-blown theory of migration has to be extremely complex.

The effects of migration on genetic differentiation of populations are seen dramatically in Figure 14. Panel (a) pertains to the moth *Biston betularia*, panel (b) to the moth *Gonodontis bidentata*. Both species have evolved melanic (black) forms in response to heavy air pollution (see discussion relating to Problem 9 in Chapter 1), and the graphs give the frequency of the melanic forms in the two species. The geographical area in (a) includes Liverpool and Manchester, as viewed from rural Wales. Note the fall-off in frequency of melanics in the nonindustrial areas toward the front of the graph. *B. betularia* occurs in low population densities and must fly relatively long distances to find a mate. The resulting high rate of migration hinders differentiation of populations, hence the smooth surface. The geographical area in (b) is smaller than that in (a), but the view is from the same perspective. *G. bidentata* occurs in high population densities and the migration rate is low, so there is substantial genetic differentiation among populations, as evidenced by the bumpy surface of the graph.

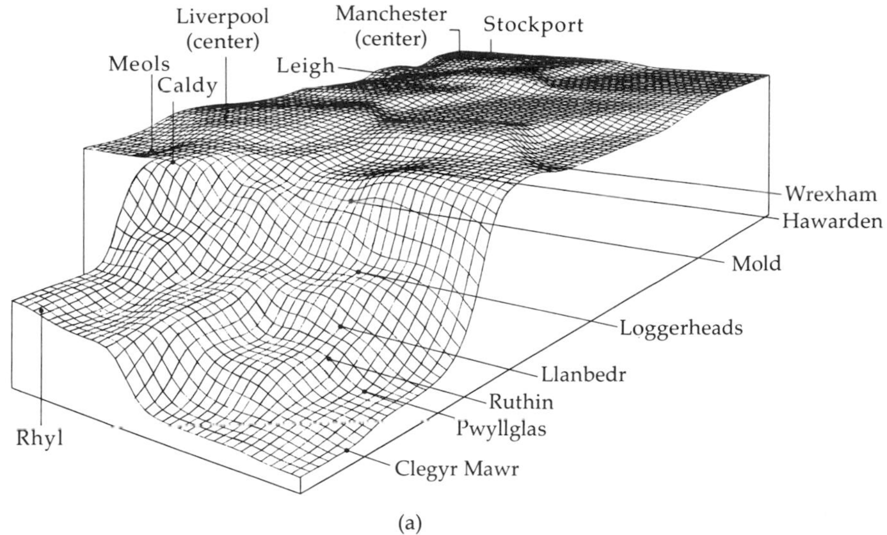

(a)

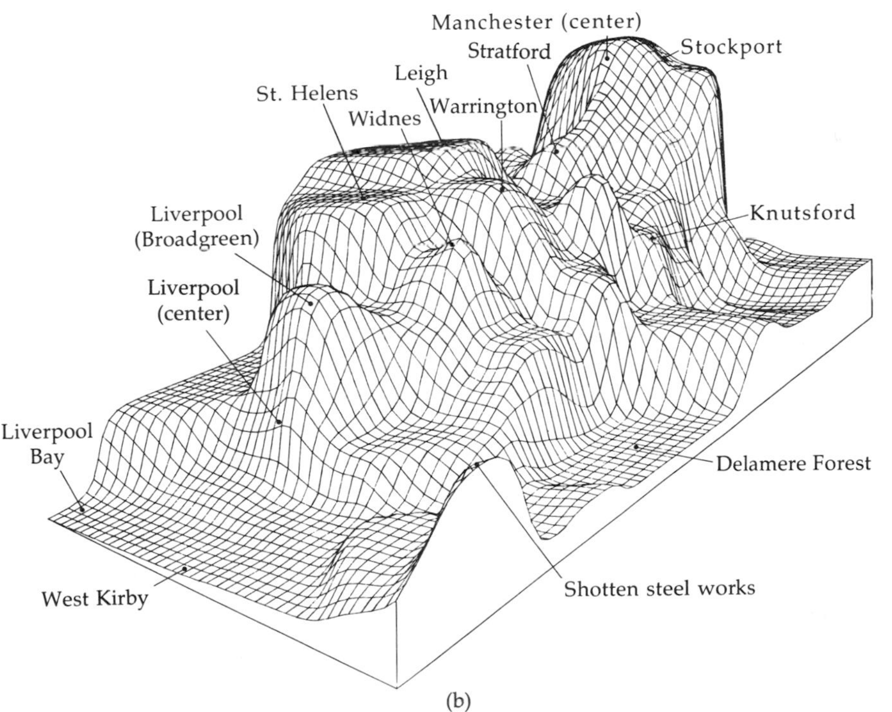

(b)

Figure 14. (a) Distribution of melanic individuals of *Biston betularia* over an area including Liverpool and Manchester, as viewed from rural Wales. (b) Distribution of melanic individuals of *Gonodontis bidentata* over a smaller area than in (a), but viewed from the same perspective. (From Bishop and Cook 1975.)

NATURAL SELECTION

So far in this book the term "natural selection" has been used in its informal, intuitive sense, the sense used by Darwin over a century ago (1872) in *The Origin of Species*:

> Owing to this struggle for life, variations, however slight and from whatever cause proceeding, if they be in any degree profitable to the individuals of a species, in their infinitely complex relations to other organic beings and to their physical conditions of life, will tend to the preservation of such individuals, and will generally be inherited by the offspring. The offspring, also, will thus have a better chance of surviving, for, of the many individuals of any species which are periodically born, but a small number can survive. I have called this principle, by which each slight variation, if useful, is preserved, by the term Natural Selection.

In this section, it is necessary to become more precise, more quantitative. First, it is important to note that natural selection is the driving force of evolution, the process that leads to greater adaptation of organisms to their environment. Subpopulations of a species are bound to live in somewhat different environments. If the environments are substantially different, natural selection, through the effect of increasing the adaptation of each subpopulation to its own environment, will promote genetic divergence of the subpopulations. If the environments are substantially similar, to consider the other extreme, natural selection, again through the effect of increasing adaptation, will tend to prevent genetic divergence of subpopulations. Random genetic drift enhances subpopulation divergence, whereas migration hinders it. Therefore, in any actual species, genetic variation produced by mutation is organized, maintained, eliminated, or dispersed among subpopulations according to the complex balance between natural selection, migration, and random genetic drift.

The Meaning of Fitness

Natural selection acts on phenotypes, not on genotypes, and it acts on the whole phenotype as determined by many genes and countless environmental factors in one of three fundamentally different ways. These are illustrated in Figure 15. When selection favors phenotypes at one extreme of the range, it is called **directional** selection (panel a). Selection favoring intermediate phenotypes is **stabilizing** or **normalizing** selection (panel b). Selection simultaneously favoring phenotypes at both extremes of the range is **disruptive** selection (panel c). Directional and stabilizing selection are discussed in detail

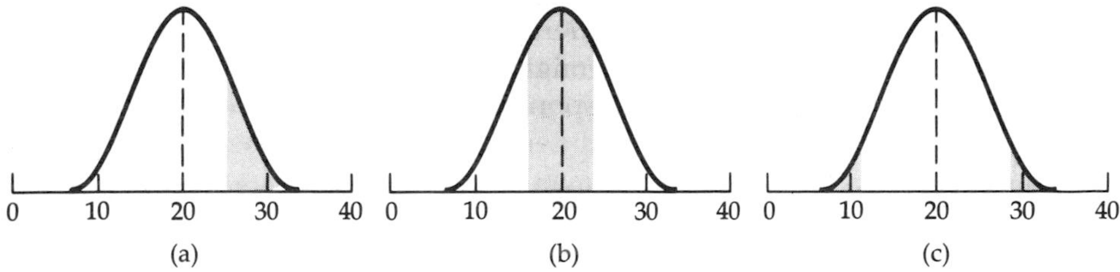

Figure 15. Three fundamentally different modes of selection. The curve represents the distribution of phenotypes in the population. In each case, the shading designates the individuals favored by natural selection. (a) Directional selection. (b) Stabilizing selection. (c) Disruptive selection.

in Chapter 4. Although selection occurs on the phenotype as determined by many genes, for purposes of exploring the consequences of selection, it is easiest to focus on how selection changes the allele frequencies of a single gene.

The simplest way to think about selection is in terms of **viability**, which refers to the probability that an individual survives from fertilization to reproductive age. For example, genotypes *AA*, *Aa*, and *aa* may have probabilities of survival (viabilities) of 0.75, 0.75, and 0.50, respectively. We call 0.75, 0.75, and 0.50 the **absolute viabilities** of the genotypes. However, selection depends only on the relative magnitudes of the viabilities, so it is often convenient to measure the viabilities relative to one another. In the above example, the relative viabilities of *AA*, *Aa*, and *aa* are (0.75)/(0.75), (0.75)/(0.75), and (0.50)/(0.75), or 1.0, 1.0, and 0.67, respectively. Of course, we could just as well write the relative viabilities as (0.75)/(0.50), (0.75)/(0.50), and (0.50)/(0.50), or 1.5, 1.5, and 1.0, respectively, the only difference being in which genotype is regarded as the standard. *Usually, the relative viabilities are calculated so that the largest relative viability equals 1.0.* Relative viabilities (such as the numbers 1.0, 1.0, and 0.67 in the preceding example) are called the **relative fitnesses** of the genotypes if the genotypes are otherwise equally capable of reproduction.

In certain instances, an approach to selection based on individual genotypes, analogous to viability differences, can also be used for fertility differences. However, this approach requires that the overall fertility of any mating pair be equal to the product of the fertilities of the individual genotypes in the mating pair. When this multiplication rule does not hold, models of selection with fertility differences become rather complex (see Ewens 1979; Clark and Feldman 1986).

Selection in Haploids

Selection in its simplest form occurs in haploid organisms such as bacteria. Consider two competing bacterial genotypes A and a with relative viabilities w_1 and w_2, respectively. Suppose that in generation t the frequencies of A and a are p_t and q_t, respectively, with $p_t + q_t = 1$. In the absence of selection, the frequencies remain the same through time, so we have $p_t/q_t = p_{t-1}/q_{t-1} = p_0/q_0$ for any value of t. With selection, a proportion w_1 of the A genotypes survive, and a proportion w_2 of the a genotypes survive, so $p_1/q_1 = p_0w_1/q_0w_2 = (p_0/q_0)(w_1/w_2)$. The relation between p_2/q_2 and p_1/q_1 is the same as that between p_1/q_1 and p_0/q_0, so $p_2/q_2 = (p_1/q_1)(w_1/w_2) = (p_0/q_0)(w_1/w_2)^2$. Continuing by generations in this manner, we have

$$p_t/q_t = (p_0/q_0)(w_1/w_2)^t \tag{2.22}$$

or

$$\ln(p_t/q_t) = \ln(p_0/q_0) + t\ln(w_1/w_2) \tag{2.23}$$

Problem 18 gives an example of the use of Equation 2.23.

PROBLEM 18 The *gnd* gene in *E. coli* codes for the enzyme 6–phosphogluconate dehydrogenase (6PGD), which is used in the metabolism of gluconate but not ribose. When otherwise genetically identical strains containing the naturally occurring alleles *gnd(RM77C)* and *gnd(RM43A)* were placed in competition in chemostats for gluconate or ribose, the data below were obtained, in which p denotes the frequency of the strain containing *gnd(RM43A)* (Dykhuizen and Hartl 1980; Hartl and Dykhuizen 1981). Estimate the fitness of the strain containing *gnd(RM77C)* relative to that containing *gnd(RM43A)* under the two growth conditions.

Genotype of strains	Medium	p_0	p_{35}
gnd(RM43A) versus *gnd(RM77C)*	gluconate	0.455	0.898
gnd(RM43A) versus *gnd(RM77C)*	ribose	0.594	0.587

ANSWER In gluconate medium, $\ln(0.898/0.102) = \ln(0.455/0.545) + [\ln(w_1/w_2)](35)$, so $\ln(w_1/w_2) = -0.0673$ and $w_1/w_2 = 1.0696$. Setting $w_1 = 1$ gives $\langle w_2 \rangle = 0.935$ as the estimated relative fitness of the *gnd(RM77C)* strain. In ribose medium, $\langle w_2 \rangle = 1.001$, which is not significantly different from 1.0.

Selection in Diploids

In diploids, the consequences of selection are most conveniently explored using the simple Hardy–Weinberg model in Chapter 1, but incorporating selection by permitting the fitnesses of the genotypes to differ. Selection is assumed to occur on the diploid genotypes, not on the gametes, and segregation is assumed to be Mendelian. For the relative fitnesses of *AA*, *Aa*, and *aa*, we use the conventional symbols w_{11}, w_{12}, and w_{22}, respectively. As noted earlier, the relative fitnesses are most easily interpreted in terms of viability, but, in certain cases, they can also be interpreted in terms of fertility.

If we assume that the allele frequencies of *A* and *a* are p_t and q_t, respectively, in generation t, then it is straightforward to derive expressions for p_{t+1} and q_{t+1}, which are the corresponding allele frequencies in the next generation. The subscripts t and $t + 1$ are cumbersome to carry along in equations, so we will use the symbols p and q for p_t and q_t, and the symbols p' and q' for p_{t+1} and q_{t+1}.

In terms of p and q, the genotype frequencies of *AA*, *Aa*, and *aa* among newly fertilized eggs in generation t are given by p^2, $2pq$, and q^2, respectively. By definition, newly fertilized eggs survive in the ratio $w_{11}:w_{12}:w_{22}$ for *AA*, *Aa*, and *aa*, so the ratio of *AA*:*Aa*:*aa* among adults is

$$p^2w_{11}:2pqw_{12}:q^2w_{22}$$

The ratio of *A*:*a* in gametes of the next generation is therefore

$$p^2w_{11} + 1/2(2pqw_{12}):1/2(2pqw_{12}) + q^2w_{22}$$

(The 1/2s enter the ratio because *Aa* heterozygotes produce 1/2 *A* and 1/2 *a* gametes from Mendelian segregation.) The ratio of *A*:*a* in gametes readily simplifies to

$$p(pw_{11} + qw_{12}):q(pw_{12} + qw_{22})$$

To obtain the *gametic frequencies* from the gametic ratio, one must divide each value in the ratio by the overall sum, conventionally denoted $\overline{w}$, which equals

$$\overline{w} = p^2w_{11} + 2pqw_{12} + q^2w_{22} \qquad \textbf{(2.24)}$$

Biologically speaking, the quantity $\overline{w}$ is simply the **average fitness** of the individuals in the population. Dividing the gametic ratio by $\overline{w}$ gives the gametic frequencies in generation $t + 1$, namely

$$\begin{aligned} p' &= \frac{p(pw_{11} + qw_{12})}{\overline{w}} \\ q' &= \frac{q(pw_{12} + qw_{22})}{\overline{w}} \end{aligned} \qquad \textbf{(2.25)}$$

It is often useful to know $p' - p$, usually symbolized Δp, which is the difference in allele frequency which results from one generation of selection. In this case,

$$\Delta p = \frac{pq[p(w_{11} - w_{12}) + q(w_{12} - w_{22})]}{\overline{w}} \qquad \textbf{(2.26)}$$

At this point, an example of the use of these equations is in order. We will use data of Teissier (1942) on the change in the frequency of the *Cy* (*Curly* wings) allele in a laboratory population of *Drosophila melanogaster*, which are plotted in Figure 16. The *Cy* allele is lethal when homozygous, so $w_{11} = 0$. The data in Figure 16 pertain to the frequency of *Cy* heterozygotes, but because *Cy/Cy* genotypes do not survive, the allele frequency p of *Cy* equals one-half the frequency of *Cy*/+ adults. The points in Figure 16 are each separated by one generation, and the initial generation has a frequency of *Cy*/+ adults of 0.67, so $p = 0.335$ and thus $q = 0.665$. Wright (1977) has studied these data and concluded that $w_{12} = 0.5$ for *Cy*/+ genotypes, relative to a value of $w_{22} = 1.0$ for +/+ genotypes. Substituting these values for p, q, w_{11}, w_{12}, and w_{22} into Equation 2.25 for p' yields

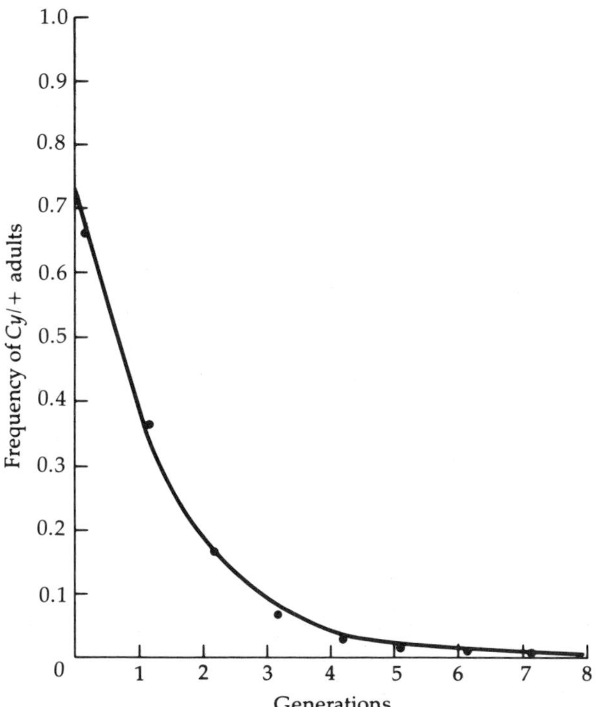

Figure 16. Change in frequency of adult *Drosophila melanogaster* heterozygous for the dominant mutation *Cy* (*Curly* wings) in an experimental population. *Cy/Cy* homozygotes are inviable, and the curve represents the theoretical change in frequency when the ratio of viabilities of *Cy*/+ to +/+ is 0.5:1. (Data from Teissier 1942. The fitness value of 0.5 was estimated by Wright 1977.)

$$p' = \frac{(0.335)[(0.335)(0) + (0.665)(0.5)]}{(0.335)^2(0) + 2(0.335)(0.665)(0.5) + (0.665)^2(1.0)} = 0.168$$

Thus, the predicted frequency of *Cy*/+ adults in generation 1 is $2p' = 0.336$, which is reasonably close to the observed value of 0.368.

PROBLEM 19

Assuming a value of $p = 0.168$ for the frequency of the *Cy* allele in generation 1 in the population in Figure 16, calculate the expected frequency of *Cy*/+ heterozygotes in generation 2.

ANSWER In this example, $p' = (0.168)[(0.168)(0) + (0.832)(0.5)]/[(0.168)^2(0) + 2(0.168)(0.832)(0.5) + (0.832)^2(1.0)] = 0.084$, so the expected frequency of *Cy*/+ adults is $2p' = 0.168$. This is very close to the observed value of 0.165. The theoretical curve in Figure 16 was calculated using this generation-by-generation procedure.

We may be excused a slight digression to point out that it is sometimes convenient to think in terms of the **marginal fitnesses** of the *A* and *a* alleles, which equals the average fitness of all genotypes containing *A* or *a*, respectively, weighed by their relative frequency and the number of *A* or *a* alleles they contain. For example, *A* alleles occur in *AA* and *Aa* genotypes with relative probabilities p and q, and therefore the marginal fitness $\overline{w}_1$ of *A*-containing genotypes equals $pw_{11} + qw_{12}$. Similarly, the marginal fitness of *a*-containing genotypes is $\overline{w}_2 = pw_{12} + qw_{22}$. Equation 2.25 for p' therefore becomes $p' = p\overline{w}_1/\overline{w}$, and Equation 2.26 becomes $\Delta p = p(\overline{w}_1 - \overline{w})/\overline{w}$. This expression makes it clear that any allele increases in frequency if the marginal fitness of genotypes containing the allele ($\overline{w}_1$) is greater than the average fitness of the population ($\overline{w}$). This approach also generalizes nicely to multiple alleles: for an allele with frequency p_i and marginal fitness $\overline{w}_i$, the change in frequency in one generation equals $\Delta p_i = p_i(\overline{w}_i - \overline{w})/\overline{w}$.

Time Required for Changes in Gene Frequency

Having derived Equation 2.25 for p_{t+1} in terms of p_t (i.e., p' in terms of p), it is an appropriate next step to express p_t in terms of p_0, as we did earlier in this chapter for the analogous equations involving mutation and migration. Unfortunately, in the case of selection, a simple formula for p_t in terms of p_0 does not exist. However, if w_{11}, w_{12}, and w_{22} are not too different, then $\overline{w}$ is close to 1.0, and the expression for Δp can be used to derive approximations. Obtaining the approximations requires calculus (the idea is to treat Δp as the derivative (dp/dt) and then to integrate), but for our purposes, only the final answers are of interest. The answers are most easily presented if we change the symbolism somewhat. For this purpose, we rewrite the fitnesses of the genotypes as follows:

$$\begin{aligned} w_{11} &= 1 \\ w_{12} &= 1 - hs \\ w_{22} &= 1 - s \end{aligned} \qquad \textbf{(2.27)}$$

Use of the h and s symbols for the fitnesses in Equation 2.27 has the advantage of making the gene effects and dominance explicit. If s is positive and h is nonnegative, the type of selection is directional selection, and A is the favored allele. In this context, s is called the **selection coefficient** against the *aa* genotype, and h is called the **degree of dominance** of the *a* allele. For example, when $h = 0$, the fitnesses of *AA*, *Aa*, and *aa* are 1, 1, and $1 - s$, respectively, and *a* is recessive to *A*. Alternatively, when $h = 1$, the fitnesses are 1, $1 - s$, and $1 - s$, respectively, and *a* is dominant to *A*. In terms of the selection coefficient and the degree of dominance, an approximate expression for Δp can be derived from Equation 2.26 as

$$\Delta p = pqs[ph + q(1 - h)] \qquad \textbf{(2.28)}$$

where the approximation in Equation 2.28 assumes that s and hs are sufficiently small that $\overline{w}$ is close to 1.0.

The following equations give p_t in terms of p_0 in three cases of importance (for methods of derivation, see Hartl 1980):

1. *A* is a **favored dominant**. In this case, $h = 0$, so $\Delta p = pq^2s$, and

$$\ln(p_t/q_t) + 1/q_t = [\ln(p_0/q_0) + 1/q_0] + st \qquad \textbf{(2.29)}$$

2. *A* is favored and the alleles are **additive** in their effects on fitness. Additive effects on fitness means that the fitness of the heterozygote is exactly intermediate between the fitnesses of the homozygotes, so in the additive case, $h = 1/2$. (Some authors refer to the additive case as *semidominance*, others use the term *genic selection*.) When $h = 1/2$, then $\Delta p = pqs/2$, and

$$\ln(p_t/q_t) = \ln(p_0/q_0) + (s/2)t \qquad \textbf{(2.30)}$$

Note that Equation 2.30 for additive alleles is similar in form to Equation 2.23 for haploid selection, when $w_1 = 1$ and $w_2 = 1 - (s/2)$ and s is small.

3. *A* is a **favored recessive**. In this case, $h = 1$, so $\Delta p = p^2qs$, and

$$\ln(p_t/q_t) - 1/p_t = [\ln(p_0/q_0) - 1/p_0] + st \qquad \textbf{(2.31)}$$

Some of the implications of these equations are explored in Problems 20 through 22. One implication should be spelled out immediately, however. The case of interest pertains to rare, harmful recessives, so the favored allele is dominant and, from Equation 2.28, $\Delta p = pq^2s$. Because the harmful allele is assumed to be rare, q is close to 0, and q^2 is therefore extremely small. Consequently, an increase in s from a value of, say, 0.5 to a value of, say, 1, has a trivial effect on the change in allele frequency because, with q^2 so small, the actual value of s matters little. The change in allele frequency of a rare harmful recessive is slow whatever the value of s. For this reason, an increase in selection against rare homozygous recessive genotypes has almost no effect in changing the allele frequency. The implication for human population genetics is that the forced sterilization of rare homozygous recessive individuals—a procedure advocated in a number of naive eugenic programs to "improve" the "genetic quality" of human beings—is not only morally and ethically questionable, it is genetically unsound.

Changes in allele frequency for Equations 2.29 through 2.31 are shown in Figure 17. Note that the change in frequency of a favored dominant allele is slow when the allele is common, which implies that the change in frequency of a harmful recessive is slow when the recessive is rare. Conversely, a favored recessive allele changes in frequency most slowly when rare, which implies that a harmful dominant allele changes most slowly in frequency when common. In general, *recessive alleles change slowly in frequency when they are rare, dominant alleles change slowly in frequency when they are common*. With a favorable additive allele, the initial increase in frequency is slower than that of a favored dominant, but the additive allele eventually overtakes and goes to fixation faster because selection continues to distinguish between the homozygous and heterozygous genotypes.

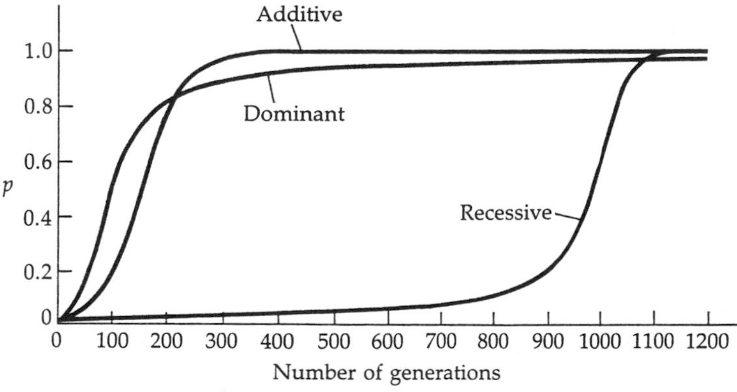

Figure 17. Changes in frequency p of favorable alleles which are dominant, additive, or recessive in their effects on fitness. The frequency of a favored dominant allele changes most slowly when the allele is common, and the frequency of a favored recessive allele changes most slowly when the allele is rare. In all cases the difference in relative fitness between the contrasting homozygous genotypes is assumed to be five percent.

PROBLEM 20

In the evolution of industrial melanism (see Problem 9 in Chapter 1), the allele resulting in black body coloration may be considered a favored dominant. In *Biston betularia*, the frequency of melanic individuals increased from 1 percent in 1848 to 95 percent in 1898. This species has one generation per year. Estimate the approximate value of the selection coefficient s against nonmelanics that would be necessary to account for the change in frequency of the melanic forms. How many generations would be required for the same change in frequency of melanic forms in a hypothetical case in which the allele for melanism is recessive, assuming the same value of s against nonmelanics?

ANSWER Let q represent the frequency of the normal (nonmelanic) allele. Then q^2 is the frequency of nonmelanic moths because the normal allele is recessive. Thus $q_0^2 = 1 - 0.01 = 0.99$, or $q_0 = 0.995$ and $p_0 = 0.005$. Also, $q_t^2 = 1 - 0.95 = 0.05$, so $q_t = 0.224$ and $p_t = 0.776$. Since the species has one generation per year, $t = 1898 - 1848 = 50$ generations. Now substitute into Equation 2.29 for a favored dominant: $\ln[(0.776)/(0.224)] + [1/(0.224)] = \ln[(0.005)/(0.995)] + [1/(0.995)] + s(50)$, or $5.707 = -4.288 + 50s$, or $\langle s \rangle = 0.20$. Thus, the relative fitnesses of melanics and nonmelanics required to account for the change are $1{:}1 - s$ or 1:0.8. In the hypothetical case of recessiveness for the melanic allele, $p_0^2 = 0.01$, so $p_0 = 0.1$ ($q_0 = 0.9$); and $p_t^2 = 0.95$, so $p_t = 0.975$ ($q_t = 0.025$); now use Equation 2.31 for a favored recessive with $s = 0.20$, which yields $t = 74.2$ generations. Over the past 20 years, smoke-control programs in industrial areas of England have resulted in a decrease in the frequency of melanic moths consistent with a selection coefficient against melanics in smoke-controlled areas of 0.12 ± 0.02 (Cook et al. 1986).

PROBLEM 21

An extensively studied isolated colony of the moth *Panaxia dominula* was cited earlier in this chapter (Problem 7) as an example of fluctuations in population size. This colony also contains a mutant allele affecting color pattern. The frequency of the mutant allele declined steadily over the period 1939 to 1968. Indeed, the corresponding steady increase in the frequency of the normal allele follows Equation 2.30 for additive genes with $s = 0.20$ (see Wright, 1978, for a graph). The species has one generation per year, and in 1965, the estimated frequency of the mutant allele was 0.008 (actually, this value is the average for the 7-year period 1962 to 1968). Estimate the frequency of the mutant allele in 1950 and in 1940.

ANSWER Here we are given $\langle q_t \rangle$ and want to use Equation 2.30 "backward" to estimate $\langle q_0 \rangle$. Between 1950 and 1965, there were $t = 1965 - 1950 = 15$ generations. We are given $\langle q_t \rangle = 0.008$, so $p_t = 0.992$, and $\ln(0.992/0.008) = 4.820$. Thus, $4.820 = \ln(p_0/q_0) + (0.20/2)(15)$, or $\ln(p_0/q_0) = 3.32$. Then $p_0/q_0 = e^{3.32} = 27.660$, or $\langle p_0 \rangle = 0.965$ and $\langle q_0 \rangle = 0.035$. For the year 1940, $t = 1965 - 1940 = 25$ generations, from which $\langle p_0 \rangle = 0.911$ and $\langle q_0 \rangle = 0.089$. (Based on actual observations made at the time, the estimates of $\langle q \rangle$ were 0.037 in 1950 and 0.111 in 1940.)

Some of the most dramatic examples of evolution in action result from the natural selection for chemical pesticide resistance in natural populations of insects and other agricultural pests. In the 1940s, when chemical pesticides were first applied on a large scale, an estimated 7 percent of the agricultural crops in the United States were lost to insects. Initial successes in chemical pest management were followed by gradual loss of effectiveness. Today, more than 400 pest species have evolved significant resistance to one or more pesticides, and 13 percent of the agricultural crops in the United States are lost to insects (May 1985).

In many cases, significant pesticide resistance has evolved in 5–50 generations irrespective of the insect species, geographical region, pesticide, frequency and method of application, and other seemingly important variables (May 1985). Equations 2.29 through 2.31 help to understand this apparent paradox, since many of the resistant phenotypes result from single mutant genes. The resistant genes are often partially or completely dominant, so Equations 2.29 and 2.30 are applicable. Prior to application of the pesticide, the allele frequency p_0 of the resistant mutant is generally close to 0. Application of the pesticide increases the allele frequency, sometimes by many orders of magnitude, but significant resistance is noticed in the pest population even before the allele frequency p_t increases above a few percent. Thus, as rough approximations, we may assume that q_0 and q_t are both close enough to 1 that $\ln(p_0/q_0) \simeq \ln(p_0)$ and $\ln(p_t/q_t) \simeq \ln(p_t)$. Using these approximations, Equation 2.30 implies that $t \simeq (2/s)\ln(p_t/p_0)$ [additive case] and Equation 2.29 implies that $t \simeq (1/s)\ln(p_t/p_0)$ [dominant case]. In many cases, the ratio p_t/p_0 may range from 1×10^2 to perhaps 1×10^7, and s may typically be 0.5 or greater. Over this wide range of parameter values, the time t is effectively limited to 5–50 generations for the appearance of a significant degree of pesticide resistance. Details in actual cases will depend on such factors as effective population number and extent of genetic isolation between local populations, and the evolution of polygenic resistance may be expected to take somewhat longer than single-gene resistance.

PROBLEM 22

Evaluate the adequacy of the approximations for t with additive and dominant genes, as compared with values calculated from Equations 2.29 and 2.30, in the following cases:

Case	p_0	p_t	s
1	1×10^{-4}	0.01	0.5
2	1×10^{-4}	0.10	0.5
3	1×10^{-4}	0.50	0.5
4	1×10^{-7}	0.10	0.5
5	1×10^{-4}	0.10	0.20

ANSWER In the following, ADDAPX = $(2/s)\ln(p_t/p_0)$, DOMAPX = $(1/s)\ln(p_t/p_0)$, ADD = Value from Equation 2.30, and DOM = Value from Equation 2.29.

Case	ADDAPX	ADD	DOMAPX	DOM
1	18.4	18.5	9.2	9.3
2	27.6	28.1	13.8	14.2
3	34.1	36.8	17.0	20.4
4	55.3	55.7	27.6	28.1
5	69.1	70.1	34.5	35.6

In all cases the approximations are acceptable.

Equilibria with Selection

An **equilibrium** value of p is a value for which $\Delta p = 0$. That is, when the allele frequency is at an equilibrium value, the allele frequency remains at that value generation after generation. There are, however, several types of equilibria depending on what happens to allele frequency when the allele frequency does *not* equal the equilibrium value. Consider first the case when the initial allele frequency is near (but not equal to) the equilibrium: if the allele frequency moves progressively farther away from the equilibrium in subsequent generations, the equilibrium is said to be **unstable**; if the allele frequency moves progressively closer to the equilibrium in subsequent generations, the equilibrium is said to be **locally stable**. A locally stable equilibrium might also be **globally stable**, which means that, whatever the initial allele frequency may be, it always moves progressively closer to the equilibrium. In cases such as those exemplified by the Hardy–Weinberg equilibrium, in which every allele frequency represents an equilibrium because, whatever the value, allele frequency does not change, the equilibria are said to be **semistable** or **neutrally stable**.

These concepts of stability can be applied to the case of directional selection for the *A* allele governed by Equation 2.26. In the case of directional selection for *A*, there are only two equilibria, namely $p = 0$ and $p = 1$. If p is close to 0, p increases in subsequent generations, so the equilibrium at $p = 0$ is unstable. On the other hand, if p is near 1, it moves still closer to 1 in subsequent generations, so the equilibrium at $p = 1$ is locally stable. Indeed, because p eventually goes to 1 whatever its initial value, the equilibrium at $p = 1$ is globally stable.

The various types of stability are important in discussing two further cases that can occur when selection involves two alleles of a single gene.

Overdominance or **heterozygote superiority** occurs when the heterozygote has a greater fitness than both homozygotes—that is, when w_{12} is greater than both w_{11} and w_{22}. When overdominance occurs, there is a third equilibrium in addition to $p = 0$ and $p = 1$ because $p(w_{11} - w_{12}) + q(w_{12} - w_{22})$ can equal 0. The third equilibrium, $\hat{p}$, can be found by solving $\hat{p}(w_{11} - w_{12}) + (1 - \hat{p})(w_{12} - w_{22}) = 0$, from which a little algebra gives

$$\hat{p} = \frac{w_{12} - w_{22}}{2w_{12} - w_{11} - w_{22}} \tag{2.32}$$

The equilibrium in Equation 2.32 is globally stable, whereas those at $p = 0$ and $p = 1$ are now unstable, as indicated in Figure 18(a) where the arrows show the direction of change in allele frequency. Figure 18(b) shows the course of change in $\overline{w}$ with overdominance, and it is of interest that the average fitness of the population is maximal when $p = \hat{p}$. Maximization of average fitness is a frequent outcome of selection in random-mating populations with constant fitnesses, but there are many exceptions to the rule when random mating does not occur, when fitnesses are not constant, or when more than one gene is involved (Ewens 1979; Curtsinger 1984).

Although overdominance might seem to be a potent force for maintaining polymorphisms in natural populations, overdominance has been documented in only a few cases. The best known case involves two alleles which code for the β chain of human hemoglobin, $Hb\beta^{+}$, the normal allele, and $Hb\beta^{S}$, the mutant allele with an amino acid substitution that causes sickle-

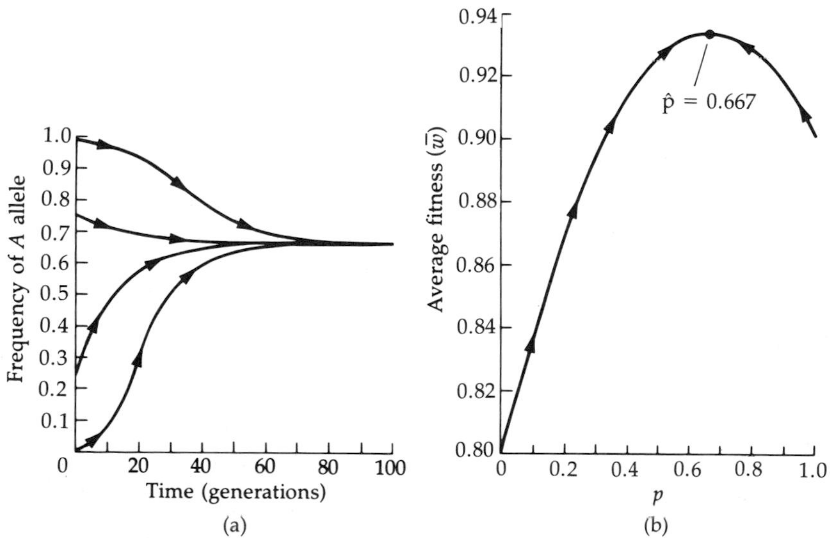

Figure 18. Selection when there is overdominance. (a) The allele frequencies converge to an equilibrium value irrespective of the initial frequency. In this example, $w_{11} = 0.9$, $w_{12} = 1$, $w_{22} = 0.8$, and the equilibrium frequency of the A allele is 0.667. (b) Average fitness $\overline{w}$ against p for the same example as in (a). Note that $\overline{w}$ is a maximum at equilibrium.

cell anemia. Homozygous $Hb\beta^S/Hb\beta^S$ individuals are severely anemic owing to sickling of red blood cells; $Hb\beta^+/Hb\beta^+$ individuals have normal hemoglobin but are sensitive to the type of malaria caused by the mosquito-borne protozoan parasite *Plasmodium falciparum*; and $Hb\beta^+/Hb\beta^S$ heterozygotes are mildly anemic but tend to be resistant to falciparum malaria, perhaps because red blood cells infested with the parasite undergo sickling and are removed from circulation. The viabilities of $Hb\beta^S/Hb\beta^S$, $Hb\beta^+/Hb\beta^S$, and $Hb\beta^+/Hb\beta^+$ individuals in high-malaria regions in Africa have been estimated as $w_{11} = 0$, $w_{12} = 1$, and $w_{22} = 0.85$, respectively (Allison 1964). Substitution into Equation 2.32 leads to a predicted equilibrium allele frequency for $Hb\beta^S$ of $(1 - 0.85)/(2 - 0 - 0.85) = 0.13$. This value is reasonably close to the average allele frequency of 0.09 observed in West Africa, but there is considerable variation in allele frequency among local populations.

PROBLEM 23 Anderson et al. (1968) established experimental populations of *Drosophila pseudoobscura* and periodically treated the flies with weak doses of the insecticide DDT. One population was initially polymorphic for five different inversions of the third chromosome, in approximately equal frequencies. After 13 generations, three of the inversions had essentially disappeared from the population. The two that remained were Standard (*ST*) and Arrowhead (*AR*). Changes in frequency of each inversion were monitored, and from the values for the first nine generations, DuMouchel and Anderson (1968) estimated the relative fitnesses of *ST/ST*, *ST/AR*, and *AR/AR* genotypes as 0.47, 1.0, and 0.62, respectively. Because the inversions undergo almost no recombination with one another, each type can be considered as an "allele." What equilibrium frequency of *ST* is predicted? What equilibrium value of $\overline{w}$ is predicted?

ANSWER Let $\hat{p}$ represent the equilibrium frequency of *ST*. Then, using Equation 2.32, $\hat{p} = (1.0 - 0.62)/(2.0 - 0.47 - 0.62) = 0.42$. (The value of $\langle p \rangle$ after 13 generations was actually 0.43.) Predicted $\overline{w}$ from Equation 2.24 equals $(0.42)^2(0.47) + 2(0.42)(0.58)(1.0) + (0.58)^2(0.62) = 0.78$.

PROBLEM 24 Warfarin is a blood anticoagulant used for rat control during and after World War II. Initially highly successful, the effectiveness of the rodenticide has gradually diminished owing to the evolution of resistance among some target populations. Among Norway rats in Great Britain, resistance results from an otherwise harmful mutation *R* in a gene in which the normal nonresistant allele may be denoted *S*. In the absence of warfarin, the relative fitnesses of *SS*, *SR*, and *RR* genotypes have been estimated as 1.00, 0.77, and 0.46, respectively. In the presence of warfarin, the relative fitnesses have been estimated as 0.68, 1.00, and 0.37, respectively (Greaves, quoted in May 1985). Calculate the equilibrium frequency $\hat{q}$ of *R* in the presence of warfarin. Noting that, in the absence of warfarin, *R* and *S* are nearly additive in their effects on fitness, estimate the approximate number of generations required for the allele frequency of *R* to decrease from $\hat{q}$ to 0.01 in the absence of the poison.

ANSWER From Equation 2.32, the equilibrium frequency $\hat{p}$ of S equals (1.00 − 0.37)/[2(1.00) − 0.68 − 0.37] = 0.66, so $\hat{q}$ = 0.34. Setting q_0 = 0.01 and q_t = 0.34 in Equation 2.30, with s = 1.00 − 0.46 = 0.54, yields t = 14.6 generations. The overdominance that occurs in the presence of warfarin appears to result from the homozygous *RR* genotypes having an excessive requirement for vitamin K.

Heterozygote inferiority refers to the opposite situation from overdominance, and it occurs when the fitness of the heterozygous genotype is smaller than that of both homozygotes (w_{12} less than w_{11} and w_{22}). A third equilibrium also exists in this case, and its value is given by Equation 2.32. However, in the case of heterozygote inferiority, the equilibrium is *unstable*, whereas the equilibria at p = 0 and p = 1 are both locally (but not globally) stable. An example of heterozygote inferiority is depicted in Figure 19(a), where the arrows again denote the direction of change in allele frequency. If the initial allele frequency is exactly equal to the equilibrium value (in this example, $\hat{p}$ = 1/3), then the allele frequency remains at that value. In all other cases,

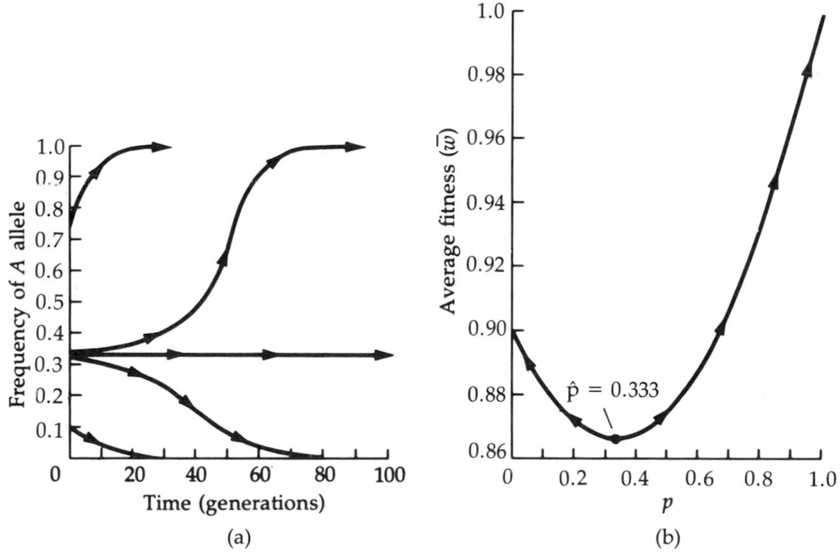

Figure 19. Selection when there is heterozygote inferiority. (a) The allele frequency goes to 0 or 1, depending on the initial frequency. In this example, w_{11} = 1, w_{12} = 0.8, w_{22} = 0.9, and there is an unstable equilibrium when the frequency of the *A* allele is p = 0.333. An infinite population with p = 1/3 maintains this frequency, but any slight upward change in the frequency of *A* results in eventual fixation of *A*, and any slight downward change in the frequency of *A* results in ultimate loss of *A*. (b) Average fitness $\overline{w}$ against p for same example as in (a). The unstable equilibrium represents the minimum of $\overline{w}$.

p goes to 1 or 0, depending on whether the initial allele frequency was above or below the equilibrium value.

Figure 19(b) shows the situation regarding average fitness when there is heterozygote inferiority. Note that the unstable equilibrium at $\hat{p} = 1/3$ represents a minimum of average fitness. The shape of the $\overline{w}$ curve has an important implication, especially in more complex examples. Imagine a population with an allele frequency near 0, where $\overline{w} = 0.9$. In terms of average fitness, the population would be better off if the allele frequency were near 1, because $\overline{w}$ would then be close to 1.0. However, as shown by the direction of the arrows, the population cannot evolve toward $p = 1$ because it cannot get through the "valley." To say the same thing in a somewhat different way, $p = 0$ represents a locally stable equilibrium. The population has no way to escape from the equilibrium even though, in doing so, it would eventually end up with a greater average fitness. This consideration would seem to limit the ability of natural selection to increase average fitness in such cases, but one way out of the impasse is suggested in the next section.

Adaptive Topographies and the Role of Random Genetic Drift

Any graph of $\overline{w}$ against allele frequency, even the simple example in Figure 19(b), is called an **adaptive topography**. In order to generalize the discussion a bit, try to imagine an adaptive topography in many dimensions, which would occur in more realistic situations in which $\overline{w}$ depended on the allele frequencies of many genes instead of only one. In many dimensions, the adaptive topography is a complex surface and, on this surface, there may be "peaks" and "pits" and even "saddle-shaped" regions. In this context, the peaks represent locally stable equilibria, and natural selection ordinarily changes the allele frequencies so as to move $\overline{w}$ to the top of some peak. However, the peak that a population perches upon may not be the highest peak that exists on the whole surface. However, as illustrated in Figure 19(b), the population may become stuck there because the peak is a locally stable equilibrium.

By what process can a population stranded on a submaximal fitness peak get off the peak, through a nearby valley, and onto a place where natural selection can carry it to the top of an even higher fitness peak? This is something that natural selection acting alone cannot accomplish because it entails a temporary reduction in fitness. There is, however, a process that can accomplish the task—random genetic drift. In a sufficiently small population, the allele frequencies can change by chance, even producing a reduction in average fitness. Theoretically, random genetic drift can shift a population from a locally stable equilibrium, through a nearby valley, and into a region where it is attracted by another locally stable equilibrium associated with a higher fitness peak. Random genetic drift can therefore play a crucial role in evolution by allowing a population to explore the full range of its adaptive topography. This role of random genetic drift has been particularly emphasized by Wright (1977 and earlier) in his proposed **shifting balance theory** of evolution. Additional discussion of the theory is found in the section below on interdemic selection, and also in Hartl (1979) and Provine (1986).

Mutation–Selection Balance

Recall from Chapter 1 that outcrossing species typically contain a large amount of hidden genetic variability in the form of recessive or nearly recessive harmful alleles at low frequencies. This situation is to be expected. Selection cannot completely eliminate harmful alleles because of their continual creation through recurrent mutation. To be specific, suppose a is a harmful allele, and consider the simple model of directional selection for A. Suppose mutation of A to a occurs at the rate μ per generation. (Because q, the allele frequency of a, remains small, reverse mutation of a to A can safely be ignored.) The calculation of p' carried out for Equation 2.25 is still valid, except that a proportion μ of A alleles mutate to a in the course of one generation. Thus,

$$p' = \frac{p(pw_{11} + qw_{12})(1 - \mu)}{\overline{w}} \qquad \textbf{(2.33)}$$

In the following, it is again convenient to use the fitness symbols defined in Equation 2.27, namely $w_{11} = 1$, $w_{12} = 1 - hs$, and $w_{22} = 1 - s$.

When selection is balanced by recurrent mutation, there is a globally stable equilibrium at an allele frequency of $\hat{p}$, where $\hat{p}$ can be found by solving Equation 2.33 with $p' = p = \hat{p}$. The equilibrium frequency of the harmful a allele is therefore $\hat{q} = 1 - \hat{p}$, and two important cases arise:

1. When the harmful allele is *completely recessive* ($h = 0$), then

$$\hat{q} = \sqrt{(\mu/s)} \qquad \textbf{(2.34)}$$

2. When the harmful allele is *partially dominant* ($h > 0$), then, approximately,

$$\hat{q} = \mu/hs \qquad \textbf{(2.35)}$$

(Equation 2.35 is an excellent approximation for realistic values of μ, h, and s.)

The Huntington disease (formerly called Huntington's chorea) serves as an example of the use of Equation 2.35. The disease is a severe degenerative disorder of the neuromuscular system that typically appears after age 35. Although the disease itself results from a "dominant" gene, Reed and Neel

(1959) estimated the fitnesses as $\langle w_{11} \rangle = 1$, $\langle w_{12} \rangle = 0.81$, and $\langle w_{22} \rangle = 0$, where w_{12} and w_{22} refer, respectively, to genotypes which are heterozygous or homozygous for the Huntington allele. Therefore, we may estimate $\langle s \rangle = 1 - \langle w_{22} \rangle = 1$ and $\langle hs \rangle = 1 - \langle w_{12} \rangle = 0.19$, so $\langle h \rangle = 0.19$. In terms of neuromuscular degeneration, the Huntington allele is dominant. In terms of fitness, however, the allele is only partially dominant owing to the late age of onset of the disease.

Because we are dealing with a partial dominant in the case of Huntington disease, Equation 2.35 is appropriate. Using Equation 2.35, we could estimate $\hat{q}$ if the value of μ were known, or we could estimate μ if $\hat{q}$ were known. In the Michigan population studied by Reed and Neel (1959), $\langle \hat{q} \rangle = 5 \times 10^{-5}$, so (assuming that the population is in equilibrium) we have $\langle \mu \rangle = (5 \times 10^{-5})(0.19) = 9.5 \times 10^{-6}$. This example illustrates one of the common indirect methods for the estimation of mutation rates in humans.

PROBLEM 25 A small amount of dominance can have a major effect in reducing the equilibrium frequency of a harmful allele. To confirm this for yourself, imagine an allele that is lethal when homozygous ($s = 1$) in a population of *Drosophila*. Suppose that the allele is maintained by selection-mutation balance with $\mu = 5 \times 10^{-6}$. Calculate the equilibrium frequency of the allele in the cases (1) complete recessive and (2) partial dominant with $h = 0.025$.

ANSWER Case 1: $\hat{q} = \sqrt{(\mu/s)} = \sqrt{[(5 \times 10^{-6})/1]} = 2.24 \times 10^{-3}$. Case 2: $\hat{q} = (\mu/hs) = [(5 \times 10^{-6})/(0.025)(1)] = 2.00 \times 10^{-4}$. The equilibrium allele frequency is reduced more than tenfold, and the frequency of homozygous recessive *genotypes* at equilibrium is reduced more than a hundredfold. It is of interest that $h = 0.025$ is near the average degree of dominance estimated for "recessive" lethals in *Drosophila* (Simmons and Crow 1977).

MORE COMPLEX TYPES OF SELECTION

We have considered only the simplest type of selection involving two alleles of a single gene with constant fitnesses determined by differences in viability. Actual mechanisms of selection in natural populations are usually more complex. As mentioned earlier, when fertilities differ among genotypes, then the simple model of selection on viability is inadequate except in special cases. Additional complications arise in actual situations because fitness is determined by many genes that may interact with each other. Simple models of selection have validity only when the genes interact in an extremely simple manner (for example, when fitness effects are additive or multiplicative across genes), or when one gene has such major effects on fitness that it overwhelms the effects of other genes. Still more complications arise because the fitnesses of the genotypes may not actually be constants, but may vary from generation to generation depending on, for example, the weather, or from place to place depending on local conditions, such as availability of food, water, or nesting sites. Hedrick (1986) has reviewed the subject of environmental heterogeneity in relation to its possible role in the maintenance of genetic polymorphisms. Realistic models of natural selection with varying fitnesses or other complications are often complex. (See Hartl 1980 for specific references.)

Allele frequencies in natural populations may vary systematically according to geographical location. The frequency of an allele may increase or decrease in a regular trend which accompanies a change in an environmental variable such as temperature, salinity (for marine organisms), population density, or some combination of environmental variables. Geographical trends in allele frequency are called **clines**.

A particularly dramatic example of a cline is found in the *hemoglobin*$-I^1$ allele in the eelpout fish *Zoarces viviparus,* the allele frequency of which drops from a value of nearly 1 in the North Sea to a value of nearly 0 in the Baltic Sea (Christiansen and Frydenberg 1974). In human aboriginal populations, there is a cline of increasing frequency of the I^B allele in the ABO blood groups from Southwest to Northeast Europe. Although clines can result from selection—for example, when one genotype is favored at one extreme of the environmental gradient but disfavored at the other extreme—clines can also result from other processes. For example, differences in allele frequency in local populations at the extremes of the range may result from founder effects, and migration of organisms from the extremes into the intermediate zone produces the cline.

Interdeme Selection and the Shifting Balance Theory

Two types of selection should be singled out for brief discussion here because they involve extended concepts of "fitness." One type is called **interdeme selection**, and it occurs between semi-isolated populations (demes) of the same species. If populations containing certain genotypes are more likely to become extinct and have their vacated habitats recolonized by migrants from other populations that are more persistent due to the particular genotypes that they contain, then the more successful populations can in some sense be considered as having a greater "fitness" than the less successful ones. Since this concept of **population fitness** is a characteristic of the entire population and not merely the average fitness of the genotypes within it ($\overline{w}$), interdeme selection is outside the realm of most conventional models of selection. Interdeme selection is one type of *group selection* (Wilson 1983).

Interdeme selection plays an essential role in the shifting balance theory of evolution (Wright 1977 and earlier). In Wright's view, subdivision of a population into small semi-isolated demes gives the best chance for the populations to explore the full range of their adaptive topography. Temporary reductions in fitness which would be prevented by selection in large populations become possible in small ones because of the random drift in allele frequencies that occurs in small populations. The lucky subpopulations which reach higher adaptive peaks on the fitness surface increase in size and send out more migrants than other subpopulations, and the favorable gene combinations are gradually spread throughout the entire set of subpopulations by means of interdeme selection. The shifting balance process includes three distinct phases:

1. An exploratory phase, in which random genetic drift plays an important role in allowing small populations to explore their adaptive topography;
2. A phase of mass selection, in which favorable gene combinations created by chance in phase (1) become rapidly incorporated into the genome of local populations by the action of natural selection; and
3. A phase of interdeme selection, in which the more successful demes increase in size and rate of migration, and the excess migration shifts the allele frequencies of nearby populations until they also come under the control of the higher fitness peak. The favorable genotypes thereby become spread throughout the entire population in ever-widening concentric circles. Where the region of spread from two such centers overlaps, a new and still more favorable genotype may occur and itself become a center for interdeme selection. In this manner, the whole of the adaptive topography can be explored, and there is a continual shifting of control from one adaptive peak to control by a superior one.

The shifting balance theory has played an important role in evolutionary thinking, in part because of the prominent role assigned to random genetic drift in the initial phase of the process. However, as a comprehensive theory of evolution, many aspects of the theory remain to be tested. For the theory to work as envisaged, the interactions between alleles must often result in complex adaptive topographies with many peaks and valleys. The population must be split up into smaller demes, which must be small enough for random genetic drift to be important, but large enough for mass selection to fix favorable combinations of alleles. While migration between demes is necessary, neighboring demes must be sufficiently isolated for genetic differentiation to occur, but sufficiently connected for favorable gene combinations to spread. Because of uncertainty about the applicability of these assumptions, the shifting balance process remains a picturesque metaphor which is still largely untested.

One implication of interdeme selection is that selection for genes that are beneficial to the group may, in some cases, override effects that are harmful to the individuals. This principle is illustrated in the model in Table 4, where the allele A' is harmful to individuals within demes but favorable to the deme as a whole. Equation 2.26 implies that, within the ith deme, $\Delta q_i = -cq_i(1 - q_i)$ (assuming that $\overline{w} \simeq 1$). Averaging across all of n subpopulations, the change in allele frequency resulting from selection within subpopulations Δq_w equals $-c\overline{q}(1 - \overline{q})(1 - F)$. At the same time within-population selection occurs, interdeme selection favors demes containing A', and the change in allele frequency in the ith population resulting from between-population selection equals $2(b - c)q_i(1 - q_i)$ (assuming $\overline{v} \simeq 1$). Averaging this across all n subpopulations gives the total change in allele frequency resulting from between-population selection as $\Delta q_b = 2(b - c)\overline{q}(1 - \overline{q})(1 - F)$. Putting both processes together, the total change in the frequency of A' is

$$\Delta q = \Delta q_w + \Delta q_b = \overline{q}(1 - \overline{q})[2F(b - c) - c(1 - F)] \tag{2.36}$$

Equation 2.36 implies that $\Delta q > 0$ if

$$\frac{b - c}{c} > \frac{1 - F}{2F} \tag{2.37}$$

This is the condition necessary for selection between demes to override selection within demes, and the formulation is quite general (Crow and Aoki 1982). (A biological interpretation of Equation 2.37 is given in the following section on kin selection.)

If there are n demes, each of size N, which exchange migrants in such a way that m is the proportion of genes in each deme that are exchanged each generation for genes chosen at random from the other demes, then the approximate value of F at equilibrium is

$$F = \frac{1}{4Nm\alpha + 1} \tag{2.38}$$

where $\alpha = [n/(n - 1)]^2$ (Crow and Aoki 1984). Note that Equation 2.38 does not depend on the mutation rate to new alleles or on the number of alleles,

Table 4. Model of interdeme selection.

Genotype	AA	AA'	$A'A'$
Frequency in deme i	p_i^2	$2p_iq_i$	q_i^2
Within-population fitness (w)	1	$1 - c$	$1 - 2c$
Between-population fitness of deme i (v)		$1 + 2(b - c)q_i$	

and when n is reasonably large, $\alpha \simeq 1$, and the equation reduces to the familiar form in Equation 2.21.

Equation 2.38 implies that, for large n, $(1 - F)/2F = 2Nm$, which equals the number of migrant diploid individuals per generation (two times the number of migrant alleles). We therefore conclude from Equation 2.37 that selection between demes overrides selection within demes only when the benefit to the group $(b - c)$, relative to the cost to the individual (c), is greater than the average number of migrant individuals per generation. This principle defines a rather stringent limit above which migration among demes cancels any possible effects of interdeme selection.

Kin Selection

A second extended concept of fitness arises in the context of **kin selection,** which refers to indirect selection for alleles that occurs through the relatives of carriers of those alleles rather than by direct selection through an increased fitness of the carriers themselves. Kin selection has been postulated in attempts to account for the evolution of altruism. **Altruism** consists of behavioral traits or other attributes that increase the fitness of other individuals at the expense of one's own fitness. Altruistic behavior is exhibited most dramatically by social insects such as termites, ants, and bees, in which certain worker castes exert their labors for the care, protection, and reproduction of the queen and her offspring but do not reproduce themselves. Other, less dramatic, examples of altruistic behavior include phenomena such as the care of offspring by their parents.

A central consideration in kin selection is that relatives have genes in common. Therefore, a gene that causes altruistic behavior can increase in frequency if the increase in the recipient's fitness as a result of altruism is sufficiently large to offset the decrease in the altruist's own fitness. The essentials of the situation can be made clear by considering the case of identical twins. Because identical twins are genetically identical, the reproduction of one's twin is genetically equivalent to reproduction by oneself. Thus, it makes no difference if an altruistic individual decreases its own fitness for the sake of an equal increase in fitness of an identical twin; from an evolutionary point of view, it is an even trade because the combined number of offspring from both twins remains unchanged. By the same token, if an altruistic act decreases the fitness of an individual by an amount less than the increase gained by an identical twin, then the altruism results in a net increase in the combined number of offspring. One would, therefore, expect altruism between identical twins to be favored by natural selection as long as the risk to the altruist is no greater than the benefit to the recipient.

These considerations of identical twins can be extended to other degrees of relationship as well, but the risk to the altruist must be correspondingly smaller than the benefit to the recipient because other types of relatives share fewer genes than identical twins. The break-even points for altruism toward various degrees of relationship have been trenchantly summarized by J. B. S. Haldane, who is said to have quipped that he would lay down his life for two brothers, four nephews, or eight cousins. In any case, fitness considerations that take into account not only an individual's own fitness but also the fitness of relatives (other than direct descendants) constitute what is called the **inclusive fitness** of the individual.

To be concrete, suppose that altruism results in a decrease in fitness c of the altruist which is offset by an increase in fitness b in the recipient. The gene for altruism increases in frequency if the ratio of cost to benefit is great enough, relative to the genetic relationship between the individuals. That is, the gene for altruism increases in frequency if

$$c/b < r \tag{2.39}$$

as shown by Hamilton (1964) and discussed in detail by Cavalli-Sforza and Feldman (1978) and Uyenoyama and Feldman (1980). The appropriate measure of genetic relationship r in this context is defined as

$$r = \frac{2F_{XY}}{(1 + F_X)} \tag{2.40}$$

where F_X is the inbreeding coefficient of X, and F_{XY} is the inbreeding coefficient of a hypothetical offspring of X and Y. As illustrated in Figure 20, r equals the probability that two gametes from X and Y contain alleles that are identical by descent (F), relative to the probability that two gametes from X contain alleles that are identical by descent [$(1 + F)/2$]. The cost–benefit

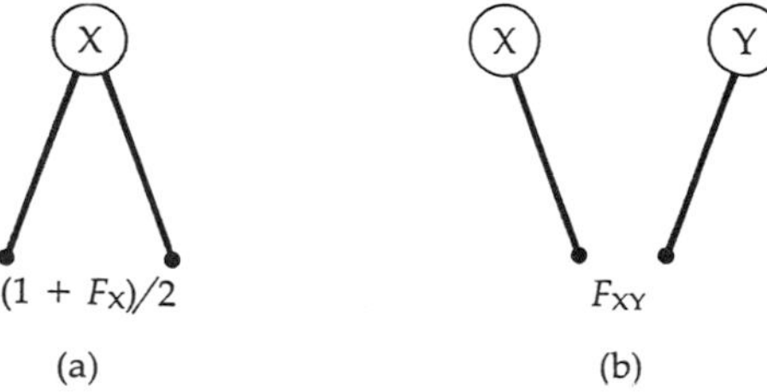

Figure 20. (a) Two genes chosen at random from an individual X are identical by descent with probability $(1 + F_X)/2$ (see Figure 23 in Chapter 1). (b) Two genes chosen at random, one from individual X and the other from individual Y, are identical by descent with probability F_{XY}, which is the inbreeding coefficient of a hypothetical offspring of X and Y. The ratio of F_{XY} to $(1 + F_X)/2$ is the appropriate measure of genetic relationship in consideration of kin selection.

tradeoff in Equation 2.39 is generally valid for weak selection when $F_X = 0$, and valid for additive alleles when $F_X \neq 0$ (Aoki 1981).

PROBLEM 26 For the illustrated pedigrees (a) and (b) of full siblings, calculate the break-even value of the benefit b to the recipient of altruism Y, relative to a cost value $c = 1$ to the donor X, in order to ensure an increase in frequency of an additive gene for altruism. Why are the answers different in the two cases?

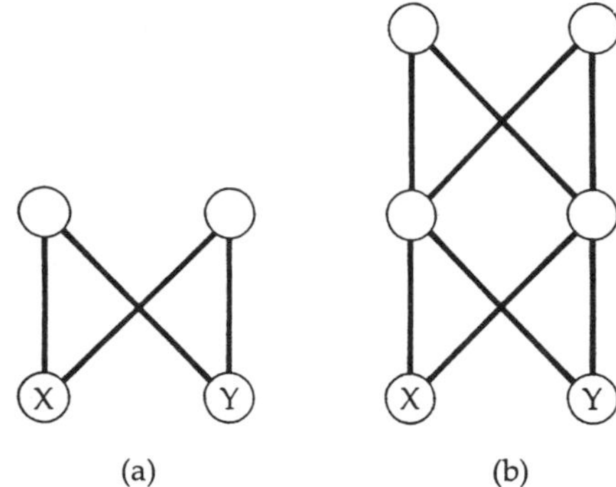

ANSWER In case (a), a hypothetical offspring of X and Y has an inbreeding coefficient of $F_{XY} = (1/2)^3 + (1/2)^3 = 1/4$, and $F_X = 0$. Therefore, $r = 2(1/4) = 1/2$, and the break-even value of $c/b = 1/2$. Hence, for $c = 1$, $b = 2$. This is the theoretical basis of Haldane's quip about laying down his life for two brothers. In pedigree (b), $F_{XY} = 4(1/2)^5 + 2(1/2)^3 = 3/8$ and $F_X = 2(1/2)^3 = 1/4$. Therefore $r = 2(3/8)/[1 + (1/4)] = 3/5$. For a cost of $c = 1$, the break-even value of b equals 5/3. The values differ in the two cases because of the differing inbreeding. In case (b), even though X is inbred, the break-even value of b is smaller because of the closer genetic relationship between X and Y.

Comparison of Equations 2.37 and 2.39 reveals a connection between kin selection and group selection of other types, which occurs because individuals within subpopulations are related. With random mating within subpopulations, the average value of r within subpopulations equals $2F/(1 + F)$, where F is given by Equation 2.38. Expressing the inequality in 2.37 in terms of r yields $c/b < r$, which is identical to Equation 2.39.

This chapter has focused on the evolutionary processes which determine the type and extent of genetic variation maintained in natural populations. It is still obscure whether the majority of genetic polymorphisms observed at the protein level are maintained by selection or a balance between the mutation and random genetic drift of selectively neutral or nearly neutral alleles. The issue is complex because the selective mechanisms by which polymorphisms are maintained may differ from gene to gene, and some polymorphic genes may experience so little selection as to be effectively neutral. Indeed, some alleles may be subject to selection, whereas other alleles of the same genes may be neutral. However, important findings regarding the neutrality hypothesis and other principles of population genetics have emerged from studies at the molecular level. The next chapter serves as an introduction to these aspects of population genetics.

SUMMARY

Random genetic drift refers to the progressive dispersion of allele frequencies among small populations and the accompanying accumulation of fixed populations. The effects of random genetic drift are conveniently measured in terms of individual heterozygosity (H_I), subpopulation heterozygosity (H_S), and total heterozygosity (H_T), and in terms of the associated F-statistics $F_{IS} = (H_S - H_I)/H_S$, which measures the effects of nonrandom mating within subpopulations, $F_{ST} = (H_T - H_S)/H_T$, which measures the effects of random genetic drift among subpopulations, and $F_{IT} = (H_T - H_I)/H_T$, which measures the combined effects of nonrandom mating and random genetic drift. When there is random mating within subpopulations, then $F_{IS} = 0$, and the genotype frequencies in any one population are given by the Hardy–Weinberg principle in terms of the allele frequencies appropriate to that subpopulation; the overall genotype frequencies, averaged across subpopulations, are given by the usual formula for inbreeding (Chapter 1) with $F = F_{ST}$. For a large group of subpopulations, each of size N, the increase in F_{ST} resulting from random genetic drift is given by $F_{ST} = 1 - (1 - 1/2N)^t$, where t represents time in generations.

The effective number of a population N_e is the size of an ideal population which has the same rate of increase in F_{ST} as the actual population. When population size varies from generation to generation, the average effective size equals the harmonic mean of the various numbers. If N_m is the number of males and N_f is the number of females, then the effective size is given by $4N_mN_f/(N_m + N_f)$. If the population is spread out uniformly in two dimensions, then the effective size is given by $4\pi\delta\sigma^2$, where δ is the density of breeding individuals per unit area and σ^2 is the one-way variance in distance between birthplace and breeding site.

F_{ST} can also be used as an index of genetic differentiation among populations. The amount of genetic differentiation associated with F_{ST} values of 0 to 0.05 is considered little, 0.05 to 0.15 is considered moderate, 0.15 to 0.25 is considered great, and above 0.25 is considered very great.

Mutation provides the raw material for evolutionary change, but, by itself, mutation pressure is a very weak force for changing allele frequency. If allele

A mutates to allele a at a rate μ per generation, and a undergoes reverse mutation at a rate ν per generation, then the equilibrium frequency of A is $\nu/(\mu + \nu)$, but the population may require tens of thousands or hundreds of thousands of generations to reach equilibrium. In finite populations, the equilibrium value of F_{ST} for genes with neutral alleles is given by $1/(4N\mu + 1)$, where μ is the mutation rate to selectively neutral alleles and $4N\mu + 1$ is called the effective number of alleles of the gene.

Wahlund's principle refers to the reduction in the frequency of rare homozygous recessive genotypes when populations become fused by extensive migration. The frequency of homozygous genotypes is reduced by the quantity $F_{ST}\overline{p}\,\overline{q}$, where $\overline{p}$ and $\overline{q}$ are the average allele frequencies in the pre-fusion populations. The reduction in frequency of homozygotes can also be expressed as σ^2, where σ^2 is the variance in allele frequency among pre-fusion populations. Thus $F_{ST} = \sigma^2/\overline{p}\,\overline{q}$.

On the whole, migration hinders genetic divergence among subpopulations. In finite populations, the equilibrium value of F_{ST} with migration is given by $1/(4Nm + 1)$, and only a few migrants per generation are sufficient to keep F_{ST} below 10 percent.

Natural selection can have many different mechanisms. The simplest case involves two alleles of one gene with constant fitnesses resulting from viability differences among genotypes. In this simple case, four outcomes are possible: (1) A becomes fixed; (2) a becomes fixed; (3) there is a globally stable equilibrium at $\hat{p} = (w_{12} - w_{22})/(2w_{12} - w_{11} - w_{22})$, where $\hat{p}$ represents the allele frequency of A and the fitnesses of AA, Aa, and aa are given by w_{11}, w_{12}, and w_{22}, respectively; or (4) there is an unstable equilibrium at the same value of $\hat{p}$. Case 3 refers to overdominance (w_{12} greater than both w_{11} and w_{22}); case 4 refers to heterozygote inferiority (w_{12} less than both w_{11} and w_{22}).

When a harmful allele is maintained in a population by a balance between selection and mutation the equilibrium frequency of the harmful allele is given by $\hat{q} = (\mu/s)^{1/2}$ when the allele is recessive and by $\hat{q} = (\mu/hs)$ when the allele is partially dominant. In these formulas, s is the selection coefficient against genotypes which are homozygous for the recessive allele, h is the degree of dominance of the harmful allele, and μ is the rate of mutation from the normal to the harmful allele.

Extended concepts of fitness can include the effects of selection acting on subpopulations or groups of relatives. Interdeme selection occurs when the differential growth or success of subpopulations results in changes in gene frequency. When interdeme selection occurs, some alleles that are harmful to their carriers within populations can nevertheless increase in frequency in the population as a whole because they have beneficial effects for selection between populations. Likewise, genes for altruism can increase in frequency if the loss in fitness of the altruist is offset by the increase in inclusive fitness to the beneficiaries. For additive alleles, the condition for the increase in frequency of an allele predisposing to altruism is $c/b < r$ where c and b are the fitness cost to altruist X and benefit to relative Y, respectively, and $r = 2F_{XY}/(1 + F_X)$.

Interdeme selection plays an important role in the shifting balance theory of evolution. According to this theory, adaptive topographies are highly complex surfaces with many peaks and valleys. In small, partially isolated subpopulations, random genetic drift promotes the random exploration of the topography. When, by chance, a subpopulation comes under the control of a higher fitness peak, mass selection occurs and rapidly multiplies the favored gene combinations. Excess migration from the successful subpopulation shifts gene frequencies in surrounding subpopulations, and through repetition of the selection process, the favored gene combinations progressively spread in waves throughout the entire population. Influential as a metaphor, the shifting balance theory has not yet been adequately evaluated as an accurate description of the principal mechanism of evolutionary change.

FURTHER READINGS

Ammerman, A. J. and L. L. Cavalli-Sforza. 1984. *The Neolithic Transition and the Genetics of Populations in Europe*. Princeton University Press, Princeton, New Jersey.

Avise, J. C. and C. F. Aquadro. 1982. A comparative summary of genetic distances in the vertebrates: Patterns and correlations. *Evol. Biol.* 15: 151–186.

Bajema, C. J. (ed.). 1983. *Natural Selection Theory: From the Speculations of the Greeks to the Quantitative Measurements of the Biometricians* (Benchmark Papers in Systematic and Evolutionary Biology, Vol. 4). Van Nostrand Reinhold, New York.

Barker, J. S. and T. Starmer (eds.). 1982. *Ecological Genetics and Evolution: The Cactus–Yeast–Drosophila Model*. Academic Press, New York.

Bishop, J. A. and L. M. Cook. 1975. Moths, melanism and clean air. *Sci. Amer.* 232: 90–99.

Bryan, G. J., J. W. Jacobson and D. L. Hartl. 1987. Heritable somatic excision of a *Drosophila* transposon. *Science* 235: 1636–1638.

Charlesworth, D. and B. Charlesworth. In press. Inbreeding depression. *Annu. Rev. Ecol. Syst.*

Christiansen, F. B. and T. M. Fenchel (eds.). 1977. *Measuring Selection in Natural Populations*. Springer-Verlag, New York.

Endler, J. A. 1977. *Geographic Variation, Speciation, and Clines*. Princeton University Press, Princeton, New Jersey.

Fisher, R. A. 1930. *The Genetical Theory of Natural Selection*. Clarendon, Oxford.

Hartl, D. L. 1980. *Principles of Population Genetics*. Sinauer Associates, Sunderland, Massachusetts.

Hartl, D. L. and D. E. Dykhuizen. 1984. The population genetics of *Escherichia coli*. *Annu. Rev. Genet.* 18: 31–68.

Hedrick, P. W. 1986. Genetic polymorphisms in heterogeneous environments: A decade later. *Annu. Rev. Ecol. Syst.* 17: 535–566.

Jacquard, A. 1974. *The Genetic Structure of Populations* (Trans. by D. and B. Charlesworth). Springer-Verlag, New York.
Jameson, D. L. (ed.). 1977. *Evolutionary Genetics* (Benchmark Papers in Genetics: Vol. 8). Academic Press, New York.
Karlin, S. and E. Nevo (eds.). 1976. *Population Genetics and Ecology*. Academic Press, New York.
Karlin, S. and E. Nevo. 1986. *Evolutionary Processes and Theory*. Academic Press, New York.
Kimura, M. 1979. The neutral theory of molecular evolution. *Sci. Amer.* 241: 98–126.
Kimura, M. and T. Ohta. 1971. *Theoretical Aspects of Population Genetics*. Princeton University Press, Princeton, New Jersey.
Krimbas, C. B. 1984. On adaptation, neo-Darwinian tautology, and population fitness. *Evol. Biol.* 17: 1–58.
Krimbas, C. B. and M. Loukas. 1980. The inversion polymorphism of *Drosophila subobscura*. *Evol. Biol.* 12: 163–234.
Lakovaara, S. (ed.). 1982. *Advances in Genetics, Development and Evolution of Drosophila*. Plenum Press, New York.
Malecot, G. 1969. *The Mathematics of Heredity* (Trans. by D. M. Yermanos). W. H. Freeman, San Francisco.
Manly, B. F. J. 1985. *The Statistics of Natural Selection on Animal Populations*. Chapman and Hall, New York.
Maruyama, T. 1977. *Stochastic Problems in Population Genetics*. Springer-Verlag, New York.
Michod, R. E. 1982. The theory of kin selection. *Annu. Rev. Ecol. Syst.* 13: 23–55.
Milkman, R. (ed.). 1983. *Experimental Population Genetics* (Benchmark Papers in Genetics: Vol. 13). Van Nostrand Reinhold, New York.
Moran, P. A. P. 1962. *The Statistical Processes of Evolutionary Theory*. Clarendon, Oxford.
Nagylaki. T. 1977. *Selection in One- and Two-Locus Systems*. Springer-Verlag, New York.
Nei, M. and A. K. Roychoudhury. 1982. Genetic relationship and evolution of human races. *Evol. Biol.* 14: 1–59.
Nevo, E. 1978. Genetic variation in natural populations: Patterns and theory. *Theoret. Pop. Biol.* 13: 121–177.
O'Donald, P. 1980. *Genetic Models of Sexual Selection*. Cambridge University Press, Cambridge.
Parsons, P. A. 1983. *The Evolutionary Biology of Colonizing Species*. Cambridge University Press, Cambridge.
Roush, R. T. and J. A. McKenzie. 1987. Ecological genetics of insecticide and acaricide resistance. *Annu. Rev. Entomol.* 32: 361–380.
Slatkin, M. 1985 Gene flow in natural populations. *Annu. Rev. Ecol. Syst.* 16: 393–430.
van Delden, W. 1982. The alcohol dehydrogenase polymorphism in *Drosophila melanogaster*: Selection at an enzyme locus. *Evol. Biol.* 15: 187–222.
Wallace, B. 1968. *Topics in Population Genetics*. W. W. Norton, New York.
Wallace, B. 1984. Adaptation, neo-Darwinian tautology, and population fitness: A reply. *Evol. Biol.* 17: 59–72.
Wijsman, E. M. and L. L. Cavalli-Sforza. 1984. Migration and genetic population structure with special reference to man. *Annu. Rev. Ecol. Syst.* 15: 279–301.
Wilson, D. S. 1983. The group selection controversy: History and current status. *Annu. Rev. Ecol. Syst.* 14: 159–187.

Wright, S. 1969. *Evolution and the Genetics of Populations*, Vol. 2. *The Theory of Gene Frequencies*. University of Chicago Press, Chicago.

Wright, S. 1977. *Evolution and the Genetics of Populations*. Vol. 3. *Experimental Results and Evolutionary Deductions*. University of Chicago Press, Chicago.

Wright, S. 1978. *Evolution and the Genetics of Populations*. Vol. 4. *Variability Within and Among Natural Populations*. University of Chicago Press, Chicago.

Wright, S. 1982. The shifting balance theory and macroevolution. *Annu. Rev. Genet*. 16: 1–19.

Wright, S. 1986. *Evolution: Selected Papers*, W. B. Provine (ed.). University of Chicago Press, Chicago.

MOLECULAR POPULATION GENETICS 3

Evolution is both process and result. The *process* of evolution includes all mechanisms of genetic change that occur in organisms through time, with special emphasis on those mechanisms that promote the adaptation of organisms to their environment or that lead to the formation of new reproductively isolated species. The *result* of evolution is observed in the evolutionary history of life on Earth and in the genetic relationships among species that exist today.

Corresponding to evolution as process and result are two principal themes that recur in evolutionary studies. The first is that of phylogenetic inference, which aims to determine the evolutionary history and genetic relatedness among a group of organisms. Phylogenetic inference emphasizes the comparative side of evolutionary studies. Traditionally it includes comparative morphology, anatomy, and embryology, but modern systematic biologists also emphasize biochemistry, biogeography, ecology, behavior, and other approaches. The second theme of evolutionary studies is that of mechanism, which is primarily within the purview of population genetics. Mechanistic studies emphasize the interplay between population structure, mating systems, migration, mutation, random genetic drift, and natural selection in the microevolutionary processes that lead to genetic change and adaptation.

Use of techniques of molecular biology, particularly those for determining amino acid or nucleotide sequences, has added a new dimension to phylogenetic inference. For example, studies of the extent of DNA relatedness among bird species, determined by measuring the thermal stability of hybrid DNA molecules containing one DNA strand from each species, have brought about several major revisions in classical bird taxonomy. One unexpected finding from DNA studies was that the tribe Corvini, which includes crows, ravens, jays, magpies, and their relatives, originally evolved as part of a larger group in Australia, which spread from there to Asia and thence throughout the world (Sibley and Ahlquist 1986).

The great power and relative ease of comparative molecular biology have given prominence to phylogenetic inference. They have not, however, eclipsed the search for evolutionary mechanisms. Indeed, the techniques of molecular biology have invigorated the study of evolutionary mechanisms through the discovery of unprecedented types of genes and DNA sequences, the evolutionary implications of which are still to be worked out. At the same time, the new molecular data provide unique opportunities for inferring mechanisms of evolutionary change and testing evolutionary hypotheses. Evolutionary biology is rapidly reaching the point at which the understanding of evolution includes not only the litany of which species begat which others, but also an appreciation of how and why such changes came to pass.

This chapter focuses on the molecular aspects of population genetics. It includes discussions of the neutral theory of molecular evolution, the evolution of multigene families, and the evolutionary implications of transposable elements ("selfish DNA"). The chapter demonstrates the important role of molecular biology in recent evolutionary studies: many discussions that are prominent today could not have been considered ten years ago because the phenomena at issue had not yet been discovered.

THE NEUTRAL THEORY OF MOLECULAR EVOLUTION

The first systematic application of molecular methods to population genetics in the 1960s, initially using the electrophoretic separation of protein molecules, revealed the existence of extensive genetic variation within most natural populations. In most species, 15–50 percent of the genes coding for enzymes were observed to include two or more widespread, polymorphic alleles. The polymorphic alleles occurred with frequencies too high to result from equilibrium between adverse selection and mutation.

In 1968, Motoo Kimura suggested that most polymorphisms observed at the molecular level are selectively neutral (Kimura 1968). Mutations that are

selectively neutral produce such small effects on the ability of their carriers to survive and reproduce that they are completely equivalent in terms of natural selection. The ultimate fate of neutral alleles in a population is determined largely by the process of random genetic drift. By extension, the hypothesis of selective neutrality would also apply to most nucleotide or amino acid substitutions that occur within a molecule during the course of evolution.

Although the neutral theory was at first thought to be in conflict with Darwinian natural selection, there is no necessary contradiction. Indeed, in *The Origin of Species,* Darwin comments:

> I am inclined to suspect that we see, at least in some [cases], variations which are of no service to the species, and which consequently have not been seized on and rendered definite by natural selection. . . . Variations neither useful nor injurious would not be affected by natural selection, and would be left either a fluctuating element, as perhaps we see in certain polymorphic species, or would ultimately become fixed. . . . We may easily err in attributing importance to characters, and in believing that they have been developed through natural selection; . . . many structures are now of no direct use to their possessors, and may never have been of any use to their progenitors. . . . [On the other hand,] we are much too ignorant in regard to the whole economy of any organic being to say what slight modifications would be of importance or not.

The neutral theory has been of considerable importance in population genetics in stimulating the collection and analysis of data in attempts to evaluate its adequacy. Mathematical investigations of its implications have resulted in what can only be regarded as one of the most complete and elegant theories in all of biology. The next section summarizes some of the theoretical implications of the neutral theory and some of the data bearing on it. Although the neutral theory of molecular evolution has gained many adherents throughout the years, it is by no means universally accepted among population geneticists (e.g., Gillespie 1978, 1986a and b), and there are doubts and inconsistencies that remain to be resolved. Definition of the adequacy and range of applicability of the neutral theory remains an important objective in both theoretical and experimental population genetics.

Theoretical Principles of the Neutral Theory

The neutral theory is more precisely called the **neutral-mutation/random-drift hypothesis** because it concerns mutations that are so nearly selectively neutral in their effects that their fate is determined largely through random genetic drift. The most common theoretical model of random genetic drift was described in Chapter 2. It assumes an ensemble of populations, each of finite size N, with nonoverlapping generations. Reproduction occurs through the N adult individuals producing an infinite pool of gametes, from which $2N$ are chosen at random to create the N zygotes of the next generation (see Figure 2 in Chapter 2). Some of the principal theoretical implications of such a population structure are summarized in the following list (Kimura 1983). Principles numbered 1, 2, 5, and 6 were already demonstrated and discussed in Chapter 2 and are included here for completeness.

1. *If a population contains a neutral allele with allele frequency p_0, then the probability that the allele eventually becomes fixed equals p_0.* In particular, a newly arising neutral mutation occurs in just one copy, so the initial allele frequency is $p_0 = 1/(2N)$, and the probability of eventual fixation of the mutation is therefore $1/(2N)$.

PROBLEM 1 Suppose a human population is founded with exactly 30 individuals and that no immigration occurs thereafter. What is the probability that a particular gene present in one of the founders ultimately becomes fixed in the population, assuming no selection?
ANSWER $1/(2 \times 30) = 0.017$. Note that the result is independent of how large or small the population might become in the meantime.

---

2. *The steady-state rate at which neutral mutations are fixed in a population equals ν, where ν is the rate of occurrence of neutral mutations per gene per generation.* The rate ν is called the **neutral mutation rate**. It is noteworthy that the equilibrium rate of fixation does not involve the population size N. The reason is that the N cancels out: the overall rate is determined by the product of the probability of fixation of new neutral mutations $[1/(2N)]$ and the average number of new neutral mutations in each generation $[2N\nu]$, hence $[1/(2N)] \times (2N\nu) = \nu$.

As you will see later in this chapter, the ultimate fate of mutations with small effects on fitness are also determined largely by random genetic drift, provided that the selection coefficients are small relative to the reciprocal of the effective population number. (The *effective population number*, defined in Chapter 2, corresponds, roughly speaking, to the number of breeding individuals in the actual population; typically, the effective size of a population is smaller, and sometimes much smaller, than the actual size.) Since the relative importance of selection and random genetic drift is determined by the relative magnitudes of the selection coefficient and the effective population number, the rate of neutral mutation ν is itself affected by the effective population number. Alleles with selection coefficients that are small in relation to effective population number are *nearly neutral*, and their ultimate fate is determined largely by random genetic drift. Alleles with equally small selection coefficients occurring in much larger populations may be under the almost exclusive control of selection.

3. *The average time that occurs between consecutive neutral substitutions equals* $1/v$. This principle follows directly from the one above. If the steady-state rate of fixation is v per unit time, then the average length of time between substitutions will be the reciprocal, or $1/v$. By way of analogy, if a Swiss clock cuckoos at the rate of 24 times per day, then the average length of time between cuckoos is 1/24th of a day, or one hour. The analogy with gene substitutions is improved if we assume a novelty clock in which the time interval between successive cuckoos is random, with a one-hour average.

4. *Among newly arising neutral alleles that are destined to be fixed, the average time to fixation is* $4N_e$ *generations; among newly arising neutral alleles destined to be lost, the average time to loss is* $(2N_e/N)\ln(2N)$ *generations*. In these equations, N_e and N refer, respectively, to the effective population number and the actual population number. The average times required for fixation or loss apply to newly arising alleles, which are necessarily present in just one copy, so $p_0 = 1/(2N)$. The symbol "ln" in the second formula is the natural logarithm, which uses the base e.

The implication of these formulas is that, on average, neutral mutations that are going to be fixed require a very long time for this to occur, but mutations destined to be lost are lost quite rapidly. This contrast is apparent in Table 1. For a population of size 1,000, for example, with $N_e = 0.8N$, the

Table 1. Average time to fixation or loss of neutral alleles.

N[a]	Average generations until loss[b]	Average generations until fixation[b]
50	7.4	160
100	8.5	320
250	9.9	800
500	11.1	1,600
1,000	12.2	3,200
5,000	14.7	16,000
10,000	15.8	32,000
50,000	18.4	160,000

[a] Tabulated values assume that $N_e = 0.8N$.
[b] Averages are for selectively neutral alleles destined to be lost or fixed, respectively.

average time to loss is 12.2 generations, whereas the average time to fixation is 3,200 generations. Table 1 also illustrates that times to loss are not markedly affected by population size. Specifically, as the population size increases from 50 to 50,000 (a 1000-fold increase), the average time to loss increases from 7.4 generations to 18.4 generations (a 2.5-fold increase).

PROBLEM 2 Suppose that the population in Problem 1 is ideal in the sense that its effective size equals its actual size (a situation not likely to be realized in practice) and that the population size remains at 30 individuals in each generation. Calculate the average time to fixation of a new neutral allele destined to become fixed, and the average time to loss of a new allele destined to become lost. Do the same calculations for a population with an effective (and actual) size of 3000.

ANSWER Average times to fixation and loss are $4 \times 30 = 120$ generations and $2\ln(2 \times 30) = 8.2$ generations, respectively, for the population with $N = N_e = 30$; and $4 \times 3{,}000 = 12{,}000$ generations and $2\ln(2 \times 3{,}000) = 17.4$ generations, respectively, in the population with $N = N_e = 3{,}000$.

---

5. *In the absence of new mutations, random genetic drift decreases the average proportion of heterozygotes in an ensemble of populations at the rate of* $1/(2N_e)$ *per generation*. Eventually each subpopulation becomes fixed for one allele or another, and at this time the average heterozygosity across the entire ensemble is zero. Nevertheless, if mating occurs at random within each subpopulation, then the genotype frequencies *within* each subpopulation are given approximately by the Hardy–Weinberg principle.

6. *If each neutral mutation creates an allele that is different from all others existing in the population in which it occurs, then, at equilibrium, when the average number of new alleles gained through mutation is exactly offset by the average number lost through random genetic drift, the expected homozygosity equals* $1/[4N_e\nu + 1]$, *where* ν *is the neutral mutation rate*. The model of mutation in which each new allele is novel is the **infinite-alleles model** of mutation. The quantity $4N_e\nu$, which shows up frequently in the neutral theory, is often denoted as θ. The equilibrium average homozygosity is therefore $1/(1 + \theta)$. Since the heterozygosity equals one minus the homozygosity, the average heterozygosity at equilibrium in the infinite alleles model equals $\theta/(1 + \theta)$.

7. *If monomorphism is defined as a gene for which the most common allele has a frequency greater than or equal to* $1 - q$, *where* q *is some small number such as 0.01 or 0.05, then, at equilibrium, the probability that a gene will be monomorphic is approximately* q^θ, *where* $\theta = 4N_e\nu$. Values of q^θ are plotted for $q = 0.01$ and $q = 0.05$ in Figure 1. Note that the scale of θ is logarithmic. In most organisms, the proportions of monomorphic genes discovered in electrophoretic surveys are consistent with values of θ in the range 0.03–0.30.

PROBLEM 3

For $q = 0.05$ and $q = 0.01$, calculate the approximate probability that a gene will be polymorphic in an equilibrium population in which $\theta = 0.09$.

ANSWER Probability of polymorphism equals $1 - q^{\theta}$, which can be calculated directly or as $1 - \exp[\theta \ln(q)]$ with $\ln(q) = -3.00$ or -4.61 as $q = 0.05$ or 0.01, respectively. Given $\theta = 0.09$, approximate probabilities of polymorphism are 0.236 ($q = 0.05$) and 0.339 ($q = 0.01$). The probabilities of monomorphism are therefore $1 - 0.236 = 0.764$ and $1 - 0.339 = 0.661$, respectively.

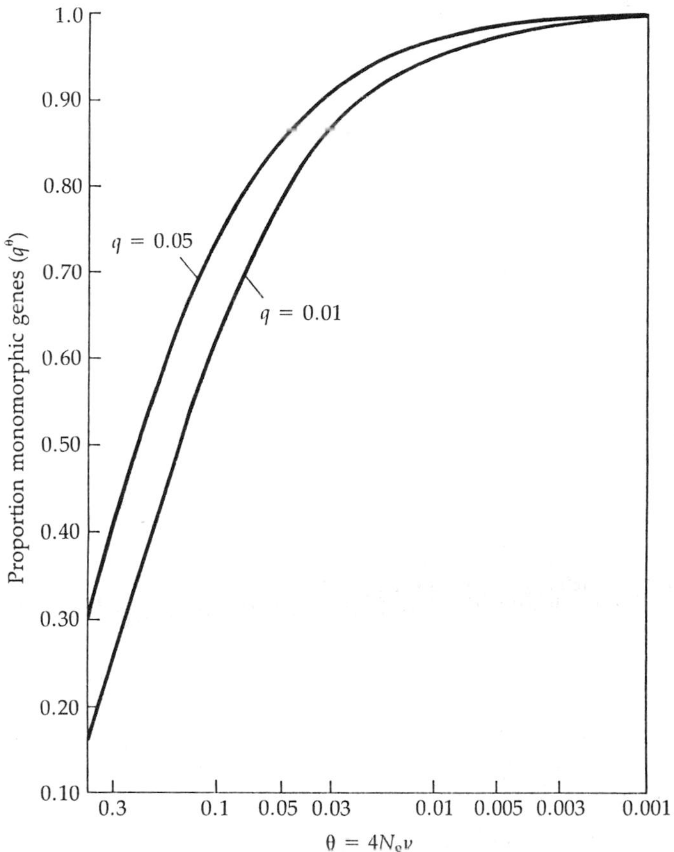

Figure 1. Approximate proportion of genes expected to be monomorphic assuming neutral mutation rate ν. A gene is considered monomorphic if the most common allele has a frequency greater than $1 - q$, where usually $q = 0.01$ or 0.05. Levels of electrophoretically detected protein polymorphism are usually consistent with values of θ in the range 0.03–0.3.

Rates of Amino Acid Replacement

The initial impetus for the neutral theory came from observations on the rate of missense (amino acid-changing) nucleotide substitutions in proteins. When extrapolated to the entire genome, the inferred rate of evolution was several nucleotide substitutions per year. This was regarded as much too high to result from natural selection, because the intensity of selection must be limited by the total amount of differential survival and reproduction that occurs in the organism. Direct DNA sequencing later revealed that rates of nucleotide substitution vary according to the function, or presumed absence of function, of the nucleotides.

The types of data that must be analyzed are illustrated in the following problem.

PROBLEM 4 The first 18 amino acids present at the amino terminal end of the human and mouse immune γ-interferon proteins constitute a signal peptide which is used in secretion of the molecules (Gray and Goeddel 1983). Calculate the proportion of amino acids which differ in the two signal sequences.

Human: Met Lys Try Thr Ser Tyr Ile Leu Ala Phe Gln Leu Cys Ile Val Leu Gly Ser

Mouse: Met Asn Ala Thr His Cys Ile Leu Ala Leu Gln Leu Phe Leu Met Ala Val Ser

ANSWER Among the 18 amino acids there are 10 differences, so the proportion different is 10/18 = 0.56.

To interpret the data in Problem 4, let us suppose that amino acid replacements occur at the rate λ per unit time. Consider two independently evolving sequences, initially identical, which at time t are found to differ in the proportion D_t of their amino acids. After the next time interval, the proportion of differences D_{t+1} is given by

$$D_{t+1} = (1 - D_t)(2\lambda) + D_t \quad \textbf{(3.1)}$$

In this equation, $(1 - D_t)(2\lambda)$ is the proportion of sites, previously identical, in which one or the other underwent an amino acid replacement during the time interval in question, which must be added to the already existing differences D_t in order to give the total. (The equation ignores the unlikely possibility of an amino acid replacement making two previously different amino acid sites identical.) The factor of 2 is present because the *total* time for evolution is $2t$ units (t units in each lineage after the split), which is illustrated in Figure 2.

Equation 3.1 suggests the differential equation

$$dD/dt = D_{t+1} - D_t = 2\lambda - 2\lambda D(t)$$

which has the solution

$$D(t) = 1 - \exp(-2\lambda t) \qquad \textbf{(3.2)}$$

An alternative argument can be used to derive Equation 3.2 without resorting to differential equations. If λ is the rate of amino acid replacement per unit time, then the probability that a particular site will remain unsubstituted for t consecutive intervals along each of two independent lineages is $(1 - \lambda)^{2t}$, which is approximately equal to $\exp(-2\lambda t)$, provided that λt is not too large. Thus, the probability $D(t)$ of one or more replacements occurring in t units of time after divergence is approximately $1 - \exp(-2\lambda t)$, which is Equation 3.2.

Since λ is the rate of amino acid replacement per unit time, the expected proportion of differences between two sequences at any time t is

$$K = 2\lambda t \qquad \textbf{(3.3)}$$

where the factor of 2 is again present because the total time for evolution is $2t$ units (Figure 2).

Substituting the K from Equation 3.3 into Equation 3.2 and rearranging yields the following estimate $\langle K\rangle$ of K:

$$\langle K\rangle = -\ln(1 - \langle D\rangle) \qquad \textbf{(3.4)}$$

where $\langle D\rangle$ is the observed proportion of sites in which two sequences differ.

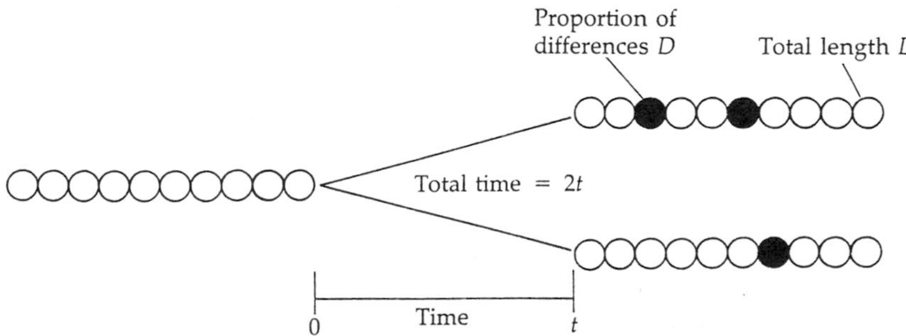

Figure 2. Two amino acid or nucleotide sequences which have each undergone independent evolution from a common ancestor for t time units are separated by a total time of $2t$ units, because there are t units in each lineage after the split. The proportion of sites which differ in the sequences is denoted $\langle D\rangle$ and the total number of sites L. In this particular example, $L = 10$ and $\langle D\rangle = 3/10$.

If the sequences under comparison are L amino acids in length, then the estimated variance $\langle Var(\langle K \rangle)\rangle$ of $\langle K \rangle$ is approximately

$$\langle Var(\langle K \rangle)\rangle = \langle D \rangle / [(1 - \langle D \rangle)L] \quad \textbf{(3.5)}$$

The rate of evolution at the molecular level is given by the amount of sequence divergence that occurs per unit of time. Thus, as suggested by Equations 3.3 and 3.4, if two sequences are compared, which are known to have diverged from a common ancestral sequence an estimated $\langle t \rangle$ time units ago, then the rate of evolution λ may be estimated as

$$\langle \lambda \rangle = \langle K \rangle / 2\langle t \rangle \quad \textbf{(3.6)}$$

The units of $\langle \lambda \rangle$ are usually expressed as replacements per amino acid site (or per nucleotide site) per year.

The quantity $\langle K \rangle$ is used in preference to $\langle D \rangle$ in estimating the rate of molecular evolution because $\langle K \rangle$ takes multiple substitutions into account. Over long periods of evolutionary time, the amino acid present at a particular site may be replaced several times, first by one alternative, then another, then still another, and perhaps, at some stage, even return to the amino acid originally present at the site. When comparing two sequences, only sites that are different can be identified. Sites that are identical at the present time may include some that were different in the past, and sites that are different at the present time might have undergone more than one substitution. The quantity $\langle D \rangle$ is determined only by the proportion of differences between the sequences observed at the present time. The estimate $\langle K \rangle$ makes a correction for multiple substitutions, but at the cost of introducing assumptions that the substitutions occur independently and at the same rate through time. For relatively short intervals of evolutionary time, during which multiple substitutions remain uncommon, the correction is minor, and the value of $\langle K \rangle$ is close to that of $\langle D \rangle$. Over longer intervals, when many multiple substitutions have occurred, the correction is important, and the assumptions on which it is based must be evaluated critically. Correction for multiple substitution events is even more important for nucleotides than it is for amino acids. With amino acids, the probability of a random replacement returning an amino acid site to its original identity is 1/20, whereas for nucleotides it is 1/4.

Application of Equations 3.4–3.5 is illustrated in the following problem.

PROBLEM 5 Use the data in Problem 4 to estimate the average rate of amino acid replacement in the signal peptide of γ-interferon during the divergence of mice and humans. Based on fossil evidence, the separation of these species occurred approximately 80 million years ago.

ANSWER For the signal peptide $\langle D \rangle = 0.56$ and $\langle K \rangle = -\ln(1 - 0.56) = 0.82$. The estimated rate of evolution is $\langle \lambda \rangle = 0.82/[2 \times (80 \times 10^6)] = 5.1 \times 10^{-9}$ amino acid replacements per amino acid site per year. The standard deviation of $\langle K \rangle$ is estimated as $[0.56/(0.44 \times 18)]^{1/2} = 0.27$.

With the sample size as small as it is in Problem 5, the estimates could ordinarily not be taken too literally. However, in this case, the average rate for the signal sequence is very close to the average rate for the molecule as a whole. For γ-interferon, among 155 amino acid sites there are 91 differences, giving $\langle K \rangle = 0.88 \pm 0.22$ and an average rate of $\langle \lambda \rangle = 5.5 \times 10^{-9}$ amino acid replacements per amino acid site per year.

Rates of amino acid replacement vary over a 500-fold range in different proteins. The rate of amino acid replacement in γ-interferon is one of the fastest rates known (Li et al.1985a). Among the slowest rates is that of histone H4, for which $\langle \lambda \rangle = 0.01 \times 10^{-9}$ per year. The average rate among a large number of proteins is very close to the rate found in hemoglobin, which is approximately 1×10^{-9} amino acid replacements per amino acid site per year.

To be concrete about the interpretation of the rate of amino acid replacement, consider a protein exactly 100 amino acids in length, in which the rate of amino acid replacement per amino acid site equals 1×10^{-9} per year. For the entire protein, the rate of replacement equals $100 \times 1 \times 10^{-9} = 1 \times 10^{-7}$ per year. In two different species, therefore, the protein would accumulate amino acid differences at the rate of one replacement every five million years since their divergence from a common ancestor (because $5 \times 10^6 \times 2 \times 1 \times 10^{-7} = 1.0$). More generally, the expected proportion of amino acid differences between the proteins at a time t years after their divergence may be calculated from Equation 3.2 with $\lambda = 1 \times 10^{-9}$.

PROBLEM 6

The evolutionary separation of kangaroos and dogs occurred approximately 135 million years ago, and in the evolution of vertebrates as a whole the α-hemoglobin molecule has been undergoing amino acid replacement at the average rate of $\langle \lambda \rangle = 1 \times 10^{-9}$ amino acid replacements per amino acid site per year. Calculate the proportion of amino acid sites that are expected to differ in the α-hemoglobin of kangaroo and dog.

ANSWER From Equation 3.2, $\langle D(t) \rangle = 1 - \exp(-2 \times 1 \times 10^{-9} \times 135 \times 10^6) = 0.237$. The observed proportion of amino acid differences is 0.234, so in this particular example the prediction is very accurate.

When protein sequences are compared between species and estimates made using Equations 3.4 and 3.6, it is usually assumed that the sequences being compared are truly representative of the species. However, as emphasized in Chapter 1, genetic variation occurs in virtually all natural populations, and the protein or gene being studied may be polymorphic. Estimation of rates of evolution become much more difficult when intraspecific genetic variation is taken into account. Indeed, intraspecific genetic variation is usually ignored in the knowledge, or sometimes only the hope, that the magnitude of intraspecific variation is small in comparison with that between the species being compared.

Rates of Nucleotide Substitution

Nucleotide sequences are analyzed in the same manner as amino acid sequences, but the analogous equation to Equation 3.1 is slightly more complicated because it has to correct for cases in which a substitution makes two previously different nucleotide sites identical. The correction is significant for nucleotide sequences because an expected 1/3 of random substitutions will make two previously different nucleotides identical. It is usually unnecessary for proteins because only 1/19 of random replacements will make two previously different amino acids identical.

Various models of nucleotide substitution have been studied, which differ primarily in the assumptions about rates of mutation between pairs of nucleotides. The simplest model is one in which mutation occurs at a rate λ per unit time, and each nucleotide is equally likely to mutate to any other (Jukes and Cantor 1969). In this case the analogous equation to Equation 3.4 is

$$\langle k \rangle = -(3/4)\ln[1 - 4\langle D \rangle/3] \qquad \textbf{(3.7)}$$

where k is the expected proportion of nucleotide sites which differ between two sequences at a time t units after their evolutionary separation. By analogy with protein evolution (Figure 2), $\langle D \rangle$ is the observed proportion of L nucleotide sites in which the sequences differ. The variance $\mathrm{Var}(\langle k \rangle)$ of the estimate can be estimated as

$$\langle \mathrm{Var}(\langle k \rangle) \rangle = \langle D \rangle(1 - \langle D \rangle)/[L(1 - 4\langle D \rangle/3)^2] \qquad \textbf{(3.8)}$$

Application of Equations 3.7 and 3.8 to nucleotide sequences is illustrated in the following problem.

PROBLEM 7 The following 60 nucleotides occur in the coding region of the *trpA* genes in strains of the related enteric bacteria *Escherichia coli* strain K12 and *Salmonella typhimurium* strain LT2 (Nichols and Yanofsky 1979). The *trpA* gene codes for one of the subunits of the enzyme tryptophan synthetase used in the manufacture of tryptophan. Estimate the amount of nucleotide divergence $\langle k \rangle$ and amino acid divergence $\langle K \rangle$ and their standard deviations. Assuming that *Escherichia* and *Salmonella* diverged at around the time of the mammalian radiation 80 million years ago, estimate the rates of nucleotide and amino acid evolution. In the sequences, nucleotides that differ are marked with asterisks and amino acids that differ are underlined.

```
     *    *  *                       *        *  *          * *  *
K12: GTCGCACCTATCTTCATCTGCCCGCCAAATGCCGATGACGACCTGCTGCGCCAGATAGCC
     Val Ala Pro Ile Phe Ile Cys Pro Pro Asn Ala Asp Asp Asp Leu Leu Arg Gln Ile Ala

LT2: ATCGCGCCGATCTTCATCTGCCCGCCAAATGCGGATGACGATCTTCTGCGCCAGGTCGCA
     Ile Ala Pro Ile Phe Ile Cys Pro Pro Asn Ala Asp Asp Asp Leu Leu Arg Gln Val Ala
```

(In K12, Val and Ile are underlined.)

ANSWER For the amino acid sequences, $L = 20$ and $\langle D \rangle = 2/20 = 0.10$; thus $\langle K \rangle = -\ln(0.90) = 0.105$ with standard deviation 0.074. For the nucleotide sequences, $L = 60$ and $\langle D \rangle = 9/60 = 0.15$; thus $\langle k \rangle = -(3/4)\ln(0.8) = 0.167$ with standard deviation 0.058. Assuming there is a divergence time of 80×10^6 years, the rates of evolution are $0.167/(2 \times 80 \times 10^6) = 1.04 \times 10^{-9}$ nucleotide substitutions per year and $0.105/(2 \times 80 \times 10^6) = 0.66 \times 10^{-9}$ amino acid replacements per year. In the gene as a whole, the values are $\langle k \rangle = 0.300$ for nucleotide substitutions and $\langle K \rangle = 0.162$ for amino acid replacements.

PATTERNS OF NUCLEOTIDE AND AMINO ACID SUBSTITUTION

The values estimated in Problem 7 serve to illustrate the general principle that nucleotide substitutions occur at a greater rate than amino acid replacements. The difference in rates, sometimes much greater than in these data, results from redundancy in the standard genetic code. As illustrated in Table 2, the codons for eight amino acids contain *N* (standing for any nucleotide) in their third position, seven terminate in *Y* (any pyrimidine, which means T or C), and five terminate in *R* (any purine, which means A or G). Coding sites that contain an *N* are called **fourfold degenerate** sites because any of the four nucleotides will do, and those containing a *Y* or an *R* are **twofold degenerate** sites (Li et al. 1985a,b). Because of degeneracies, nucleotides in a gene can change without affecting the amino acid sequence. These changes are called **synonymous** or **silent** nucleotide substitutions. Nucleotide substitutions that do change amino acids are **nonsynonymous** substitutions.

Paralleling the evolutionary rates for amino acid-changing substitutions, the rates of nonsynonymous nucleotide substitution vary tremendously among different proteins. Among the slowest rates is that of histone H4, for which $\langle \lambda \rangle = 0.004 \times 10^{-9}$ substitutions per nonsynonymous nucleotide site per year, and among the fastest is that of γ-interferon, for which $\langle \lambda \rangle = 2.80 \times 10^{-9}$ substitutions per nonsynonymous nucleotide site per year. The average rate among a large number of proteins is very close to the rate found in hemoglobin, which is 0.87×10^{-9} substitutions per nonsynonymous nucleo-

Table 2. Degeneracy in the genetic code.

First nucleotide in codon	Second nucleotide in codon: T	C	A	G
T	TT*Y* Phe TT*R* Leu	TC*N* Ser	TA*Y* Tyr TA*R* End	TG*Y* Cys TGA End TGG Trp
C	CT*N* Leu	CC*N* Pro	CA*Y* His CA*R* Gln	CG*N* Arg
A	AT*H* Ile ATG Met	AC*N* Thr	AA*Y* Asn AA*R* Lys	AG*Y* Ser AG*R* Arg
G	GT*N* Val	GC*N* Ala	GA*Y* Asp GA*R* Glu	GG*N* Gly

Note: In this representation of the standard genetic code, the symbol *N* stands for any nucleotide (T, C, A, or G), the symbol *Y* for any pyrimidine (T or C), and the symbol *R* for any purine (A or G). The *H* in the set of codons for isoleucine (Ile) stands for "not-G" (T, C or A). Degeneracies are as follows: *N* represents a fourfold degenerate site, *Y* and *R* represent twofold degenerate sites. The *H* in the set of codons for isoleucine is considered as twofold degenerate, as are the first nucleotides in four leucine codons (TTA, TTG, CTA, and CTG) and four arginine codons (CGA, CGG, AGA, and AGG). All other nucleotides are nondegenerate.

tide site per year (Figure 3a). As in the examples given here, rates of nonsynonymous nucleotide substitution are usually quite similar to the rates of amino acid replacement in the same genes.

In contrast with the highly variable rates of nonsynonymous nucleotide substitutions among proteins, the rates of synonymous substitution are much more uniform. For example, in mammalian genes, the fastest rate of synonymous substitution is only 3–4 times greater than the slowest rate (Figure 3b). However, the average rate, $\langle\lambda\rangle = 4.7 \times 10^{-9}$ substitutions per synonymous site per year, is not only greater than the average rate of nonsynonymous substitutions, but is greater than the fastest known rate of nonsynonymous substitutions (γ-interferon).

In calculations involving synonymous and nonsynonymous nucleotide sites, the total number of synonymous sites is calculated as the number of fourfold degenerate sites plus one-third of the number of twofold degenerate sites. The total number of nonsynonymous sites in a coding region is defined as the number of nondegenerate sites (nucleotides in which any change results in an amino acid substitution), plus two-thirds of the number of twofold degenerate sites (the latter because, with random mutation at twofold

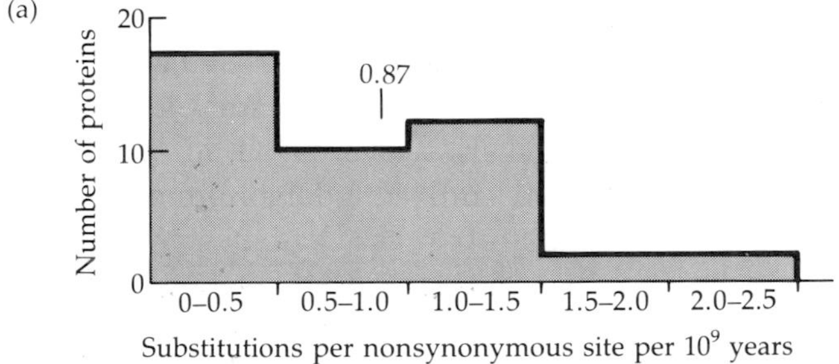

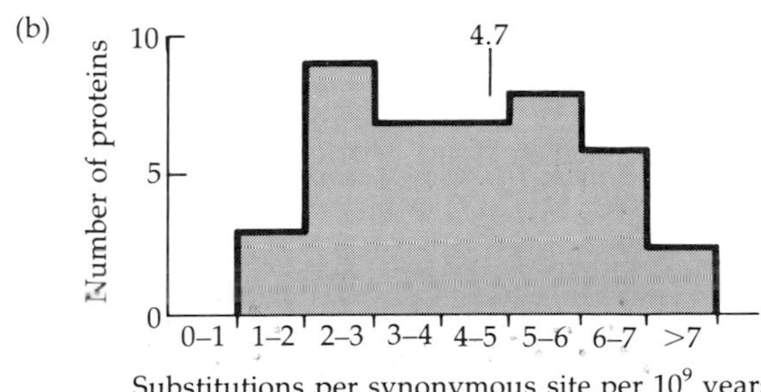

Figure 3. Distributions of rates of nucleotide substitution in protein-coding regions in animal genes. (a) Nonsynonymous substitutions. (b) Synonymous substitutions. Note that the scales in (a) and (b) are different. (Data from Li et al. 1985.)

degenerate sites, two-thirds of the mutations are expected to result in amino acid changes). These conventions are illustrated in Problem 8.

PROBLEM 8

Use the degeneracy values in Table 2 to determine the degeneracy of each nucleotide in the region of the *E. coli trpA* gene given in Problem 7. For each difference between *E. coli* and *Salmonella*, consider the difference as synonymous if the site is fourfold degenerate or if it is twofold degenerate and the change is a transition (that is, A to G or the reverse, or T to C or the reverse). Consider the difference as nonsynonymous if the site is nondegenerate or if it is twofold degenerate and the change is a transversion (that is, A or G to T or C). Use Equation 3.7 to calculate the proportion of nonsynonymous nucleotide substitutions per nonsynonymous site and the proportion of synonymous substitutions per synonymous site.

ANSWER Degeneracy numbers 0, 2, and 4 are as follows:

```
     *    *  *                       *        *  *         **   *
     004004004002002002002004004002004002002002204204004002002004
K12: GTCGCACCTATCTTCATCTGCCCGCCAAATGCCGATGACGACCTGCTGCGCCAGATAGCC
     Val Ala Pro Ile Phe Ile Cys Pro Pro Asn Ala Asp Asp Asp Leu Leu Arg Gln Ile Ala

LT2: ATCGCGCCGATCTTCATCTGCCCGCCAAATGCGGATGACGATCTTCTGCGCCAGGTCGCA
     N    S  S                       S        S  S         NN   S
```

The stars above indicate differences with *Salmonella* and the letters below indicate which changes are nonsynonymous and which synonymous. Altogether there are 38 nondegenerate sites, 12 twofold degenerate sites, and 10 fourfold degenerate sites. The total number of nonsynonymous sites is 38 + (2/3)12 = 46, and the total number of synonymous sites is 10 + (1/3)12 = 14. There are 3 nonsynonymous changes ($\langle D \rangle$ = 3/46 = 0.065) and 6 synonymous changes ($\langle D \rangle$ = 6/14 = 0.429). Using Equation 3.7, the number of nonsynonymous nucleotide substitutions per nonsynonymous site is $\langle k \rangle$ = 0.068, and the number of synonymous nucleotide substitutions per synonymous site is $\langle k \rangle$ = 0.635.

The great variability among proteins in the rate of nonsynonymous nucleotide substitution, when contrasted with the much smaller variability found in the rate of synonymous substitutions, is illustrated graphically in Figure 4. This disparity has been used as evidence in favor of the neutral theory. Interpreted according to the neutral theory, the variation in rates occurs because there are selective constraints on amino acid substitutions which do not operate as strongly on synonymous nucleotide substitutions.

Not just any amino acid will serve at a particular position in a protein molecule, because each amino acid must participate in the chemical interactions which fold the molecule into its three-dimensional shape and give the molecule its specificity and ability to function. The need for proper chemical interactions and folding constrains the acceptable amino acids that can oc-

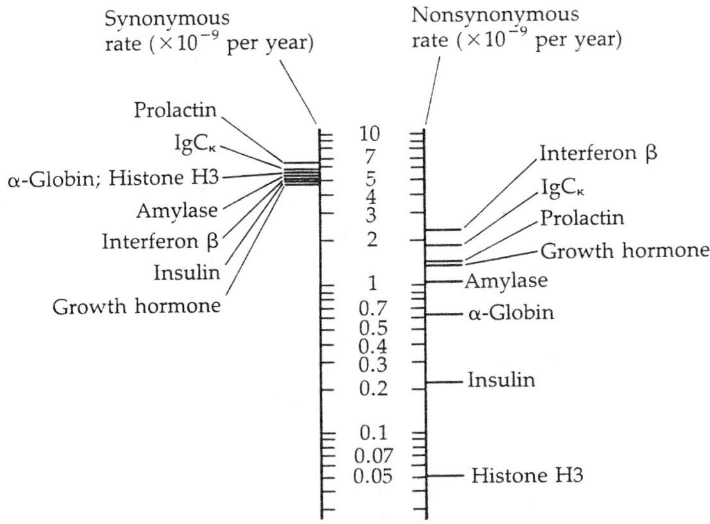

Figure 4. Comparison of rates of synonymous and nonsynonymous nucleotide substitutions. Synonymous rates are generally much faster and much more uniform than nonsynonymous rates. (From Kimura 1986.)

cupy each site. Although some amino acid substitutions may be functionally equivalent or nearly equivalent, many more are expected to impair protein function to such an extent that they reduce the fitness of the organisms that contain them. Thus, the constraints on acceptable amino acids are **selective constraints** because unacceptable amino acid substitutions are eliminated by selection.

With synonymous substitutions, however, the selective constraints are weaker and, for some substitutions, may be negligible. One type of constraint on synonymous substitutions occurs through possible secondary structures that the RNA might form, in which certain nucleotides must undergo base pairing. A second, and perhaps more important, type of constraint occurs through codon preferences, which are correlated with the relative abundance of tRNA molecules that interact with and translate the codons. In bacteria and yeast, for example, highly abundant proteins tend to use codons for abundant tRNA molecules, whereas proteins produced in small amounts tend toward codons for less abundant tRNA molecules (Ikemura 1985a,b).

The rate of nucleotide substitution not only differs between nonsynonymous and synonymous sites, but it also differs in different regions of genes. Figure 5(a) illustrates the average rates of nucleotide substitution found in different parts of genes. The different parts are the 5′ (upstream) flanking region, the 5′ transcribed but untranslated region, the coding region (nonsynonymous substitutions only, and both the slowest and fastest rates are shown), intervening sequences, the 3′ (downstream) transcribed but untranslated region, and the 3′ untranscribed region. On the average, the fastest rates of substitution occur in intervening sequences and the 3′ flanking regions, but the average rates in the 5′ flanking regions and the 3′ untranslated region are all substantially faster than 0.87×10^{-9} substitutions per nucleotide site per year, which is the average rate of nonsynonymous substitution in coding regions (see Figure 3a). Neutralists would argue that the high rates of substitution in noncoding regions, and variation among different parts of the noncoding regions, result from varying degrees of selective constraints on different parts of the gene. It is to be emphasized that Figure 5(a) depicts the averages of rates that are variable among genes, and in individual instances, especially in comparisons of closely related species, there may be fewer substitutions observed in flanking sequences than in coding sequences, or fewer changes in synonymous sites than in nonsynonymous sites.

For comparison of rates, it would be useful to study rates of nucleotide substitution in stretches of DNA wholly devoid of function and therefore subject exclusively to the whims of mutation and random drift. A likely candidate is found in a class of genes called **pseudogenes**, which are DNA sequences that are homologous to known genes but which have undergone

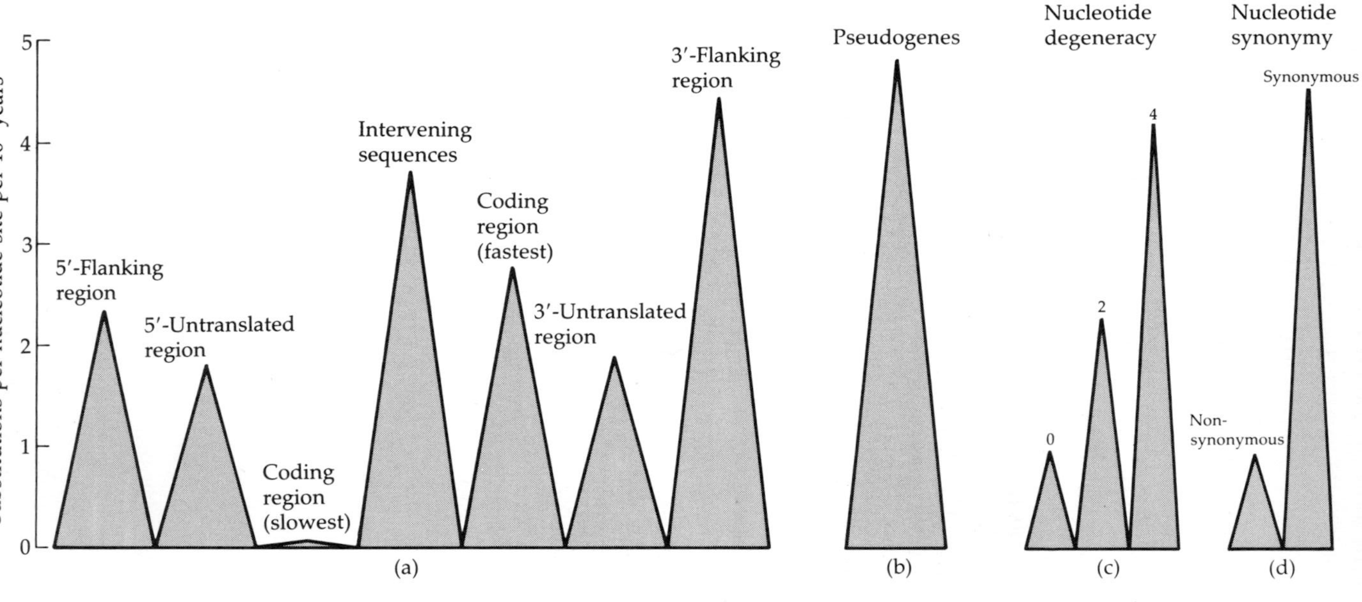

Figure 5. Comparison of average rates of nucleotide substitution in different parts of genes, types of genes, and nucleotide sites. (a) Different parts of genes. (b) Average substitution rate in pseudogenes. (c) Average substitution rates in nondegenerate (0), twofold degenerate (2), and fourfold degenerate (4) nucleotide sites. (d) Average rates of nonsynonymous and synonymous substitution. (Data from Li et al. 1985.)

one or more mutations eliminating their ability to be expressed. Pseudogenes are thought to be completely nonfunctional relics of mutational inactivation, and, in fact, their extremely rapid rate of nucleotide substitution is offered in support of this view. The average rate of nucleotide substitution in pseudogenes is shown in Figure 5(b). It is faster than the average rate found in intervening sequences and 3′ flanking regions. It is indeed faster than the rate of substitution occurring at fourfold degenerate sites in coding regions (Figure 5c), and faster than the rate of synonymous substitution in coding regions (Figure 5d). Pseudogenes evolve at the fastest rates known, which may correspond to rates of substitution when DNA is completely (or minimally) constrained by natural selection.

Rates of nucleotide substitution also vary *within* protein molecules. Human insulin is a good illustration. The A and B polypeptide chains found in the mature insulin molecule are created by posttranslational cleavage of a longer polypeptide known as preproinsulin (Figure 6). Preproinsulin contains a signal peptide for secretion, and an internal C-peptide, which are not present in the active molecule. The rates of nucleotide substitution in these three regions are as follows, where the rates are expressed in terms of nonsynon-

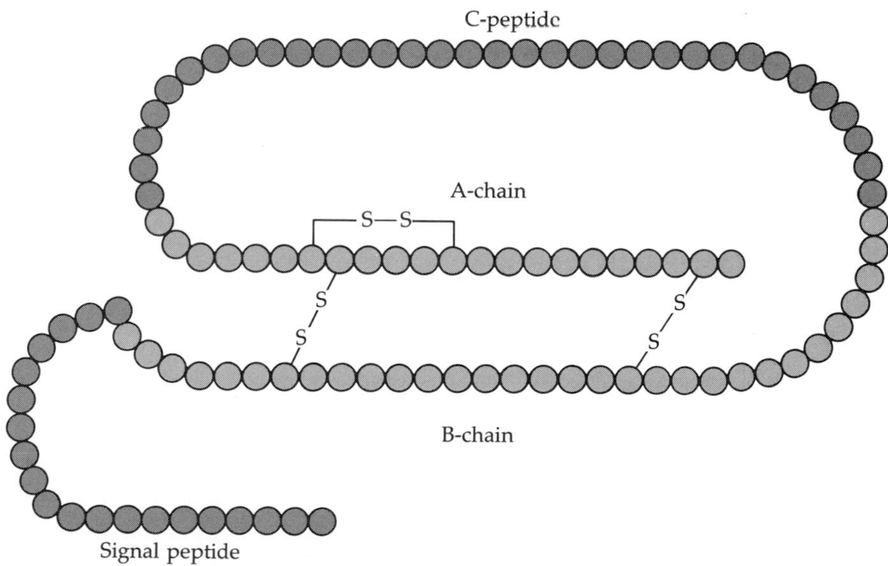

Figure 6. Functional specialization of regions within the insulin preproinsulin protein. The signal peptide functions in secretion, and the C-peptide is cleaved from the molecule in maturation. The mature insulin polypeptide consists of the A and B peptides connected through disulfide bonds (—S—S—).

ymous nucleotide substitutions per nonsynonymous site per year (Li et al. 1985a):

Insulin A and B chains:	$0.16 \pm 0.09 \times 10^{-9}$
Insulin C-peptide:	$0.99 \pm 0.26 \times 10^{-9}$
Insulin signal peptide:	$1.16 \pm 0.36 \times 10^{-9}$

In insulin, while there is a sevenfold difference between the maximum and minimum rates of nonsynonymous substitution in different regions of the molecule, the rates of synonymous substitution differ only twofold. Moreover, there is a negative correlation between functional importance and rate of nonsynonymous substitution within the insulin molecule. Many diverse amino acid sequences can serve as signal peptides provided they are hydrophobic, which suggests that selective constraints on signal peptides may be reduced in comparison with sequences in mature polypeptides. In insulin, as expected, the rate of nonsynonymous substitution is fastest in the signal peptide and slowest in the functional subunits of the mature molecule. This kind of negative correlation between selective constraint and substitution rate has also been observed in several other proteins (Li et al. 1985a).

From the selectionist viewpoint, while granting that substitutions in pseudogenes may be neutral, and synonymous substitutions may be constrained by natural selection only weakly, it is nevertheless maintained that nucleotide substitutions that change amino acid sequences are inevitably subject to the action of natural selection of an intensity which is sufficient to counteract the effects of random genetic drift. Thus, selectionists would argue that the amino acid substitutions which have occurred in a protein during the course of evolution became fixed by natural selection because they increased the fitness of the carriers through improvement in function of the molecule. However, neutralists argue back, the selectionist viewpoint cannot easily explain the negative correlation between functional importance and rate of substitution within proteins.

Mitochondrial and Chloroplast DNA Evolution

The rates of evolution summarized in Figure 5 pertain to genes located in the germ cell nucleus. Mitochondria and chloroplasts contain their own DNA molecules. These molecules undergo an evolutionary process of their own, and in mammalian mitochondria, the DNA undergoes nucleotide substitutions at a rate faster than occurs even in pseudogenes in the nucleus.

In animals, mitochondria are usually inherited through the egg cytoplasm (**maternal inheritance**) and are genetically uniform within an individual (**homoplasmic**). The mitochondrial genome consists of a single circular DNA molecule, denoted **mtDNA**, the size of which varies over a remarkably nar-

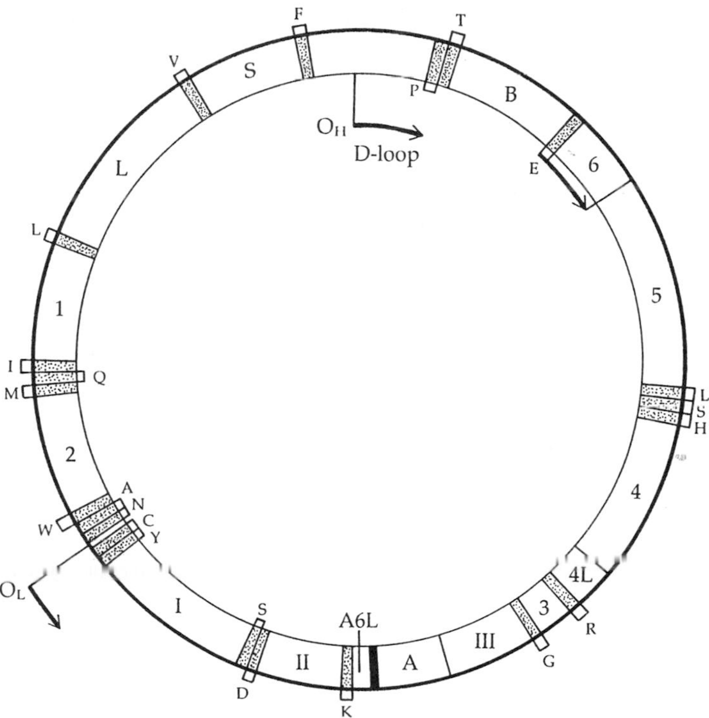

Figure 7. Organization of human mitochondrial DNA is typical of that in other mammals. It contains 13 protein-coding regions, 12 of which are essential in electron transport and ATP synthesis, plus genes for 22 transfer RNA molecules and two ribosomal RNA molecules. The typical length is 16,569 nucleotide pairs. In this diagram, the tRNA genes are indicated with stippling, L and S code for ribosomal RNA molecules, and the other labeled regions between the heavy and light lines code for proteins. The heavy and light lines represent the complementary strands of the mitochondrial DNA molecule. The arrows at O_H and O_L indicate the origin and direction of replication of each strand. H is the coding (template) strand for both rRNA genes and all proteins except number 6, and the coding strand for each tRNA gene is the strand nearest the label. See Brown (1983) for details. (From Brown 1983.)

row range in different species (15.7–19.5 kb), averaging about 16 kb. Human mtDNA, illustrated in Figure 7, is fairly typical. It contains control regions for the initiation of DNA replication, plus genes for two ribosomal RNA molecules, 22 transfer RNA molecules, and 13 proteins—12 of which code for subunits of enzyme complexes that carry out electron transport and ATP synthesis. The genetic code of animal mitochondria differs from the standard code (Table 2) in that ATA codes for Met, TGA codes for Trp, and AGR codes for End (termination of protein synthesis); thus, every codon in the mitochondrial code can be written as either *NNY* or *NNR*. Animal mitochondria also contain several hundred enzymes used in metabolic functions, but these are coded for by nuclear genes, and the enzymes are transported into the mitochondria.

At the nucleotide level, the rates of substitution in mammalian mtDNA are typically 5–10 times greater than occur in single-copy nuclear genes, averaging approximately 10×10^{-9} substitutions per nucleotide site per year. In protein-coding mitochondrial genes, the rate of synonymous substitution is about five times greater than the rate of nonsynonymous substitution, which is comparable with the ratio found in nuclear genes. Mitochondrial tRNA genes evolve approximately 100 times as rapidly as their nuclear counterparts (Brown 1985; Avise 1986).

The extraordinary rate of substitution in mammalian mtDNA is thought to result from the occurrence of a high rate of mutation during mtDNA replication, either through a high rate of nucleotide misincorporation or a low efficiency of repair. Support for the latter view comes from the observation that the mitochondrial DNA polymerase lacks the proofreading function for error correction that occurs in the nuclear DNA polymerase.

Because mammalian mtDNA evolves so rapidly and tends to be highly polymorphic within species, it provides a useful tool with which to study the geographical structure of subpopulations within a species, or genetic differentiation between closely related species. For example, one restriction-enzyme study of 87 pocket gophers (*Geomys pinetis*) yielded 23 distinct mtDNA types and revealed a strong geographical pattern in the occurrence of the types (Avise et al. 1979).

PROBLEM 9 The mtDNA molecules of 21 humans of diverse geographic and racial origin were digested with 18 restriction enzymes, 11 of which exhibited one or more fragments in which size polymorphism occurred (Brown 1980). All fragment polymorphisms could be explained by single-nucleotide differences, thus there was no evidence for insertions, deletions or other mtDNA rearrangements. Altogether, 868 nucleotide sites were assayed for differences among individuals, and the average number of differences per nucleotide site per individual was estimated at 0.0018. Assuming that mammalian mtDNA undergoes sequence divergence at the rate of $5\text{–}10 \times 10^{-9}$ nucleotide substitutions per site per year, and that the rate is uniform in time, calculate the length of time since all of the 21 contemporary mtDNA molecules last shared a common ancestor.

ANSWER With an average number of differences per nucleotide site per individual of 0.0018, and given an average rate of divergence of $5\text{–}10 \times 10^{-9}$ per site per year, the time of the last common ancestor would be between $0.0018/(10 \times 10^{-9})$ and $0.0018/(5 \times 10^{-9})$, or 180,000–360,000 years. That is, the present mtDNA genetic diversity in this sample of individuals could have originated from a single mitochondrion present in a female as recently as 9,000 to 18,000 generations ago, assuming 20 years per generation. One possible explanation of this finding is that the human population might have undergone a bottleneck in population size at about this time. In the popular press, this conclusion has been interpreted as evidence for the existence of Eve (and, by implication, Adam). However, in the list of principles of neutral evolution discussed earlier in this chapter, Principle 4 says that, among alleles destined to be fixed, the average time to fixation equals $4N_e$ generations. Assuming that a single mitochondrial genome became fixed in 9,000–18,000 generations, and assuming that these values bracket the average, the corresponding range of estimates of $\langle N_e \rangle$ (for mitochondrial DNA) is 2250–4500 individuals. This range appears not unreasonable for small societies of the sort that have characterized most of human evolution. Moreover, the average effective size of human populations in regard to mitochondrial DNA may not be representative of the average effective size in regard to nuclear DNA.

Mammalian mtDNA is unique in undergoing a rapid rate of nucleotide substitutions but a slow rate of structural changes such as insertions and deletions. The DNA of other organelles shows different patterns of evolution, which are summarized qualitatively in Table 3 (Palmer 1985). All four possible patterns of fast and slow evolution are found.

Chloroplast DNA (**cpDNA**) ranges in size from 135–160 kb, and it occurs in multiple copies in each chloroplast. Its structural organization is conserved in higher plants, and the rate of synonymous nucleotide substitution is approximately 1×10^{-9} substitutions per site per year. Thus, the evolution of cpDNA is conservative in regard to both sequence and structure. The rate of evolution of fungal mtDNA is at the other extreme—rapid in both sequence and structure.

The mtDNA of angiosperm plants has the opposite pattern of evolution as in animals. In sequence evolution the rate is slow, but in structural evolution, fast. In plants, the mtDNA genome is large and highly complex. In some instances a single molecule can resolve itself into smaller circles and even linear molecules. For example, in the turnip (*Brassica campestris*), a 218-

Table 3. Rates of evolution in organelle genomes.

Genome	Rate of nucleotide substitution	Rate of structural evolution
Angiosperm cpDNA	Slow	Slow
Angiosperm mtDNA	Slow	Rapid
Mammalian mtDNA	Rapid	Slow
Fungal mtDNA	Rapid	Rapid

(From Palmer 1985, p. 214)

kb molecule undergoes an internal recombination event which produces smaller circles of 135 kb and 85 kb. Maize mtDNA contains six pairs of repeated sequences which can undergo recombination and create a variety of structural derivatives. In addition, many plant mitochondria contain autonomously replicating plasmid DNA molecules, and they are also capable of incorporating segments of cpDNA. Why plant mtDNA genomes are so large, complex and variable in size is not understood.

In Problem 9, the assumption was made that the rate of sequence evolution was uniform in time. In the next section we discuss some of the evidence supporting the hypothesis that the rate of gene substitution at the molecular level is approximately constant.

Molecular Clocks

In spite of variation in rates of evolution which occur within different parts of molecules and among different molecules, the average rate of molecular evolution often manifests a remarkable uniformity throughout long periods of evolutionary time. Such uniformity in the rate of amino acid or nucleotide substitution is known as a **molecular clock**.

An example of clocklike uniformity in amino acid substitutions is illustrated in the evolution of the α-globin gene in the organisms depicted in Figure 8. The data are summarized in Table 4. The numbers above the

Table 4. Rate of evolution in the α-globin gene.

	Shark	Carp	Newt	Chicken	Echidna	Kang	Dog	Human
Shark		59.4	61.4	59.7	60.4	55.4	56.8	53.2
Carp	**0.90**		53.2	51.4	53.6	50.7	47.9	48.6
Newt	**0.95**	**0.76**		44.7	50.4	47.5	46.1	44.0
Chicken	**0.91**	**0.72**	**0.59**		34.0	29.1	31.2	24.8
Echidna	**0.93**	**0.77**	**0.70**	**0.42**		34.8	29.8	26.2
Kang	**0.81**	**0.71**	**0.64**	**0.34**	**0.43**		23.4	19.1
Dog	**0.84**	**0.65**	**0.62**	**0.37**	**0.35**	**0.27**		16.3
Human	**0.76**	**0.67**	**0.58**	**0.28**	**0.30**	**0.21**	**0.18**	
Avg *K*	0.87	0.71	0.63	0.35	0.36	0.24	0.18	
Time	450	410	360	290	225	135	80	

(Percentage data from Kimura 1983.)
Note: Values above the diagonal are the observed percent amino acid differences (D) between the α-globin sequences in the species, values in **boldface** are the expected amino acid differences per site [$\langle K \rangle = -\ln(1 - D)$]. Average values of $\langle K \rangle$ and the estimated times of divergence (in millions of years) are given at the bottom of the table. Abbreviation: Kang, kangaroo.

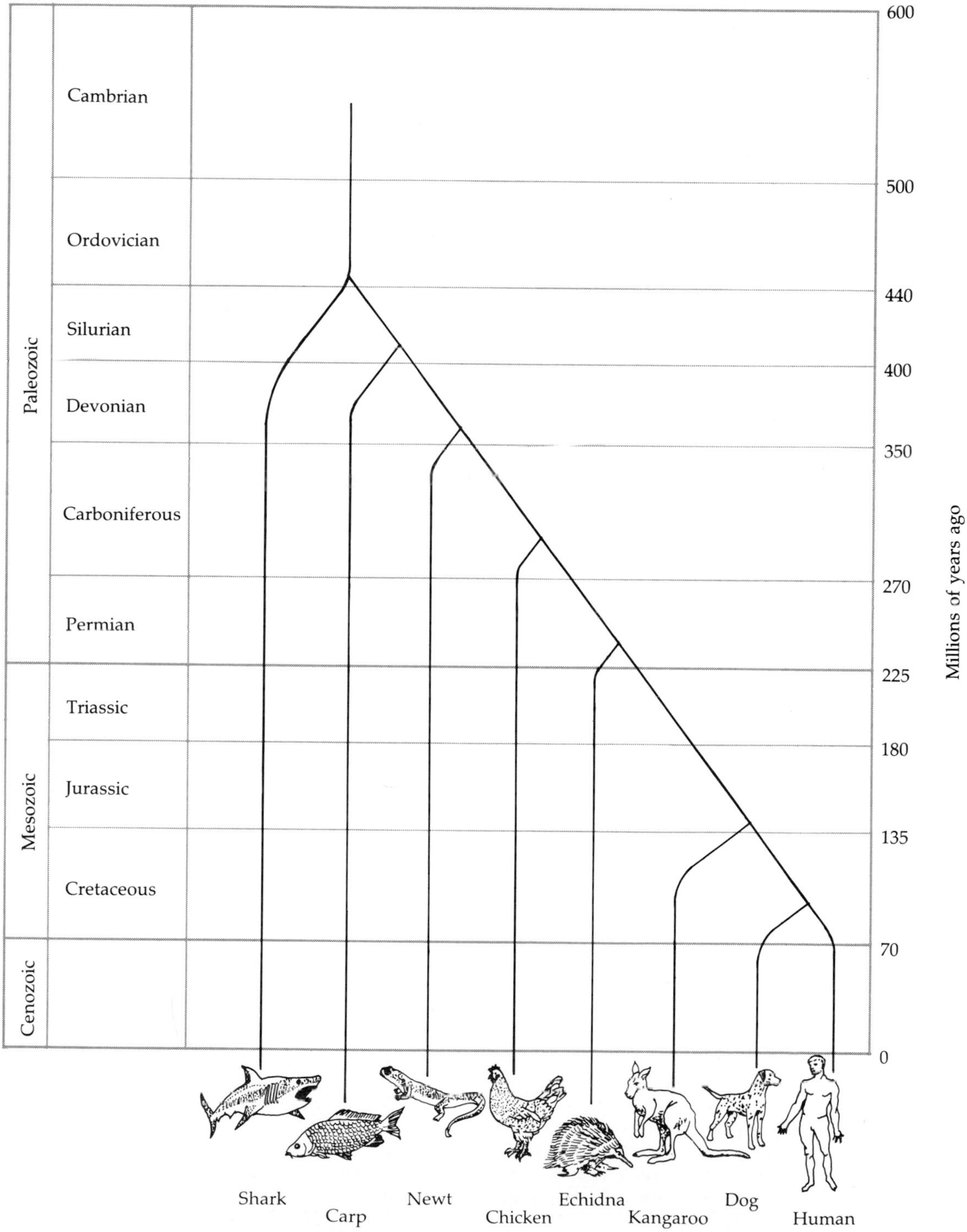

Figure 8. Phylogenetic relationships among eight vertebrate species and their approximate times of evolutionary divergence. (From Kimura 1983.)

diagonal are the percent amino acid differences ($\langle D \rangle \times 100$) between the α-globin sequences. For example, the α-globin genes of dog and human differ in 16.3 percent of their amino acid sites; since mammalian α-globin contains 141 amino acids, this percentage corresponds to 23 sites in which the amino acids differ. The percentages exclude differences that result from the insertion or deletion of amino acids, which are called **gaps** in sequence comparisons. For example, the comparison between human and shark α-globin is based on 139 amino acid sites that are homologous, and excludes gaps amounting to 11 additional amino acid sites.

Beneath the diagonal in Table 4 are the estimated proportions of differences per amino acid site, calculated from Equation 3.4 as $\langle K \rangle = -\ln(1 - \langle D \rangle)$. Table 4 also gives the average value of $\langle K \rangle$ in all comparisons with the shark, carp, newt, chicken, echidna, kangaroo, and dog, respectively, and the divergence times from the bifurcations in Figure 8.

The average proportion of differences per site is plotted against divergence time in Figure 9. The very close fit to a straight line is evident. Since the divergence time is exactly half of the total time available for evolution (Figure 2), the rate of evolution λ can be estimated as one-half times the slope of the line in Figure 9. For these data, the slope is 1.8×10^{-9}, and therefore $\langle \lambda \rangle = 0.9 \times 10^{-9}$ amino acid substitutions per amino acid site per year. The good

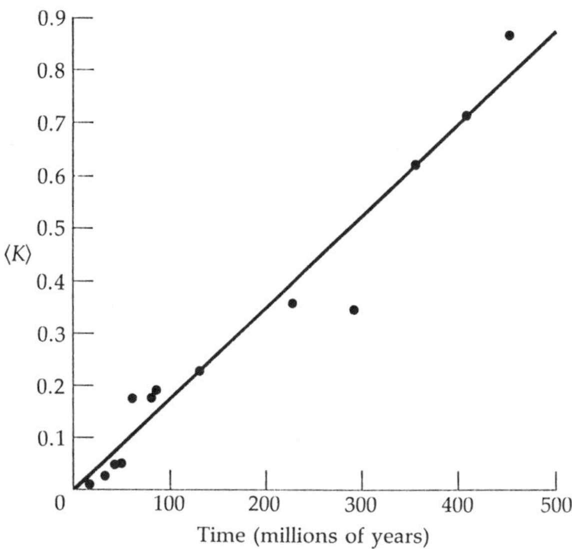

Figure 9. Relation between estimated number of amino acid substitutions in α-globin ⟨K⟩ between pairs of the vertebrate species in Figure 8, against time since each pair diverged from a common ancestor. The straight line is expected based on a uniform rate of amino acid substitution during the entire period. (From Kimura 1983.)

fit of the points to the straight line indicates that the actual rate of α-globin evolution has deviated little from the average for the past 450 million years.

The data in Problem 10 pertain to amino acid substitutions in the β-globin gene during primate evolution. They indicate a rate of evolution of the β-globin gene not markedly different that that observed for the α-gene.

PROBLEM 10

Calculate the average rate of evolution of β-globin molecule in primates using the data below on the average number of amino acid differences and the times of divergence. The β-globin molecule in primates contains 146 amino acids. (Data from Kimura 1983.)

Time of divergence (millions of years)	Average number of amino acid differences
85	25.5
60	24.0
42	6.25
40	6.0
30	2.5
15	1.0

(Note: the slope of the regression of y on x is given by

$$\frac{\Sigma(xy) - \Sigma(x)\Sigma(y)/n}{\Sigma(x^2) - [\Sigma(x)]^2/n}$$

where n is the number of pairs of points. In this application, x is the time of divergence and y is the average value of $\langle K \rangle$ estimated from Equation 3.4.)

ANSWER $\langle D \rangle$ values are obtained by dividing each number of amino acid differences by 146, and average values of $\langle K \rangle$ are estimated as $-\ln(1 - \langle D \rangle)$. The average $\langle K \rangle$ values, from top to bottom, are 0.192, 0.180, 0.044, 0.042, 0.018, 0.007, respectively. These are the y values in the linear regression, and the x values are the divergence times, as in Figure 9. In total there are $n = 6$ points. In this case, $\Sigma(xy) = 3.1263 \times 10^7$, $\Sigma(x) = 2.72 \times 10^8$, $\Sigma(y) = 0.482$, and $\Sigma(x^2) = 1.5314 \times 10^{16}$. The slope of the regression is 3.15×10^{-9}, and the rate of evolution is half of this, or 1.58×10^{-9} amino acid replacements per amino acid site per year. This estimate is reasonably close to the value of 0.9×10^{-9} per year calculated for α-globin. (Note: Rather than calculate $\langle K \rangle$ from the average number of amino acid differences, it would be more accurate to calculate $\langle K \rangle$ for each species comparison and then take the average; however, in this example, it makes very little difference.)

Although the rate of evolution is approximately constant for a given gene, the rates vary among genes. The extremes are represented by histone H4 at the low end, and γ-interferon at the high end, with globin genes near the middle of the spectrum. In short, different molecular clocks "tick" at different rates. Some exceptions to the molecular clock are well documented. For example, the insulin gene evolved much more rapidly in the evolutionary line leading to the guinea pig than in other evolutionary lines (King and Jukes 1969), and the C-type viral sequences integrated into the primate genome evolved at twice the rate in Asian primates as in African primates (Benveniste 1985).

Furthermore, molecular evolutionary clocks are "sloppy" clocks in that they demonstrate an enhanced variance in the evolutionary rate. In a theoretically perfect molecular clock driven by a random process identical to that of radioactive decay, the variance in the rate of ticking would be equal to the average rate of ticking. However, in molecular evolutionary clocks, the estimated variance is typically about twice the magnitude of the mean. Gillespie's (1986a,b) analysis suggests that the ratio of the variance to the mean might be even greater.

The possibility that gene substitutions might occur at an approximately constant rate gave some credence to the simplest version of the neutral theory. Theoretical Principle 2 discussed earlier in this chapter states that the expected rate of substitution of neutral alleles equals the rate of mutation ν to neutral alleles. Therefore, on the face of it, the occurrence of molecular clocks would seem to support the neutral theory. However, one feature of molecular evolutionary clocks is that their rate is approximately constant in a time scale measured in *years*. This is quite unexpected, because mutation rates are thought to be more nearly constant when measured in generations. However, the appropriate time scale of molecular evolution is not completely settled, as there is some evidence that the rate of synonymous substitution in genes in the rodent lineage (short generation time) might be about two times as rapid as occurs in the same genes in the human lineage.

Even if true, a nearly constant rate of gene substitution per year is not necessarily in conflict with a constant rate of neutral mutation per generation. The reason is that organisms with short generation times tend to be small and to maintain large population sizes. In such organisms, the proportion of nearly neutral mutations will be reduced because effective neutrality requires that $Ns << 1$, where s is the selection coefficient against the mutation. However, the smaller proportion of nearly neutral mutations in these organisms is balanced against the occurrence of more mutations per unit time than in larger organisms, because the generation time is shorter. Thus, the effects of short generation time and larger population size act in opposite directions and tend to cancel each other out (Crow 1985).

It has been argued that the apparent clocklike regularity of gene substitutions is actually an artifact of averaging evolutionary rates over long periods of time. A continuous record of gene substitutions might indicate that substitutions occur in clusters rather than at nearly regular intervals. This kind of pattern is suggested by the analyses of Gingerich (1986) and Gillespie (1986a,b).

The explanation of molecular clocks as an artifact of averaging has been examined with data on the occurrence of gene substitutions in short-lived organisms. An example occurs in the NS genes of human influenza A virus. Influenza is an RNA-containing virus, and the NS genes code for nonstructural proteins that are not involved in immune recognition by the host. Strains of influenza have been isolated over the past 50 years, and, except for the earliest isolates, have been maintained in refrigeration.

Figure 10 shows the number of nucleotide differences observed in the influenza NS genes, plotted against the year of isolation of the virus containing them. The rate of gene substitution averages $\langle\lambda\rangle = 1.94 \pm 0.09 \times 10^{-3}$ nucleotide substitutions per nucleotide site per year. Although the rate of gene substitution is about 10^6-fold faster than observed in germline genes in eukaryotes, it is nevertheless approximately constant during the period available for study. The extraordinary rate of evolution in influenza virus is thought to be related to a high rate of spontaneous mutation resulting from errors in replication (Holland et al. 1982). As in many other RNA-based viruses, the RNA replicase enzyme which replicates the influenza genome lacks a proofreading function.

The existence of molecular evolutionary clocks, in spite of their greater

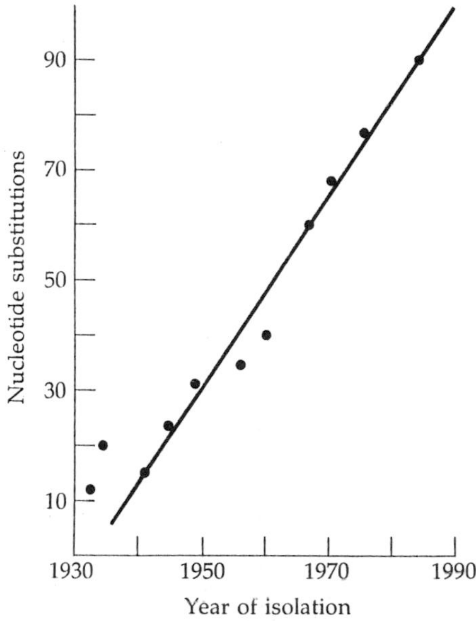

Figure 10. Molecular evolution in the NS genes of influenza virus determined from strains isolated and stored during the past 60 years. The total rate of evolution in the 890-nucleotide sequence averages 1.73 ± 0.08 nucleotide substitutions per year, and the rate is remarkably uniform. (From Buonagurio et al. 1986.)

variation than theoretically perfect clocks, provides a valuable tool in inferring the phylogenetic relationships among species and in estimating their times of divergence. Molecular data thereby come to bolster, and in some cases refine, relationships and divergence times inferred from the fossil record. In a few cases, molecular data have demanded changes in the established views of evolutionary history (e.g., Sibley and Ahlquist 1986).

Determining the evolutionary relationships between organisms that are closely related is often difficult because the phylogenetic inferences can be affected by even relatively small inhomogeneities in rates of divergence or sampling errors in the data. This difficulty is illustrated in the case of higher primates. Figure 11 shows the two most popular phylogenetic hypotheses for the relationships between the human, chimpanzee, and gorilla. Phylogeny (a) groups humans and chimpanzees, phylogeny (b) groups chimpanzees and gorillas.

Based on restriction-fragment differences in mtDNA, Templeton (1983) rejected phylogenetic hypothesis (a) in favor of (b). However, the validity of Templeton's statistical analysis was contested by Nei and Tajima (1985), who concluded that neither phylogeny could be rejected based on the mtDNA data. On the other side, analysis of differences in nuclear DNA sequences as judged by their ability to form thermally stable hybrid molecules, led Sibley and Ahlquist (1984) to reject phylogeny (b). These data were reanalyzed by Templeton (1985), who concluded that rejection of phylogeny (b) was unwarranted. His statistical methods were challenged by Saitou (1986) and Ruvolo and Smith (1986), and defended by Templeton (1986). The situation is therefore very much in dispute, but identifying the correct phylogeny is important in determining whether human traits such as the bipedal gait and human limb proportions should be considered as evolutionarily ancestral or derived (Templeton 1983). The essence of the problem is that the species are so closely related that neither the presently available data nor the methods of analysis can make the critical distinctions.

Misconceptions About the Neutral Theory

Some objections to the neutral theory are based on misunderstandings or oversimplifications of what the theory implies. Several of the more common of these misconceptions are as follows.

Fallacy 1: Only genes that are unimportant can undergo neutral mutations. The fallacy here stems from failing to understand that neutral mutations are assumed to be *equivalent* in function, not *lacking* in function. By way of analogy, a dollar bill is monetarily equivalent to four quarters, but this hardly implies that money has no economic function.

A biochemical basis for the functional equivalence of certain enzyme var-

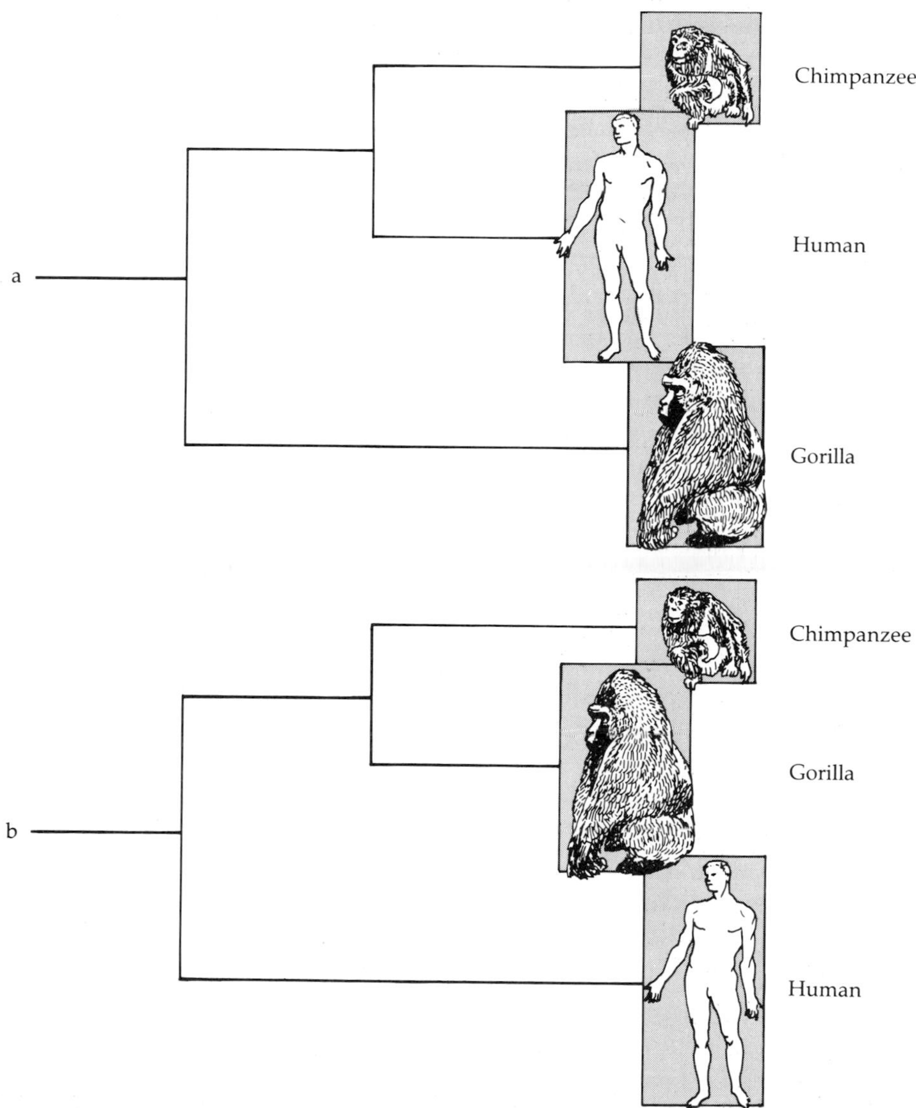

Figure 11. Two popular models for the phylogenetic relationship among the human, chimpanzee, and gorilla. Neither model can be convincingly rejected based on currently available data.

iants has been suggested, which is based on studies of the control of the rate of flux of metabolites through metabolic pathways (Kacser and Burns 1973; Hartl et al. 1985). This concept is illustrated by the observations in Figure 12, where relative fitness (f) is given as a function of the relative enzyme activity (a) of β-galactosidase in cells of *E. coli* competing for growth in lactose medium. The shape of the curve is a hyperbola, like that of first-order enzyme kinetics, with $f = a/(1 + a)$, and the curve implies that, for cells with enzyme activities near that of wildtype (circle), a small change in enzyme activity results in a disproportionately smaller change in fitness. For example, in terms of the arbitrary enzyme units used in the figure, a change from 22 to 21 units represents a decrease of 4.5 percent in enzyme activity, but it results in only a 0.2 percent decrease in relative fitness. Of course, not all enzymes obey a hyperbolic fitness–activity relation of the type shown in Figure 12 (Burton and Place 1986; Hartl et al. 1986), but for those that do, there is certainly scope for mutations that change activity to such a small extent that the resulting change in fitness is negligible. The kind of hyperbolic relation between fitness and the activity of an enzyme that is exemplified in Figure 12 is generally to be expected when fitness is proportional to the flux of substrate across the metabolic pathway in which the enzyme occurs (Kacser and Burns 1973; Hartl et al. 1985).

PROBLEM 11 In populations of *E. coli* undergoing competition for lactose in chemostats, the fitness w of a mutant, relative to the wildtype K12 strain, is given approximately by

$$1/w = 0.130/a_0 + 0.004/a_1 + 0.866$$

where a_0 is the activity of the β-galactoside permease and a_1 is that of the β-galactosidase, both measured relative to their counterparts in the wildtype K12 strain (Dykhuizen et al. 1987). Consider two mutations, one of which decreases the relative permease activity by 5 percent, and the other of which decreases the β-galactosidase activity by 5 percent. How much is relative fitness decreased by each of the two mutations? How much would relative fitness be decreased in a strain containing both mutations?

ANSWER For the permease mutation,

$$1/w = (0.130/0.95) + (0.004/1.00) + 0.866 = 1.00684$$

or $w = 0.9932$, for a decrease of 0.68 percent. For the β-galactosidase mutation,

$$1/w = (0.130/1.0) + (0.004/0.95) + 0.866 = 1.00021$$

or $w = 0.99979$, for a decrease of 0.02 percent. Note that the permease mutation has a 34-fold greater affect on fitness. Also, the double mutant has a fitness of $w = [(0.130/0.95) + (0.004/0.95) + 0.866]^{-1} = 0.9930$, which is essentially equivalent to that of the permease mutation alone.

Fallacy 2: The neutral theory of molecular evolution diminishes the role of natural selection. The premier role of natural selection in promoting adaptive changes in developmental, morphological and behavioral traits is accepted even by adherents to the neutral theory. However, the neutral theory asserts that, at the molecular level, the majority of gene substitutions which occur result

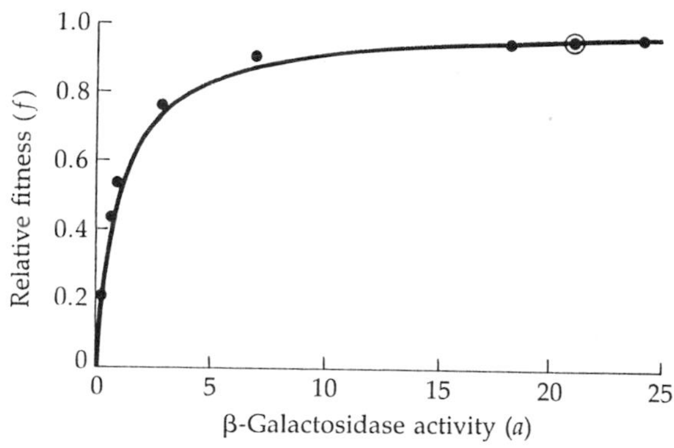

Figure 12. Hyperbolic relation between relative fitness f of strains of *Escherichia coli* undergoing competition for lactose in chemostats, against relative activity of the β-galactosidase enzyme (a). The circled point represents the wildtype laboratory strain *E. coli* K12; the other points are from mutants. The curve is arbitrarily scaled so that $f = 1.0$ at $a = 30$. (From Hartl et al. 1985.)

from the random fixation of nearly neutral mutations. Although some molecular changes certainly do promote increased adaptation at the molecular level (for example, the amino acid substitution of valine for glutamic acid at position number 6 in human β-globin, which confers increased resistance to malaria), gene substitutions that are correlated with adaptive changes are much less common than random fixations.

Fallacy 3: Nucleotides or amino acids that can undergo neutral mutation are not subject to selective constraints. The fallacy in this case is to infer that what is true of some mutations at a site must be true of all of them. It is entirely possible that some nucleotide or amino acid replacements at a site may be neutral, but others affecting the same site may be selected.

Some insight into the nature of selective constraints on particular amino acids may be gained from mutational studies of an *E. coli* class A β-lactamase, which is an enzyme responsible for ampicillin resistance. Position 71 in the enzyme, adjacent to a key serine in the active site, is occupied by threonine in all class A β-lactamase enzymes. Thus, the threonine at position 71 appears to be functionally quite important. However, when site-directed mutagenesis was used to replace the threonine-71 with each of the other 19 possible amino acids, only five substitutions (thr → tyr, trp, asp, lys, or arg) eliminated activity altogether (Schultz and Richards 1986). A total of 14 mutants continued to demonstrate appreciable resistance to ampicillin. Although the mutant proteins were less stable to degradation than wildtype, suggesting a possible reason why the threonine-71 is conserved, the example demonstrates that different types of substitutions at a site can have dramatically different consequences.

Different selective constraints on different substitutions also occur at the nucleotide level. For example, many codons that terminate in a pyrimidine (Y: T or C) or a purine (R: A or G) [see Table 2] can undergo transition mutations (T → C, C → T, A → G or G → A) without changing the amino acid. However, a transversion mutation ($Y \rightarrow R$ or $R \rightarrow Y$) at the same site often changes the amino acid, which may significantly impair, or much less likely improve, the function of the protein.

Selective constraints on nucleotide sequences also occur at the level of secondary structure. A familiar example occurs in the foldback structures formed by base pairing in transfer RNA molecules, of which a typical example is illustrated in Figure 13. Duplex RNA structures can accomodate GU base pairs, so a molecule can theoretically undergo the following sequence of substitutions without undergoing any significant change in conformation: GC → GU → AU.

Fallacy 4: Neutral mutations have a selection coefficient defined by $s = 0$. This fallacy fails to recognize the fact that the real populations are finite in size, and the ultimate fate of a mutation is determined jointly by the selection coefficient and the population size. In a diploid species, the key quantity proves to be $4N_e s$, where N_e is the effective population number and s is the

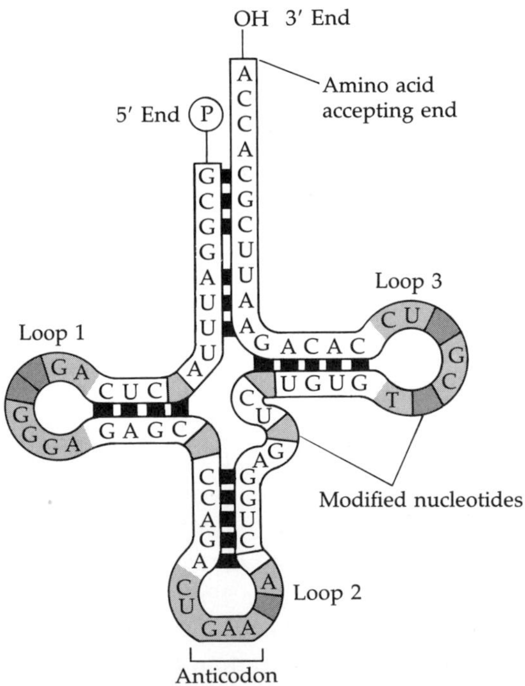

Figure 13. Secondary structure resulting from foldback base pairing (heavy black bars) in a typical tRNA molecule. In this example, the phenylalanine tRNA is from yeast. (From Alberts et al. 1983.)

selection coefficient of the mutation, assuming no dominance in fitness (Kimura 1964). The following guidelines serve as approximations for defining the relative roles of selection and random genetic drift in determining the ultimate fate of a mutation:

1. If $4N_e s > 10$, then selection is predominant.
2. If $4N_e s < 0.1$, then random genetic drift is predominant.
3. If $0.1 < 4N_e s < 10$, then selection and random genetic drift both play a role.

Thus, small effective population size promotes the relative importance of random genetic drift by increasing the magnitude of the selective effect that is necessary to make selection paramount. Therefore, the rate of occurrence of neutral mutations, mentioned earlier in this chapter in Principle 2, is itself a function of the effective population number.

Ohta (1973, 1974) emphasized that many gene substitutions and polymorphisms, particularly those that change amino acids, may involve mutations with slightly deleterious effects on fitness, but effects so small that random genetic drift still plays an appreciable role in governing their frequencies and fate. In a statistical analysis of the occurrences of amino acid polymorphisms in the gene for 6-phosphogluconate dehydrogenase among natural isolates of *E. coli*, as compared with the occurrences of silent polymorphisms in the same gene, Sawyer et al. (1987) concluded that no more than half of the observed amino acid polymorphisms in the gene could be truly selectively neutral, and they estimated that the average selection coefficient against the polymorphic amino acids might be approximately 1.6×10^{-7}.

Fallacy 5: Neutral mutations are neutral for all time. The environment of natural populations changes with time, and the essence of evolution is that the genotypes within populations also change. Thus, it is perfectly conceivable that a mutation could be selectively neutral for a time, but later, owing to a change in environment or genetic background, become subject to selection. Indeed, a selectively neutral mutation ($4N_e s < 0.1$) could even become subject to selection because of an increase in the effective population number.

The possibility that the selective effect of alleles can differ dramatically depending on the environment or genetic background has been called the **potential for selection** (Hartl and Dykhuizen 1981, 1985; Dykhuizen and Hartl 1983). An example is illustrated in the following problem.

PROBLEM 12

The following data were provided by D. E. Dykhuizen on the selective effect of a *gnd* (6-phosphogluconate dehydrogenase) mutation in *E. coli* undergoing competition in chemostats with the isogenic parental strain. The *gnd* mutation, *gnd*(862), contains a small deletion upstream from the promoter sequence and results in a three-fold increase in enzyme activity (Miller et al. 1984). In the data, p represents the frequency of mutant cells, and time is given in hours. The selection coefficient of the mutant, relative to wildtype, is given by the slope of the regression of $\ln[p/(1-p)]$ against time, and the units of the selection coefficient are therefore hr^{-1}. In environment A the cells were undergoing competition for gluconate, in environment B for a mixture of 75 percent gluconate and 25 percent ribose, and in environment C for ribose. Calculate the selection coefficient per hour of the mutant in each of the three environments.

A	B	C	Time
0.1205	0.0937	0.0851	0
0.1937	0.0639	0.0742	17.0
0.2524	0.0498	0.0785	23.5
0.3918	0.0296	0.0777	40.5
0.4479	0.0254	0.0799	47.5
0.4590	0.0247	0.0877	64.5

ANSWER Let y represent $\ln[p/(1-p)]$ for the values in each of A, B, and C, and let x represent time in hours. Then $n = 6$, $\Sigma(x) = 193.0000$, $\Sigma(x^2) = 8898.0000$, and the other sums needed in the regression, and the slopes, are as follows:

	A	B	C
$\Sigma(y)$	−5.31298	−18.71550	−14.62150
$\Sigma(xy)$	−88.10887	−666.61250	−468.11620
Slope (s)	0.0308	−0.0240	0.0008

In appropriate statistical tests, the estimated selection coefficient s is significantly different from zero only in chemostats A and B. The important point, in terms of potential for selection, is that the *gnd*(862) mutation is selectively favored in gluconate, disfavored in the mixture of gluconate and ribose, and without detectable effect in ribose.

The Neutral Theory: An Appraisal

In the recent history of population genetics, protagonists in the selection-neutrality controversy have been at times detached, bemused, animated, disagreeable, or occasionally churlish, depending on mood and personality. However, as data have accumulated, the neutral theory has been enlarged to accomodate slightly deleterious and other types of mutations, and most selectionists have become comfortable with the idea of the near-neutrality of nucleotide substitutions in regard to pseudogenes or synonymous sites, if not in regard to those which change amino acids. Some commentators suppose that there is little left to debate. Granting the existence of many shades of opinion, most contemporary population geneticists seem to agree that the neutral theory, truth or falsehood aside, has been valuable as a scientific theory and important as a stimulus for experimental investigation.

Crow (1985, p. 11) has provided a list of some of the reasons why the neutral theory has been good as a scientific theory, and his summary serves as a fitting appraisal of the current status of the theory:

> 1. The theory provides the best explanation so far of the gross difference in the pattern of rates of change between molecular and morphological evolution.
> 2. The neutral theory brings together the superficially unrelated rates of change of synonymous codon positions, introns, and pseudogenes.
> 3. The neutral theory predicted that pseudogenes would have rapid evolutionary rates and that there would be as much molecular differentiation and polymorphism in living fossils [like the horseshoe crab] as in rapidly changing species.
> 4. The neutral theory has been remarkably heuristic in two ways. As for theory, it has stimulated development of formulas for the mean time until fixation or loss of an allele, the mean time to fixation conditional on fixation, the total number of individuals and the total number of heterozygotes along a sample path, and it has provided not only means but [also other attributes], or the entire distribution. As for experiments, it has stimulated a much more systematic and relevant study of molecular polymorphisms and has provided guidance for gathering data on molecular evolutionary rates.
>
> [The neutral] theory depends on assumptions and parameter values that cannot at present be adequately tested. Only the future will tell whether it is correct.

EVOLUTION IN MULTIGENE FAMILIES

Some genes exist in multiple related copies within the genome of a single species. Within individuals of the species, the discrete copies of the multigene family usually bear a greater resemblance to each other than they do to the members of homologous gene families present in the genomes of closely related species.

Genes increase in number through **duplication**. Several successive rounds of duplication result in a family of homologous genes with related functions, a **multigene family**, the members of which are often tandemly arranged within the chromosome, as illustrated in Figure 14.

In this section we summarize some of the principal processes that govern the evolution of multigene families.

Mechanisms of Concerted Evolution

Members of multigene families tend to retain more similarity than would be expected. They evolve together, in concert, because of mechanisms that

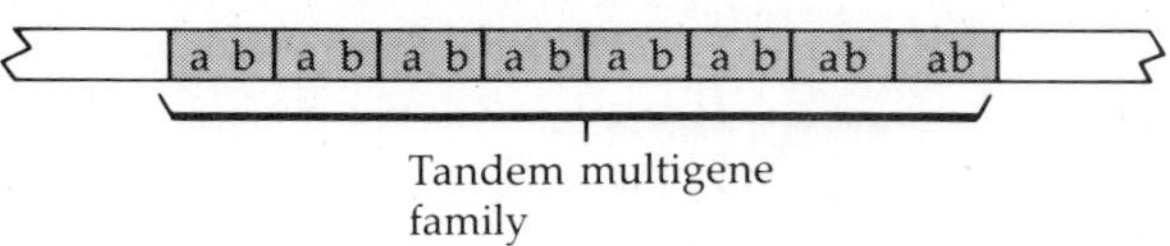

Figure 14. Organization of genes in a tandemly repeated multigene family. Some multigene families contain hundreds of members and may not be adjacent but separated by "spacer" DNA. The small letters show the identical orientation of units within the array.

operate to homogenize the sequences within the family. This tendency toward homogenization is known as **concerted evolution**. Two important mechanisms of concerted evolution are illustrated in Figure 15. Figure 15(a) shows **gene conversion**, a process in which nucleotide pairing between two sufficiently homologous genes is accompanied by the excision of all or part of the nucleotide sequence in one gene and its replacement by a replica of the nucleotide sequence in the other gene. Formally, the result is that the sequence in one gene "converts" the sequence in the other gene to be exactly like itself.

Figure 15(b) illustrates concerted evolution by means of **unequal crossing-over**. In tandem multigene families, the pairing of the genes during meiosis can be offset, and crossing-over results in one type of sequence within the family becoming over-represented, and an increase in copy number overall. However, a subsequent crossing-over can correct the copy number without changing the over-representation of the one sequence. Formally, the net result is genetically equivalent to gene conversion, because the nucleotide sequences in some members of the multigene family have been replaced with

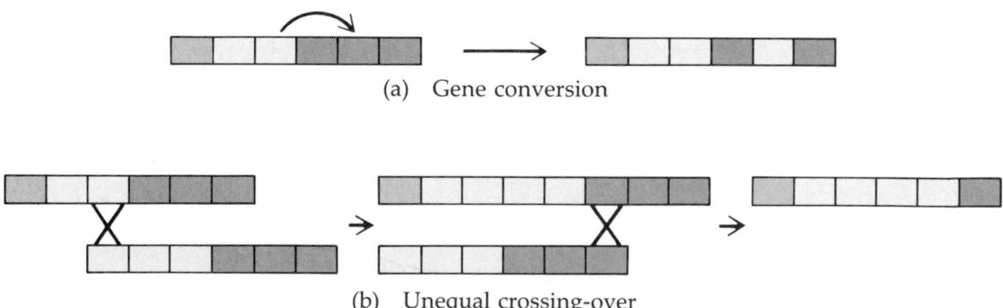

Figure 15. Two mechanisms of concerted evolution that can homogenize the sequences of members of multigene families. (a) Gene conversion, in which genetic information in one gene is replicated in the place of another. (b) Unequal crossing-over. The increase in the number of members of the gene family because of one unequal crossing-over can be corrected by a subsequent unequal crossing-over, resulting in identical DNA sequences among several adjacent members of the multigene family.

the nucleotide sequences in other members.

Gene conversion and unequal crossing-over operate to maintain similarity between copies of genes in multigene families. A theoretical study of the processes has been carried out by Ohta (1982). In Ohta's model, a tandemly arranged multigene family consists of a fixed number n of members, and λ is the probability that a particular member of the gene family becomes converted by another member in any one generation. (Equivalently, λ is the probability of completion of a cycle of unequal crossing-over resulting in the replacement of one gene in the family by another.) The mutation rate per gene is v, and the effective population number is N. The probability that equal crossing-over occurs within the gene cluster in any generation is given by β.

In this model there are three types of gene identity by descent that can occur between members of the multigene family, which are illustrated in Figure 16:

1. The probability c_1 that two genes in the same chromosome are identical by descent.
2. The probability c_2 that two genes in nonhomologous positions in different chromosomes are identical by descent.
3. The probability f that two genes in homologous positions in different chromosomes are identical by descent.

Complex formulas for the equilibrium values of c_1, c_2, and f have been derived by Ohta (1982), but they are greatly simplified when $4Nv << 1$ and when there is no equal crossing-over within the gene cluster ($\beta = 0$). In this case, to a good approximation,

$$c_1 = c_2 = \frac{\lambda}{\lambda + (n - 1)v} \tag{3.9}$$

$$f = \frac{4N\lambda c_2 + 1}{4N\lambda + 4Nv + 1} \tag{3.10}$$

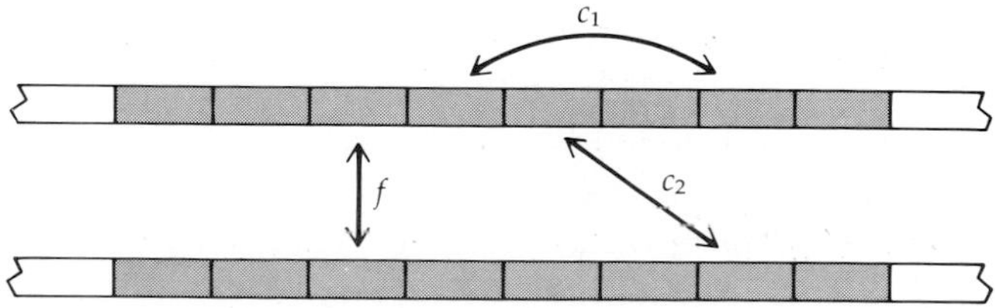

Figure 16. Three types of identity by descent in multigene families. They are the identity between genes at homologous sites (probability f), between genes at nonhomologous sites in the same chromosome (probability c_1), and between genes at nonhomologous sites in different chromosomes (probability c_2). (After Ohta 1982.)

Equation 3.10 is valid for all values of $4Nv$. In this model, the quantity nv equals the *total* rate of mutation in the entire multigene family, and this is very nearly $(n - 1)v$ when n is reasonably large. Thus, the implication of Equation 3.9 is that, when the rate of gene conversion λ is substantially greater than the total mutation rate, then the probability of identity by descent of nonhomologous genes within the family (c_1 and c_2) will be close to 1.0; and when λ is substantially smaller than the total mutation rate, then the probability of identity by descent of nonhomologous genes will be close to zero. The random conversion process is analogous to that of random genetic drift in that, if a single chromosome is followed through successive generations, the continual replacement of individual members of the linked family corresponds to the repeated sampling of gametes during reproduction.

PROBLEM 13 Using the approximations in Equations 3.9 and 3.10, and the parameter values $n = 50$, $N = 5 \times 10^4$, and $v = 1 \times 10^{-8}$, calculate c_1, c_2, and f for the two cases $\lambda = 1 \times 10^{-5}$ and $\lambda = 1 \times 10^{-7}$.

ANSWER For the case $\lambda = 1 \times 10^{-5}$, the equilibrium values are $c_1 = c_2 = 0.953$ and $f = 0.968$; for $\lambda = 1 \times 10^{-7}$ the corresponding values are 0.169 and 0.982. With the larger rate of gene conversion, the identities differ little between genes at homologous (f) or nonhomologous positions (c_1, c_2). With the smaller rate of gene conversion, genes in homologous positions are much more likely to be identical than genes in nonhomologous positions. Incidentally, a small amount of equal crossing-over in the gene cluster does not affect the qualitative conclusions. For example, using the same parameters as above except with $\beta = 1 \times 10^{-3}$, the solutions to Ohta's formulas are $c_1 = c_2 = 0.88$ and $f = 0.92$ when $\lambda = 1 \times 10^{-5}$, and $c_1 = c_2 = 0.17$ and $f = 0.98$ when $\lambda = 1 \times 10^{-7}$.

Duplication and Divergence in Multigene Families

Duplicate genes can evolve in separate ways under the influence of natural selection, and also mutation and random genetic drift. In time, some members of a multigene family may diverge to a greater or lesser degree in their function. This process of **duplication and divergence** is thought to be the major mechanism by which genes with novel functions are created.

Figure 17 illustrates the result of duplication and divergence in two related multigene families in humans—those coding for the β-like and α-like polypeptide chains of hemoglobin. The genes are specialized for different periods of life. The ϵ (epsilon) and ζ (zeta) genes are expressed in embryos; the γ (G-gamma and A-gamma) genes and the α genes in the fetus; and the β, δ, and α genes in the adult. The time of expression and function of the θ1 gene is unclear (Marks et al. 1986).

The inference from differences in nucleotide sequence is that the original

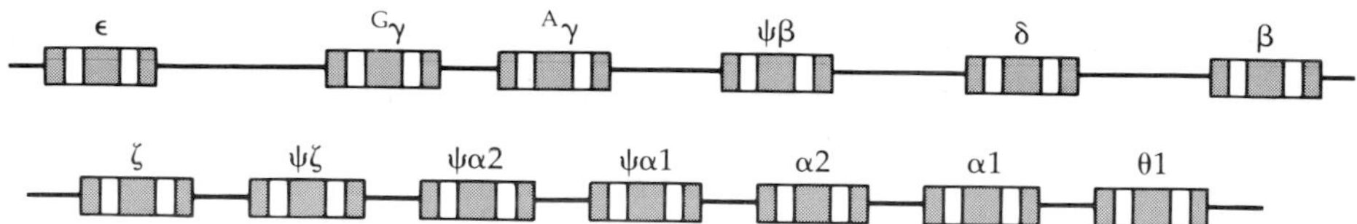

Figure 17. Organization of the β-globin and α-globin gene families in humans (located in chromosomes 11 and 16, respectively). The designation ψ denotes a pseudogene. The functional adult α-globin gene copies α1 and α2 are identical, and the fetal globin genes $^{G}\gamma$ and $^{A}\gamma$ are nearly identical. The genes ε and ζ code for embryonic hemoglobins. Judged by its nucleotide sequence, the α-like gene θ1 is potentially functional and may be expressed during restricted times or in restricted tissues during development. (Organization of the α-gene cluster after Marks et al. 1986.)

β–α duplication took place approximately 500 million years ago, when vertebrates were represented by the bony fishes, and the β–γ duplication took place about 80 million years ago, during the mammalian radiation. More recent duplications have also occurred, for example those leading to the two functional α genes, the cluster of three α-like pseudogenes, and the two γ genes.

The evolutionary history of the fetal globin genes in humans reveals that the $^{G}\gamma$ and $^{A}\gamma$ genes originated as part of a relatively recent 5 kb tandem duplication (Shen et al. 1981). Furthermore, evidence from nucleotide sequences strongly suggests that a gene conversion event also occurred, which converted part of one particular $^{A}\gamma$ allele into a $^{G}\gamma$ allele. The converted $^{A}\gamma$ allele is very similar to a $^{G}\gamma$ allele for about 1550 bp on the upstream (5′) side of a putative recognition signal for gene conversion (a stretch of repeating TG and CG dinucleotides); but on the downstream (3′) side of the putative signal, the converted $^{A}\gamma$ allele is typical of other $^{A}\gamma$ alleles in the human population.

The $^{A}\gamma \rightarrow {}^{G}\gamma$ gene conversion occurred much more recently than the duplication resulting in the $^{A}\gamma$ and $^{G}\gamma$ genes, as indicated by the analysis in the following problem.

PROBLEM 14

The nucleotide sequence data are that, on the upstream side of the putative gene-conversion signal, seven out of 1550 nucleotides differ between $^{G}\gamma$ and the converted $^{A}\gamma$ allele; but on the downstream side, 145 out of 1550 nucleotides differ. In both cases the majority of the 1550 nucleotides are noncoding. Use Equations 3.7 and 3.8 to estimate the amount of nucleotide substitution and its standard deviation, on the upstream and downstream sides of the gene-conversion signal. What evidence is there that the amounts are actually different in the two regions? Assuming that the rate of sequence divergence in the α-globin multigene family accumulates at the rate of one percent per 2.2 million years (an average based mainly on noncoding sequences), use the upstream data to estimate the time since the gene conversion event occurred, and the downstream data to estimate the time since the $^{G}\gamma$–$^{A}\gamma$ duplication occurred.

ANSWER With $\langle D \rangle = 7/1550$ differences upstream from the signal, $\langle k \rangle = 0.00453 \pm 0.0017$; and with 145/1550 differences downstream from the signal, $\langle k \rangle = 0.0999 \pm 0.0084$. The difference seems to be real, as its magnitude ($0.0999 - 0.00453 = 0.095$) is more than 11 times greater than the larger standard deviation. Assuming one percent sequence divergence per 2.2 million years, the time of the gene conversion is set at $0.00453 \times 100 \times 2.2 = 1.0$ million years; the time of the duplication itself is set at $0.0999 \times 100 \times 2.2 = 22.0$ million years.

The estimate of the time of occurrence of the $^{A}\gamma$–$^{G}\gamma$ duplication in Problem 14 can be improved by using the nucleotide sequence data from the entire whole duplicated 5 kb region. In the entire region there are 14 percent nucleotide differences, which translates into $\langle k \rangle = 0.155 \pm 0.006$; this suggests a time for the duplication of $0.155 \times 100 \times 2.2 = 34$ million years (Shen et al. 1981).

PROBLEM 15 In the regions of the $^{A}\gamma$ and $^{G}\gamma$ fetal globin genes that were unaffected by the gene conversion analyzed in Problem 14, there were 34 gaps, resulting from addition/deletion events, among 4146 nucleotides (Shen et al. 1981). The length distribution of the gaps, in number of nucleotides, was as follows:

Length	No.	Length	No.	Length	No.
1	12	6	1	13	1
2	3	7	2	23	1
3	2	8	0	122	1
4	4	9	0	Sum	34
5	6	10	1		

Assuming that each gap results from a single addition/deletion event, calculate the $\langle k \rangle$ for the gaps in the sequences, and the standard deviation. Assuming a divergence time of 34 million years between the sequences, what is the rate of incorporation of gaps? How does this number compare with the rate of 1 percent nucleotide sequence divergence per 2.2 million years?

ANSWER For the gaps, $L = 4146$ and $\langle D \rangle = 34/4146 = 0.0082$. These imply that $\langle k \rangle = 0.00825 \pm 0.00142$. The rate of gap incorporation is therefore $0.00825/(2 \times 34 \times 10^6) = 0.121 \times 10^{-9}$ gap substitutions per nucleotide site per year. Also, a rate of divergence in nucleotides of one percent per 2.2 million years equals $0.01/(2.2 \times 10^6) = 4.55 \times 10^{-9}$ nucleotide substitutions per nucleotide site per year. In the fetal globin genes, the rate of nucleotide substitution in noncoding regions is approximately 40-fold greater than the rate of incorporation of gaps.

Unequal crossing-over in multigene families can result in a decrease in the number of genes as well as an increase (see Figure 15). It is therefore not surprising that deletions of one or more of the hemoglobin genes are found in most parts of the world. Although usually very rare, in a few places the frequency of the deletions reaches levels too great to be accounted for by chance, especially in view of the observation that the carriers are mildly to severely anemic, or worse. For example, deletion of the β-gene results in death when homozygous, but it, and other mutations that decrease the abundance of the β-hemoglobin chain, are relatively common in the Mediterranean Sea basin where malaria is endemic. For this reason, the decreased β-chain diseases are called **β-thalassemias** (literally translated as "sea anemias"). The well-established link between sickle-cell anemia and malaria (see Chapter 2), along with the geographical correlation between the β-thalassemias and malaria, provides a strong circumstantial case for malarial parasites being an important selective agent.

Deletion of one or more of the α-globin genes results in **α-thalassemia**. There are two α genes (Figure 17), and deletion of one of them results in a mild form of anemia. For example, when a normal chromosome (α α) and a chromosome containing an α-deletion (− α) are combined in heterozygous condition (α α/− α), the individual is often indistinguishable from normal; on the other hand, the − α/− α homozygote exhibits mild anemia.

The frequency of α-deletion chromosomes is significant throughout the malarial regions of the world, suggesting a causal link. Figure 18 strengthens the argument in demonstrating a direct correlation between the frequency of chromosomes containing the α-deletion and the prevalence of malaria. The study was carried out in Melanesia in the southwest Pacific, including the islands of Papua New Guinea, the Solomon Islands, Vanuatu, and New Caledonia (Flint et al. 1986). In north coastal Papua New Guinea, where malaria is most severely endemic, the frequency of the α-deletion chromosome reaches 68 percent. This figure may represent an approximate equilibrium in which the mildly harmful effects of the deletion are offset by the beneficial effects that supposedly result from increased resistance to malaria.

Multigene Superfamilies

In some cases, several sets of multigene families and single-copy genes may share recognizable homology, implying a common ancestry, but they have undergone major divergence in function and relocation of position within the genome. These sets of historically related but functionally distinct genes constitute a **multigene superfamily**.

One extensively studied multigene superfamily which serves diverse functions in immunity is illustrated in Figure 19 (Hood et al. 1985; Hunkapiller

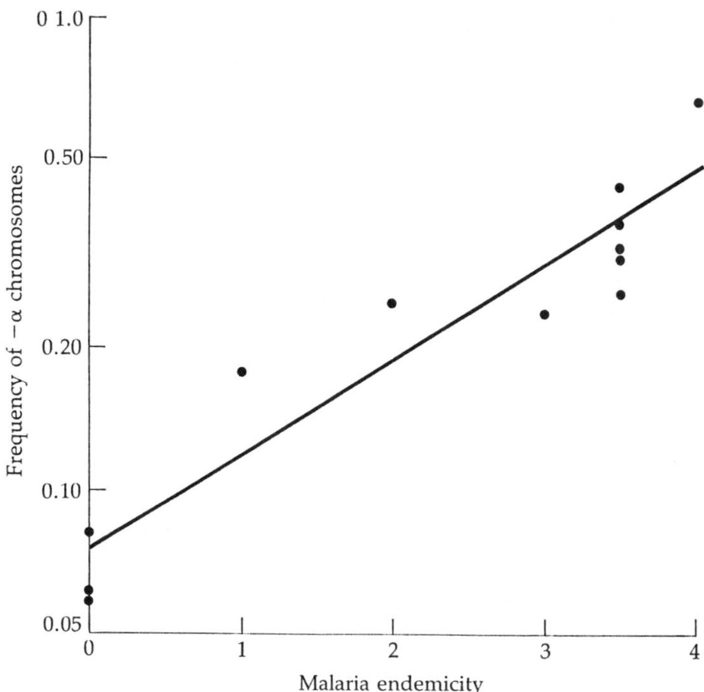

Figure 18. Association of α-thalassemia deletion (−α) with prevalence of malaria in Melanesia. The scale of gene frequency is logarithmic, and that of malaria endemicity is proportional to the occurrence of the disease in different geographical regions. (Data from Flint et al. 1986.)

Figure 19. Proposed evolution of the immunoglobulin multigene superfamily from ➤ a primordial gene coding for a cell-surface receptor. Details of the evolutionary relationships are speculative. The superfamily has diversified into 12 single-gene representatives (all of those at the left, plus β_2-microglobulin—β_2-m—at the right), and eight multigene families (remaining representatives at the right). These include genes for antibodies, T-cell receptors, major histocompatibility antigens, and other functions. The single-gene members include T-cell molecules implicated in MHC recognition (CD4 and CD8) and possibly ion channel formation (T3δ, T3ε), an immunoglobulin-transport protein (poly-Ig), a plasma protein (α1B-glycoprotein), two molecules restricted to lymphocytes and neurons (Thy-1, OX-2), two brain-specific proteins (N-CAM, NCP3), and β_2-microglobulin. The multigene families include the heavy (H) and light (κ, λ) components of antibody molecules, the α, β and γ chains of T-cell receptors, and the Class I and Class II molecules from the major histocompatibility complex (MHC). (Adapted from Hood et al., 1985, and Hunkapiller and Hood, 1986.)

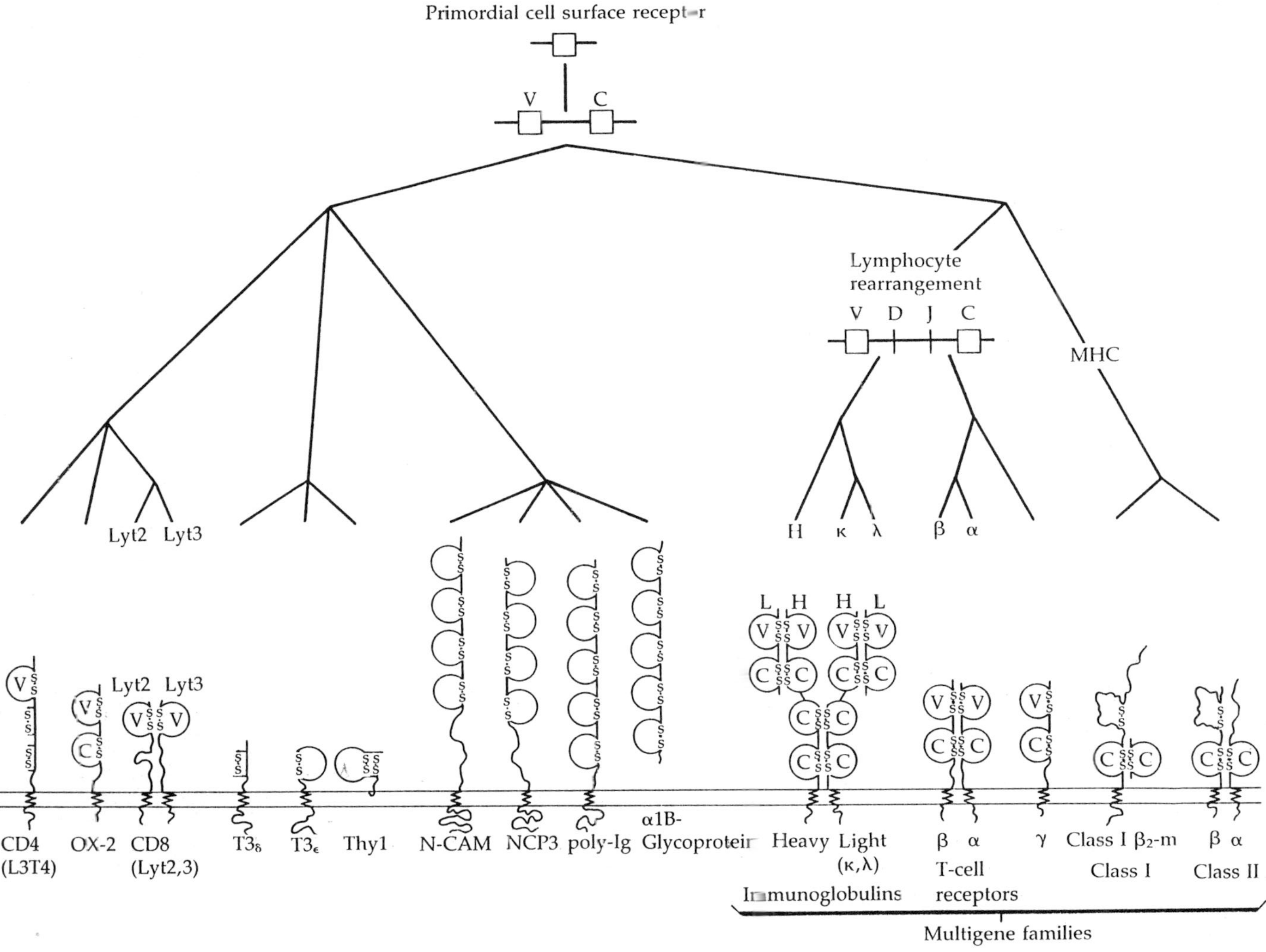

Primordial cell surface recept-r
V
C
Lymphocyte
rearrangement
V D J C
MHC
Lyt2 Lyt3
H κ λ
β α
Lyt2 Lyt3
CD4
(L3T4)
OX-2
CD8
(Lyt2,3)
T3δ
T3ε
Thy1
N-CAM
NCP3
poly-Ig
α1B-
Glycoprotein
L H H L
Heavy
Light
(κ,λ)
Immunoglobulins
β α
T-cell
receptors
γ
Class I β2-m
Class I
β α
Class II
Multigene families

and Hood 1986). The primordial single-copy gene may have coded for a cell-surface receptor containing the basic homology unit of the superfamily, which is about 110 amino acids in length with a strategically placed disulfide bridge and folding characteristics enabling it to combine with other similar units. An early duplication and divergence of the primordial gene resulted in the variable (V) and constant (C) domains that have been so versatile in their diversification for specialized immune functions. In some members of the immunoglobulin superfamily, shown at the left of Figure 19, the functional products are usually individual polypeptide chains, sometimes containing internal duplications of the primordial folding unit. These products include the poly-Ig receptor, which mediates the transport of immunoglobulin molecules across cell membranes.

In the other main branch of the superfamily, shown at the right of Figure 19, the functional products are usually aggregates of polypeptide chains. In this branch, there occurred multiple duplications of the V regions and specialization of D (diversity) and J (joining) regions during the evolution of the DNA splicing mechanism in lymphocytes, which today results in the tremendous diversity of antibodies and T-cell receptors. During the formation of heavy-chain antibody genes in the lymphocytes, any one of a large number of DNA sequences coding for the variable part of the molecule can become spliced with any one of a small number of DNA sequences coding for the constant part, with diversity and joining regions incorporated in between. The many possible V–D–J–C combinations enables enormous numbers of different possible antibodies to be formed, which is increased still further by slight variation in the exact positions of the splice junctions. An analogous type of splicing process occurs in the formation of antibody light-chain genes and T-cell receptor genes.

Quite a different evolutionary pathway for formation of antibody heavy-chain genes occurred in the primitive shark (Gilbert 1986; Hinds and Litman 1986). In the horned shark *Heterodontus francisci,* which diverged from mammals approximately 450 million years ago (Figure 8), the unit that became duplicated was the complete V–D–J–C unit. Although the sequences within each duplicated unit are joined by splicing, and some antibody variability results from variation in the splice junctions, the enormous number of possible V–D–J–C splicing combinations has been forfeited. As a result, the horned shark has a diminished repertoire of possible antibodies (Hinds and Litman 1986).

In yet another offshoot of the immunoglobulin superfamily, shown at far right in Figure 19, the C region underwent duplication and specialization to form molecules of the **major histocompatibility complex** (**MHC**), which, among other functions, are necessary for the T cells of the immune system to recognize foreign antigens.

Although many aspects of the immunoglobulin superfamily tree in Figure 19 are speculative, the molecules are undoubtedly related because comparison of the relevant units gives 15–40 percent homology at the amino acid level, and at the DNA level each homology unit is encoded in a separate exon. The immunoglobulins thus demonstrate the immense evolutionary potential of repeated rounds of duplication and divergence through specialization of function.

Localized Highly Repetitive DNA Sequences

Most eukaryotic genomes contain tandemly arranged, highly repetitive DNA sequences near the centromeres or telomeres of chromosomes. In some cases the localized highly repetitive DNA accounts for a significant fraction of the genome. For example, in the American kangaroo rat *Dipodomys ordii*, more than half of the DNA consists of tandem arrays of variants of the simple sequences AAG, TTAGGG, and ACACAGCGGG, which comprise 2400, 2200 and 1200 *million* nucleotide pairs, respectively, and altogether account for more than 50 percent of the entire genome (Miklos 1985). The simple sequences are often not transcribed, and they are quite variable. For example, only about 25 percent of the sequences in the TTAGGG cluster in *Dipodomys* actually read TTAGGG; most of the others differ from TTAGGG in one or two nucleotides. Often called **satellite DNA**, localized highly repetitive DNA sequences provide abundant opportunities for gene conversion and unequal crossing-over (Charlesworth et al. 1986).

The functions of localized highly repetitive DNA sequences, if any, are completely obscure. Miklos (1985) has classified the functional hypotheses under four headings. Localized highly repetitive DNA sequences are important for their role in:

1. Chromosome structure, e.g., determining centromere strength.
2. Chromosome pairing and recombination.
3. Determining the three-dimensional architecture of the nucleus.
4. Chromosome changes (genome reorganization), which may accompany the process of speciation.

An alternative hypothesis is that differences among organisms in localized highly repetitive sequences result in minimal phenotypic effects, and hence the sequences are not subject to the usual forces of natural selection acting at the level of phenotype.

Nonfunctional DNA sequences might have their own evolutionary dynamics determined by gene conversion and unequal crossing-over, and the number of copies of each sequence could be amplified or diminished through unequal crossing-over or other phenomena. When the evolution of DNA sequences is determined mainly by their own ability to replicate and undergo changes, and the course of events is affected minimally if at all by natural selection acting at the phenotypic level, the sequences are called **selfish DNA** (Doolittle and Sapienza 1980; Orgel and Crick 1980). The molecular mechanisms that determine the evolution of selfish DNA are sometimes known collectively as **molecular drive** (Dover 1982).

All four functional hypotheses listed earlier may seem somewhat unattractive because of the variability that occurs in localized highly repeated DNA within species. For example, the genome of *Drosophila melanogaster* contains 20 percent satellite DNA, localized in the heterochromatic regions around the centromeres, and composed of variants of the simple sequences AATAT, AATATAT, AAGAG, AAGAGAG and AATAACATAG. However, approximately 3 percent sequence divergence in the satellites occurs among individuals, and up to 30 percent sequence divergence occurs in related species. Furthermore, the amount of heterochromatic DNA in *D. melanogaster* can be altered drastically without apparent effects on viability or fertility.

Functional hypotheses are also weakened by the sometimes peculiar location of satellite DNA in the genome. For example, approximately 60 percent of the genome in *Drosophila nasutoides* consists of satellite sequences, but unlike *D. melanogaster*, all of the satellite DNA is localized to one giant chromosome (Figure 20). At the other end of the spectrum are species such as *D. littoralis* and *D. ezoana*, which seem to be perfectly good species but lack detectable satellite DNA (Miklos 1985).

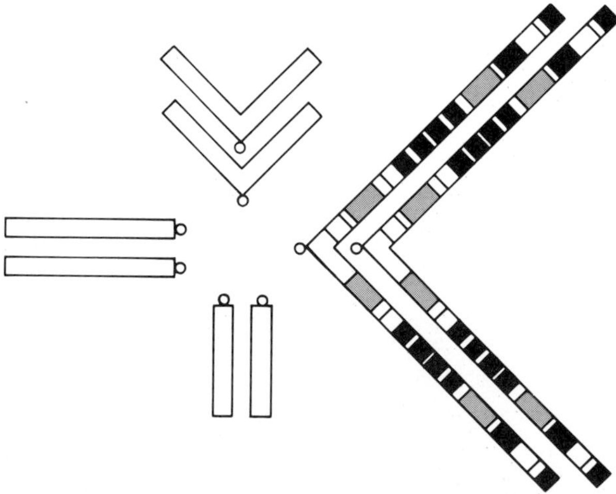

Figure 20. ***Drosophila nasutoides*** **contains several localized highly repetitive DNA families (indicated by different shaded boxes), all of which are located in a single very large chromosome. (From Miklos 1985.)**

The chromosome-reorganization hypothesis runs into difficulty because, while chromosome structure is sometimes drastically different between related species, this does not appear to be correlated with the amount of satellite DNA. For example, chromosomes in the rat species *Rattus villosissimus* and *R. sordidus* have undergone extensive centromere rearrangement, with the result that many acrocentric chromosomes in the former species occur as metacentric chromosomes in the latter, yet the species contain negligible quantities of localized highly repetitive centromeric DNA. On the other hand, the *Drosophila* of Hawaii contain large amounts of satellite DNA, but the species are not differentiated by major chromosome reorganizations.

The whole question of the function of localized highly repetitive DNA has been reviewed by Miklos (1985), who cites the inadequacies of all of the functional hypotheses. He concludes that satellite DNA is categorically selfish DNA, and that the principal mechanisms of its evolution are those of mutation, gene conversion, unequal crossing over, and random genetic drift.

One possible difficulty with the selfish DNA hypothesis is that, in some cases, simple-sequence DNA is transcribed in a developmentally specific manner. Figure 21 shows such a sequence from *D. melanogaster*, which is located in the X chromosome. Note the many repetitions of the GATA motif (underlines). The DNA strand is transcribed into RNA from lower right to upper left, and the boldface region indicates a hypothetical product of 93 amino acids derived from translation of the RNA, starting with Met (codon AUG) at lower right (arrows). Related GATA sequences are found in the W sex chromosome in snakes and in the male-determining region of the Y chromosome in mice, raising the possibility that the sequence may play some unknown role in sex development (Singh et al. 1984).

Not all localized highly repeated DNA consist of variants of very short

5′—ATATAGATAGATGTATAGATGGTTAGATAGATAGGTAGATAGATAGATAGATAGATAGATGT
AAGATAGATAGATAGATAGATAGATAGATAGATGGTTAGATAGATAGATAGATAGATAGATAGAT
AGATAGATAGATAGATAGATAGATATATAGATTATAGACAGCTAGATAGATAGATAGATAGATAT
ATAGATATATATATATAGATAGATAGGTAGATATATAGATAGATATGTAGATATATAGATATATA
TATATAGATTGATAGGTAGATAGATAGATAGATAGATGGATAGATAGGTAAATAGATAGATATAT
AGATAGATAGATAGATAGATAGATAGACAGATAAATAGATATAAAGATAGATATAAAGATAGATA
GATAGATCAACCAATAGATAGATGTTTAGATAGATAGATAGTTTGATTGCAACCTGCAGCAT—3′
←←←

Figure 21. Localized repetitive DNA sequence containing multiple GATA units (underlined) from *Drosophila melanogaster* X chromosome. The arrows at the lower right define an ATG sequence in the opposite DNA strand which potentially initiates an open translational reading frame of 93 amino acids. The function of the sequence is unknown. (Data from Singh et al. 1984.)

sequences, such as those occurring in *Dipodomys ordii* and *Drosophila melanogaster*. The genome of the killer whale (*Orcinus orca*) contains a highly repeated sequence of 1,579 base pairs, present in about 500,000 copies and accounting for approximately 15 percent of the total genomic DNA (Widegren et al. 1985). The sequence is organized in tandem repeats and is located in the heterochromatin of 8 of the 21 autosomes. Repeated sequences related to the one in the killer whale are found in all cetaceans (whales, dolphins, and porpoises), testifying to the conservation of the sequence during evolution. Unfortunately, just as with simple-sequence highly repetitive DNA in other organisms, the function of the cetacean sequence is completely unknown. Indeed, it is not even certain that it has a function.

Dispersed Highly Repetitive DNA Sequences

A second major class of highly repetitive DNA in eukaryotes is not localized in clusters of tandemly repeating units, but is dispersed throughout the genome with single-copy sequences. In vertebrates, this **dispersed highly repetitive DNA** occurs in two categories, denoted SINEs and LINEs (Singer 1982).

SINEs (**S**hort **IN**terspersed **E**lements) are sequences typically shorter than 500 base pairs which occur in 10^5 or more copies in the genome. Like tRNA genes, they contain internal transcriptional start sites and are transcribed by RNA polymerase III. **LINEs** (**L**ong **IN**terspersed **E**lements) are sequences typically greater than 5,000 base pairs which occur in 10^4 or more copies in the genome. They are processed pseudogenes (see below) and, when transcribed, are transcribed by RNA polymerase II. Marked differences in the particular array of subfamilies of SINEs and LINEs or both are frequently observed among even closely related species. As in the case of localized highly repetitive DNA sequences, the mechanisms and possible significance of such massive and rapid changes in repetitive DNA in the genome are very obscure.

One example of SINEs in human DNA is the ***Alu*** family, named because the sequence contains a characteristic restriction site for the restriction enzyme *Alu*I. The *Alu* sequence is about 300 nucleotides in length, and members of the family show an average sequence divergence of about 15–20 percent from the *Alu* consensus sequence. *Alu* sequences are present in more than 500,000 copies in the human genome and constitute approximately seven percent of the total DNA. Sequences closely related to *Alu* are found in other primates, and more distantly related sequences occur in rodents and probably in all placental mammals. The length of the rodent *Alu*-related sequence is only 130 nucleotides. The human *Alu* sequence consists of a head-to-tail dimer of the rodent sequence, with an additional 30 base pairs of DNA inserted into one of the monomers (Schmid and Shen 1985).

Two randomly chosen human *Alu* sequences differ on the average at 15–20 percent of their nucleotide sites. Assuming a rate of sequence evolution of 5 substitutions per nucleotide site per 10^9 years (approximately the pseudogene rate; see Figure 5), estimate the average time of divergence of two random *Alu* sequences.

PROBLEM 16

ANSWER $\langle D \rangle = 0.15$ to 0.20 with corresponding $\langle k \rangle = 0.167$ to 0.233. The times are given by $\langle k \rangle/(2 \times 5 \times 10^{-9})$, which calculate to between 16.7 and 23.3 million years. To the extent that gene conversion between *Alu* sequences is significant in keeping the sequences more similar than they would otherwise be, the divergence times are underestimated.

The abundance of *Alu* sequences in the genome implies that many genes might have one or more *Alu* sequences in the same neighborhood. For example, the cluster of β-globin genes contains several copies of *Alu* sequences, which are of some significance in undergoing rare unequal crossing-over resulting in hemoglobin disorders. A benign condition known as the hereditary persistence of fetal hemoglobin (high F-disease) is associated with a deletion of the β and δ genes caused by unequal crossing-over between two *Alu* repeats within the gene cluster; the reciprocal recombination product results in one form of β-thalassemia.

An example of LINEs in the human genome is the ***Kpn*** family of sequences (also called *LINE*-1 or *L*1), where the designation *Kpn* refers to a characteristic restriction site. The *Kpn* sequences average about 2,000 nucleotides, and the 50,000 copies of the sequence in the human genome account for about four percent of the total DNA. As with the *Alu* family, sequences related to *Kpn* are found in other mammals, including the mouse (Hardies et al. 1986).

The abundance of dispersed repeated gene families is evident. Approximately 20 percent of the human genome consists of such sequences, and 11 percent comes from the *Alu* and *Kpn* families themselves. The profligacy of such sequences is matched only by our ignorance of their function. No function of highly repeated dispersed DNA sequences has yet been demonstrated. As with the localized highly repeated sequences, the possibility must be considered that the dispersed class of sequences is essentially functionless selfish DNA.

Some clues to the origin and evolution of LINE sequences come from analysis of their molecular structure. In molecular organization they resemble processed RNA molecules. At the 5′ end they contain transcriptional start signals, at the 3′ end they often contain many consecutive adenine nucleotides (poly-A tail), and they contain no intervening sequences. Dispersed repeated DNA sequences also feature a short direct (not inverted) duplication of 10–20 nucleotides at the termini.

In their molecular organization, LINE sequences strongly resemble a class of pseudogenes known as **processed pseudogenes**. Processed pseudogenes are thought to result from the reverse transcription of an RNA molecule into DNA, followed by insertion of the DNA into the genome. The reverse transcription and integration process can be carried out by an enzyme called **reverse transcriptase**, which is coded in the genome of a class of RNA-containing viruses called **retroviruses**. In cells infected with retrovirus, the reverse transcriptase makes a DNA copy of the viral RNA, and another enzyme inserts the DNA into the chromosome. When reverse transcription and integration happen to a processed RNA molecule, the result is a processed pseudogene. Many genes are known to have processed pseudogene counterparts, including the genes for human λ-immunoglobulin and β-tubulin, rat α-tubulin and cytochrome *c*, and mouse α-globin.

One hypothesis for the origin of LINE sequences is that they originate from genes that are actively transcribed at times in development that are particularly favorable for reverse transcription and integration (Schmid and Shen 1985). This hypothesis has the virtue that the family of sequences could be maintained even if only a small minority of the members were capable of transcription on their own. And, indeed, most dispersed repeated sequences are not transcribed, or are transcribed only because of accidental read-through transcription from upstream promoters. In this hypothesis, as pseudogene copies of the sequence are lost through deletion and other processes, new ones are created by reverse transcription and integration of transcripts from functional members of the gene family. Walsh (1985) has estimated that, at equilibrium, 1.7–15 percent of the members of a repeated gene family may retain their ability to function.

The processed pseudogene model of dispersed repeated DNA evolution is illustrated in Figure 22 (Hardies et al. 1986). The functional, transcribed copies of the gene family are shown at the top, and the dotted lines represent gene conversion which promotes concerted evolution of the functional genes. The gene in the center is a preferred donor for gene conversion (**biased gene conversion**).

Emanating from the functional genes are numerous copies of processed pseudogenes distributed throughout the genome. These copies are essentially functionless and undergo sequence divergence promoted by mutation and random genetic drift, which is offset in part by gene conversion and other homogenizing processes among the pseudogenes. Eventually the pseudogene sequences are cleared from the genome by deletion or extreme sequence rearrangement or divergence.

One implication of the model in Figure 22 is that, eventually, a balance is reached in which the clearance of old pseudogenes from the genome is equalled by the creation and insertion of new ones. In the equilibrium state there is a steady turnover among sequences in the family, but the total number neither grows nor shrinks. Studies of a dispersed repeated sequence

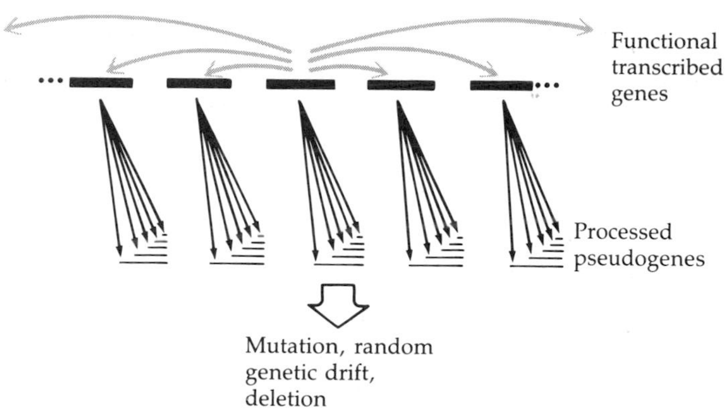

Figure 22. Model for the evolution of dispersed highly repetitive family of processed pseudogenes. A small number of functional genes (top), which undergo concerted evolution by means of gene conversion, are transcribed under conditions that favor reverse transcription and integration into numerous dispersed chromosomal locations. The resulting nonfunctional genes undergo mutation and random genetic drift, and are ultimately eliminated by deletion or other mechanisms. (From Hardies et al. 1986).

in the mouse related to human *Kpn* suggest a turnover with a half-life of approximately two million years. That is, after two million years, half the members of the gene family will have been removed and replaced with new ones. However, the *Kpn* family may evolve more rapidly than is typical.

Hardies et al. (1986) have emphasized that the model of sequence evolution in Figure 22 allows ample scope for the occurrence of natural selection at the level of phenotype, not only through the direct effects of the functional members of the family, but through indirect effects of the pseudogene copies. The very abundance of the pseudogenes implies that many unrelated genes may have pseudogenes in the same vicinity, as is the case with *Alu* sequences interspersed in the β-globin cluster. Some fraction of these linked pseudogenes may alter the level, timing, or tissue distribution of transcription of the genes to which they are linked, or they may have subtle effects on chromatin structure that affect gene expression. Through any of a diversity of mechanisms, pseudogene copies of dispersed highly repeated gene families could, in principle, have effects on phenotype and thus be subject to the influence of natural selection. While true in principle, such effects have not yet been demonstrated. To the extent that such effects can safely be ignored, the evolutionary mechanism of highly dispersed repeated DNA sequences is that of selfish DNA subject to the conflicting forces of neutral mutation/random drift and the diverse homogenizing processes of concerted evolution.

PROBLEM 17 If the disappearance of dispersed repetitive DNA sequences from the genome is like radioactive decay, in that a constant fraction θ of the sequences in the genome are eliminated per unit time, then the fraction of sequences present at time 0 that still exist at time t is proportional to $e^{-\theta t}$. The rate constant θ can be estimated by linear regression using the equation $\ln[1 - F(t)] = C - \theta t$, where $F(t)$ is the fraction of observed sequences having an age $\leq t$. Estimates of t can be obtained from the observed values of $\langle k \rangle$, using $\langle t \rangle = \langle k \rangle / \langle \lambda \rangle$, assuming an estimate of λ is available.

The data below are from Hardies et al. (1986) on the evolution of the rodent *L1* sequence related to the human *Kpn* family. In this case $\langle \lambda \rangle = 4.1 \times 10^{-9}$ substitutions per nucleotide site per year. Carry out the regression to estimate θ, and calculate the half-life of the turnover process and the average persistence of a particular sequence. The half-life equals $-\ln(1/2)/\theta$, and the average persistence equals $1/\theta$. (In the table, $f(t)$ is the probability density function and $F(t)$ is the cumulative distribution.)

$f(t)$	$F(t)$	$\langle k \rangle$	$f(t)$	$F(t)$	$\langle k \rangle$
0.312	0.312	0.00334	0.061	0.819	0.02027
0.196	0.508	0.00670	0	0.819	0.02370
0.031	0.539	0.01007	0.027	0.846	0.02715
0.126	0.665	0.01345	0.061	0.907	0.03062
0.093	0.758	0.01685	0.061	0.968	0.03410

ANSWER Values of $\langle t \rangle$ corresponding to $\langle k \rangle$ are, in the same order, 0.814×10^6, 1.63×10^6, 2.46×10^6, 3.28×10^6, 4.11×10^6, 4.94×10^6, 5.78×10^6, 6.62×10^6, 7.47×10^6, and 8.32×10^6. The slope of the regression of $\ln[1 - F(t)]$ on t estimates θ; in this case $\langle \theta \rangle = 3.40 \times 10^{-7}$ per year. The half-life is 2.04×10^6 years and the average persistence of a sequence is 2.94×10^6 years.

TRANSPOSABLE ELEMENTS

DNA sequences that can change their location within the genome are called **transposable elements**. The process of transposition requires a protein, called **transposase**, which is usually coded within the sequence of the transposable element itself. Most transposable elements undergo transposition through a replicative process with DNA or RNA intermediates. In most cases, when transposition to a new location occurs, one copy of the transposable element remains behind in its original location. Transposable elements can thereby increase in copy number in the genome. Most transposable elements are also able to regulate their own rate of transposition. Generally speaking, the rate of transposition is regulated in such a way that the rate of transposition per element decreases as the number of copies of the element increases, which prevents the occurrence of runaway transposition. Regulation of transposition is thought to be an evolved characteristic rather than being intrinsic to the biochemistry of the process. Several different types of transposable elements can be distiguished by their nucleotide sequence organization or by the details of their mechanisms of transposition or regulation.

Factors Controlling the Population Dynamics

Transposable elements were originally discovered in maize through genetically unstable mutations resulting from insertion near the affected genes. They are now known to be ubiquitous among prokaryotes and eukaryotes. Insertions of transposable elements can, in some cases, alter the regulation of genes in their vicinity, which prompted their original designation in maize as *controlling elements*. The propensity of transposable elements to transpose and increase in copy number reveals a dynamic, fluid aspect of the genome not previously recognized. Furthermore, many transposable elements are related to viruses, or they use other mechanisms which enable their dissemination among organisms or even among species. Transposable elements are thus examples of selfish DNA. Realistic models for the population dynamics of transposable elements usually incorporate several features:

1. A rate of infection, in which genomes previously lacking the transposable element become infected with it.
2. A rate of transposition, which determines how rapidly the copy number increases; the effects of regulation are taken into account by assuming that the rate of transposition is a decreasing function of copy number.
3. A mechanism, or combination of mechanisms, for eliminating elements from the population, otherwise the copy number would increase indefinitely. The usual assumption is that the presence of transposable elements in the genome decreases fitness, resulting in the elimination of some elements by means of natural selection, or that elements can be eliminated from the genome by means of genetic deletion.

Through the study of such models, the diversity and novel attributes of transposable elements are coming to be incorporated into the concepts of population genetics (e.g., Langley et al. 1983; Montgomery and Langley 1983; Kaplan and Brookfield 1983; Sawyer et al. 1987; Sawyer and Hartl 1986).

Insertion Sequences and Composite Transposons in Bacteria

Bacteria contain several types of transposable elements, the simplest of which are insertion sequences. **Insertion sequences** are transposable sequences, typically about 1000–2000 nucleotides in length, which contain short nucleotide sequences repeated at each end in inverted orientation, and at least one long open translational reading frame thought to code for the transposase protein. A typical type of insertion sequence from intestinal bacteria, denoted IS50, is diagrammed in Figure 23(a). IS50 is 1500 nucleotides in length and produces two overlapping mRNA molecules, translated in the same reading frame, one for the transposase protein and the other for a repressor protein which regulates the rate of transposition. The intestinal bacterium *Escherichia coli* contains six well-characterized insertion sequences, the numbers of which have been determined among 71 natural isolates (Sawyer et al. 1987). These distributions have been used to estimate the parameters in population models.

Models of transposable elements in *E. coli* are greatly simplified because of asexual reproduction, a low rate of recombination among strains, and a low rate of deletion of insertion sequences. The "state" of a strain may be described in terms of the number of copies n of an element it contains, and in accord with the controlling factors discussed in the previous section, let

1. u equal the rate at which uninfected cells become infected (that is, go from $n = 0$ to $n = 1$),
2. T equal the rate of transposition in infected strains [that is, the rate at which cells in state n ($n \geq 1$) go to state $n + 1$], and
3. S equal the decrease in fitness of infected cells, relative to uninfected cells.

The most general models of this type allow T and S to be functions of n, but here we will assume that they are constant. However, assumption (2) does incorporate regulation of transposition because the rate of transposition *per element* in a strain equals T/n, which is a decreasing function of n.

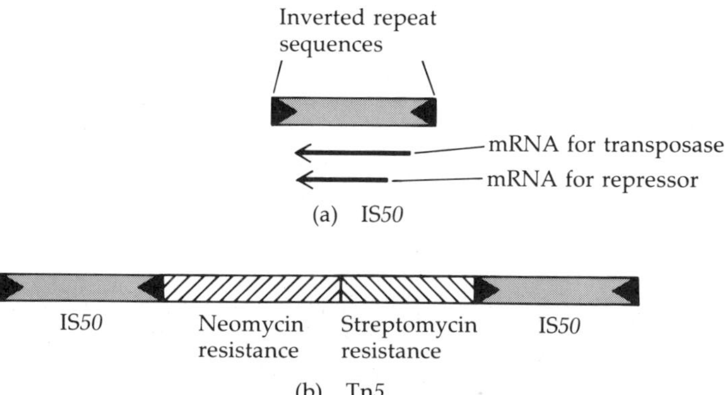

Figure 23. Examples of two types of transposable elements in bacteria. (a) Insertion sequence IS50, which is flanked by short inverted repeats and codes for its own transposase protein and repressor. (b) Composite transposon Tn5, which contains two copies of IS50 flanking a central region containing genes for neomycin and streptomycin resistance.

With these assumptions, the population attains an equilibrium distribution of numbers of elements, in which the probability p_i that a strain contains exactly i elements is given by

$$\begin{aligned} p_0 &= \alpha \\ p_i &= (1 - \alpha)(1 - \Theta)\Theta^{i-1} \quad (i \geq 1) \end{aligned} \tag{3.11}$$

where $\alpha = 1 - (u/S)$ and $\Theta = T/(T + S - u)$ (Sawyer and Hartl 1986, Sawyer et al. 1987).

Equation 3.11 can be applied to the concrete case of insertion sequence IS30 in *E. coli*, in which the distribution of numbers among 71 strains fits a model with $\alpha = 1/2$ and $\Theta = 1/2$. With these parameters the distribution simplifies to the remarkably simple formula $p_i = (1/2)^{i+1}$ for $i \geq 0$. Among 71 strains, therefore, the expected numbers of strains containing i elements are as follows (the observed numbers are also given):

No. elements	Expected no. strains	Observed no. strains
0	35.5	36
1	17.8	16
2	8.9	13
3	4.4	2
4	2.2	2
≥5	2.2	2

The strains with five or more elements have been grouped in order to carry out a χ^2 test of goodness of fit. This χ^2 test has three degrees of freedom because α and Θ were estimated from the data. The value of χ^2 equals 3.48, which has an associated probability level of about 0.35. Thus, the simple model for IS30 fits the observed data very well. Although the χ^2 test cannot be completely trusted in this case because of the small expected numbers in some of the categories, the conclusion is supported by a more exact statistical test (Sawyer et al. 1987).

The following problem fits the model in Equations 3.11 to three other insertion sequences in *E. coli*.

PROBLEM 18

The distribution of IS1 fits Equations 3.11 with $\alpha = 1/5$ and $\Theta = 5/6$; IS2 fits the same model with $\alpha = 2/5$ and $\Theta = 2/3$; and IS4 fits the same model with $\alpha = 2/3$ and $\Theta = 3/4$. Calculate the expected numbers for 71 strains and carry out a χ^2 test. The observed numbers below are from Sawyer et al. 1987.

No. elements	0	1	2	3	4	≥5
IS1	11	14	8	6	7	25
IS2	28	8	12	5	5	13
IS4	43	5	5	3	5	10

ANSWER For IS1 the distribution is given by $p_0 = 1/5$, $p_i = (4/25)(5/6)^i$ for $1 \le i \le 4$, and $p_{\le 5} = 1 - (p_0 + p_1 + p_2 + p_3 + p_4)$; for IS2 $p_0 = 2/5$ and $p_i = (3/10)(2/3)^i$ for $(1 \le i \le 4)$; and for IS4, $p_0 = 2/3$ and $p_i = (1/9)(3/4)^i$ for $(1 \le i \le 4)$. Expected numbers, χ^2 values and associated probabilities are as follows:

No. elements	0	1	2	3	4	≥5	χ^2	Prob.
IS1	14.2	9.5	7.9	6.6	5.5	27.4	3.58	0.35
IS2	28.4	14.2	9.5	6.3	4.2	8.4	6.31	0.10
IS4	47.3	5.9	4.4	3.3	2.5	7.5	4.00	0.28

As with IS30, more exact statistical tests confirm the conclusion that the model fits. However, the distribution of IS1 has a very long tail, with nine strains containing 15–20 copies and six strains containing 21–30 copies; this distribution is approximated even better by a model in which the regulation of transposition decreases more gradually than as T/n (Sawyer et al. 1987).

It should not be assumed that insertion sequences and other transposable elements are necessarily always harmful to their hosts. The following problem demonstrates that insertion sequence IS50 can result in an *advantage* in fitness in *E. coli* cells cultured in chemostats.

PROBLEM 19 Two strains A and B of *E. coli* were subjected to competition for glucose in chemostats. The strains were genetically identical except that strain A contained one copy of the insertion sequence IS50. Use the method in Problem 12 to estimate the selection coefficient favoring the IS50-bearing strain. The data are from Hartl et al. (1983).

Hours of competition	Percent strain A
0	56.2
14	71.5
24	81.0
32	86.3
50	94.5

ANSWER From Problem 12, the selection coefficient per hour favoring strain A is estimated as the slope of the regression of $\ln(A/B)$ against time (in hours), where A and B are the relative frequencies of strains A and B in the chemostat. The logarithms, from top to bottom, are 0.249, 0.920, 1.450, 1.840, and 2.844; and the slope of the regression is 0.052 per hour. This is a very large selective advantage, but it persists for only about 50–75 hours in the chemostat. The mechanism of the selective advantage is unknown, but it does not result from transposition (Hartl et al. 1983).

Aside from their own evolutionary dynamics, insertion sequences are important because they can mobilize other sequences in the genome. When two copies of an insertion sequence are on flanking sides of an unrelated sequence, the inverted repeats that are used in transposition are preferentially those at the extreme ends. This kind of insertion sequence "sandwich" constitutes a composite transposable element or **transposon**, which transposes as a single unit. Transposon Tn5, an example of a transposon mobilized by IS50, is illustrated in Figure 23(b). The central sequence of Tn5 contains two regions coding for resistance to the antibiotics neomycin and streptomycin. Possession of Tn5 would obviously confer a major selective advantage on a bacterial cell in an environment containing either or both antibiotics.

Mobilization of genes for antibiotic resistance, heavy-metal resistance, and other functions is one of the principal evolutionary implications of transposable elements in bacteria. They enable the piecewise assembly of specialized, infectious plasmids. **Plasmids** are autonomously replicating circular molecules of DNA that exist in bacterial cells, which in many cases contain genes that promote their transfer between different organisms. Many plasmids contain genes, such as genes for antibiotic resistance, which are highly advantageous to their hosts in certain environments. These genes are often contained in transposons, and they undoubtedly entered the plasmid through transposition from a different plasmid or from the genome of a previous host. Infectious plasmids containing multiple antibiotic-resistance genes are called **resistance transfer factors**. Transposon Tn5 (Figure 23) was originally found in a resistance transfer factor in a species related to *E. coli*. Testifying to the ability of transposons to get around is the fact that the streptomycin-resistance gene in Tn5 is not expressed in intestinal bacteria but in the soil bacterium *Rhizobium*.

Transposable Elements in Eukaryotes

Transposable elements can have important genetic consequences as mutagenic agents, through effects on the expression of genes they become inserted into or lie near. Their consequences in population genetics and evolution are less well understood. Transposable elements of several different types are widespread in both animals and plants. For example, *Drosophila melanogaster* contains multiple copies of each of 50–100 different transposable elements (Rubin 1983). Although few of these elements have been studied in detail from the standpoint of population genetics, indirect evidence suggests that most of the elements, like insertion sequences in bacteria, are mildly harmful to the host (Golding et al. 1986).

Two of the most thoroughly studied types of transposable elements are diagrammed in Figure 24. Although both examples occur in *D. melanogaster*, other tranposable elements with analogous types of sequence organization occur in other organisms. Figure 24(a) illustrates the **P element**, in which the coding region in the middle is flanked by short inverted repeated sequences.

The coding region of many P-like elements is interrupted by one or more intervening sequences, but some P-like elements contain a single uninterrupted reading frame (for example, the *Drosophila* element *mariner* discussed in Jacobson et al. 1986). In the case of the P element, transposition of the element is normally restricted to cells in the germ line because, in somatic cells, one of the introns in the transcript of the P element is not excised (Laski et al. 1986).

Figure 24(b) illustrates the sequence organization of the ***copia*** transposable element. *Copia*-like elements are flanked by long direct repeats (that is, not inverted), and their coding region often contains a single uninterrupted open reading frame. *Copia* exemplifies transposable elements that are closely related to, or identical with, the RNA retroviruses discussed earlier, which reverse-transcribe their RNA into DNA and integrate it into the genome.

The next section focuses on a phenomenon occurring in *D. melanogaster* that results from the presence of the P element but serves more generally to illustrate the potential importance of transposable elements in evolution. The evolutionary implications of retroviral-like elements are discussed in the following section.

Hybrid Dysgenesis in *Drosophila*

Severe genetic abnormalities occur in the germ line of certain hybrids of *Drosophila melanogaster*, which are referred to collectively as **hybrid dysgenesis** (Kidwell et al. 1977; Engels 1983). The abnormalities include partial or complete sterility in either or both sexes, increased mutation, chromosome breakage, and nondisjunction. These diverse phenomena result from mobilization of the P element in the germ line of the hybrids (Bingham et al. 1982).

The occurrence of hybrid dysgenesis is related to the distribution of the P element among strains. The element occurs in multiple copies in most strains of *D. melanogaster* recently collected from natural populations, but it is absent

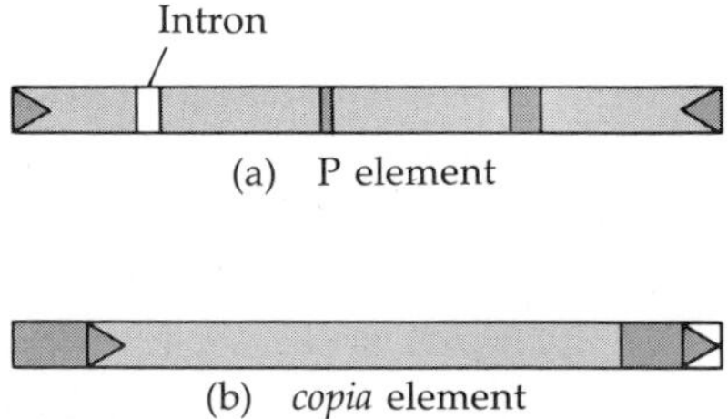

Figure 24. Examples of two types of transposable elements in *Drosophila*. (a) The P element, flanked by short inverted repeats and containing four exons interrupted by three introns. (b) The *copia* element, a retrovirus-like element, which contains a single uninterrupted reading frame between long direct repeats.

from old laboratory strains. Recently collected strains containing P elements are called **P strains**, and old laboratory strains and others lacking the P element are called **M strains**. Hybrid dysgenesis occurs when females of an M strain are crossed with males of a P strain.

Hybrid dysgenesis does not occur in the reciprocal cross of P females with M males, because of substances that P females transmit through the egg cytoplasm (Engels 1979). The nature of the suppressing cytoplasm, known as **P cytotype**, and its relation to the genomic P elements themselves, remains obscure (Engels 1983). However, recent evidence suggests that the P element codes for its own repressor of transposition, and that maternal transmission of the repressor accounts for P cytotype (Robertson and Engels, personal communication).

Although the hybrid animals in hybrid dysgenesis exhibit reduced fertility, the mobilization of P element in the germ line means that most of the chromosomes transmitted by the hybrids will have become infected with P elements. Thus, provided that the reduction in fertility in the hybrids is not too great, the P element can spread throughout the entire population. This conclusion has been confirmed for P elements both theoretically (Uyenoyama 1985) and experimentally (Kiyasu and Kidwell 1984; Scavarda and Hartl 1987). Apparently, the P element is a relative newcomer to the genome of *D. melanogaster*, having invaded the genome just in the past 50 years since the old laboratory strains were collected. Since another example of hybrid dysgenesis in *D. melanogaster* is well documented (Bucheton et al. 1984), which results from the mobilization of a transposable element completely unrelated to the P element, it appears that the infection, spread, and turnover of transposable elements may be a very rapid and frequent process, at least in this species.

Endogenous Retroviruses in Mammals

Mammals and some other organisms contain, as a normal constituent of their cellular DNA, multiple copies of the genome of several unrelated classes of retroviruses. DNA sequences that are homologous to retroviruses and that have become fixed in the genome of a species are known as **endogenous** retroviral sequences, or **virogenes**. Several distinct types of endogenous sequences occur in the vertebrate genome. Present in 10–100 copies per genome, each type of virogene had its origin in reverse transcription and integration of a different RNA retrovirus. Under suitable conditions of cell culture, endogenous viruses can give rise to mature retrovirus particles. Although not ordinarily infectious in cells of their species of origin, the released particles are infectious in cells of related species. The released viruses are typical retroviruses in that a small number of reverse transcripts of the viral RNA become incorporated into the host cell DNA. These observations and others suggest that endogenous retrovirus sequences are advantageous to the host in conferring some degree of immunity to viral infection (Benveniste 1985).

Endogenous viruses are retained in the genome for long periods. For example, the Type C endogenous retrovirus of baboons is found in all species of Old World monkeys, but not in New World monkeys. The sequence therefore invaded the germ line subsequent to the split between Old World and New World monkeys, and it remained within the Old World lineage through multiple speciation events.

The Type C endogenous retrovirus of baboons also illustrates the rapid rate of evolution characteristically found in virogenes. The evidence is shown in Figure 25, which summarizes data from DNA hybridization experiments. In **DNA hybridization**, DNA molecules from two sources are separated into individual strands at high temperature and then mixed to promote reannealing of the strands to form **heteroduplex** molecules that contain one strand from each source. The heteroduplex molecules normally contain nucleotide mismatches, and the degree of mismatch is measured by the difference in melting temperature (ΔT_m), which is the temperature at which the heteroduplex molecules come apart into their individual strands. In practice, the DNA from one source is rendered radioactive, and the melting temperature is determined from the temperature at which radioactive strands are released from the heteroduplex molecules. As a rule of thumb, a change in ΔT_m of one degree centigrade results from one percent mismatches in the heteroduplex molecules.

Figure 25 summarizes data obtained with two types of probe DNA from the baboon *Papio cynocephalus*, when hybridized with cellular DNA from other primate species. The first probe was baboon *cellular* DNA, which gave the solid line plotted with equal values on both axes; note that the DNA from all species falls on the line, which is evidence for a molecular clock with a constant rate of nucleotide substitution, reflected in ΔT_m, when averaged over the entire nonrepetitive genome.

The second baboon probe was the Type C virogene probe, which gave the lines shown in long or short dashes. For each point on these lines, the x-coordinate is the ΔT_m found with the baboon cellular DNA probe, and the y-coordinate is the ΔT_m found with the baboon virogene probe. These data define two straight lines rather than one, which are distinguished by the geographical origin of the species. The upper line includes species of Asian origin, the lower one species of African origin.

The two virogene lines in Figure 25 imply that virogene sequences have evolved at different rates in Asian and African primates (Benveniste 1985). Comparison of the slopes indicates that the Asian virogene sequence has evolved at approximately two times the rate of the African virogene sequence. An argument pointing to the relaxation of natural selection as a possible reason for the difference has been suggested by Benveniste (1985). In the first place, baboons and their close relatives, which range throughout most of Africa, are unusual in their ability to release infectious retroviruses. These

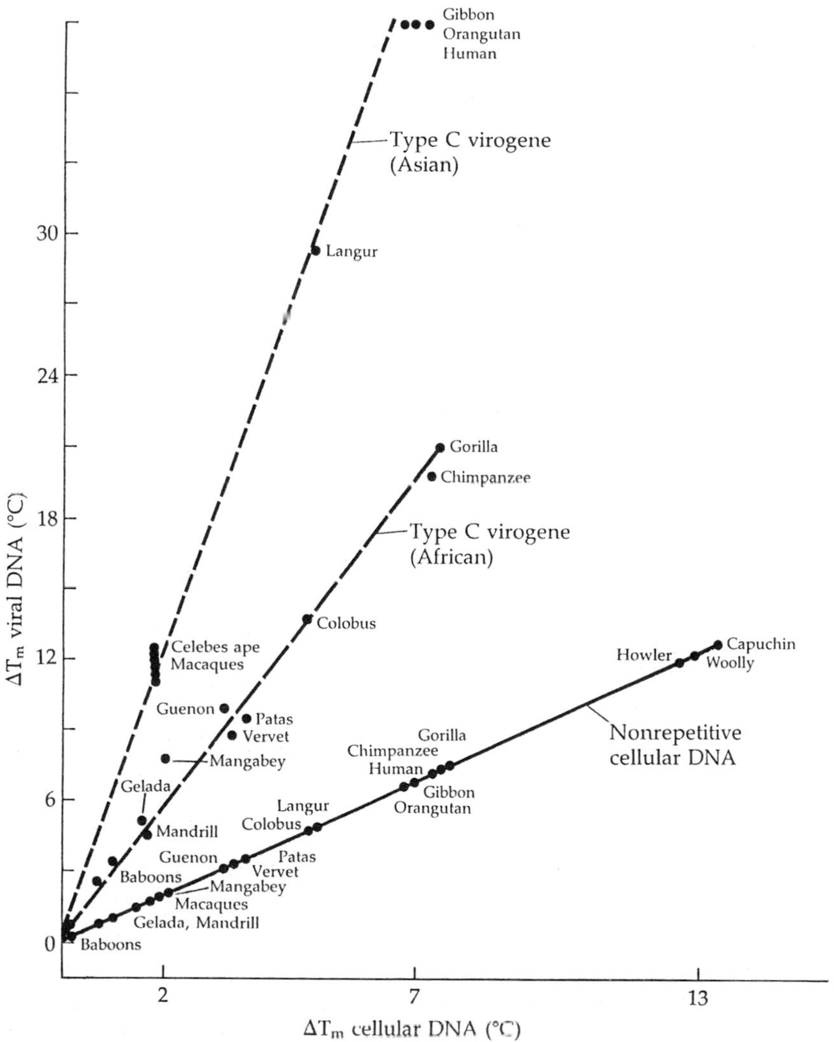

Figure 25. Evolutionary rate of baboon Type C endogenous retrovirus (dashed lines) as compared with nonrepetitive cellular DNA (solid line). In African primates, these virogenes have evolved almost threefold faster than nonrepetitive cellular DNA, and in Asian primates they have evolved about sixfold faster. (After Benveniste 1985.)

viruses can infect other organisms, and there is convincing evidence, discussed later, that one virus was transmitted to an ancestor of the domestic cat and became integrated into DNA of the cat germ line. However, while baboon retroviruses replicate well in cells of Asian primates, they do not infect cells of African primates. It appears, therefore, that close nucleotide sequence homology of virogene sequences provides some immunity to retrovirus infection. African primates, frequently exposed to the retroviruses released from baboons, would have been under selection to maintain the virogene sequence, whereas Asian primates, not exposed to baboons, would not have been subjected to selection pressure for conservation.

PROBLEM 20 Given (1) that a ΔT_m of 1°C results from one percent nucleotide mismatches and (2) that a ΔT_m of 1°C occurs in 10 million years of evolution, what rate of nucleotide substitution per nucleotide site per year corresponds to a ΔT_m of 1°C? The slopes of the Asian and African lines in Figure 25 are 6.0 and 2.8, respectively, relative to a value of 1.0 for nonrepetitive cellular DNA. What are the corresponding rates of evolution of Type C virogene sequences in terms of nucleotide substitution per site per year, using the rate just calculated for nonrepetitive cellular DNA? In Figure 25, the ΔT_m of the baboon–human comparison is 7.3°C. What is the predicted length of time of evolutionary divergence between these species?

ANSWER One percent nucleotide mismatches equals 0.01 mismatches per nucleotide site, a number which is sufficiently small that it need not be corrected for multiple substitutions. A value of ΔT_m of 1°C represents 0.01 mismatches per site divided by 10×10^6 years, or 1.0×10^{-9} substitutions per nucleotide site per year. The Type C virogene rates are $2.8 \times 1.0 \times 10^{-9} = 2.8 \times 10^{-9}$ substitutions per nucleotide site per year for the African sequences, and $6.0 \times 1.0 \times 10^{-9} = 6.0 \times 10^{-9}$ for the Asian sequences. In the baboon–human comparison, a ΔT_m of 7.3°C would correspond to a divergence time of $7.3 \times (1/2) \times 10.0 \times 10^6 = 36.5 \times 10^6$ years. The factor of 1/2 enters because a total evolutionary time of ten million years represents five million years since divergence (Figure 2).

It is of some interest in Figure 25 that the sequence of the human Type C virogene lies on the line characteristic of Asian primates. Taken at face value, this observation implies that a significant portion of human evolution may have occurred in an Asian setting (Benveniste 1985), which is contrary to the popular belief favoring Africa.

Horizontal Transfer of Endogenous Retroviruses

That endogenous retroviruses can undergo **horizontal transfer** between species that are only remotely related phylogenetically, and subsequently incorporated into the germ line, is now well documented. Examples include the transfer and germ-line integration of retroviral sequences from ancestors of modern primates to ancestors of carnivores, other primates, and birds; and the horizontal transfer from ancestors of modern rodents to ancestors of carnivores, primates, and artiodactyls (cleft-hoofed mammals).

Figure 26 provides evidence for the horizontal transfer of an endogenous retrovirus (named RD-114) from an ancestor of modern baboons (genus *Papio*) to an ancestor of the domestic cat (Benveniste 1985). The baboon-type virogene is not found in the lion, leopard or cheetah, but it is in species closely related to the domestic cat. Thus, the baboon-type retrovirus must have been horizontally transferred after the divergence of the leopard cat but prior to

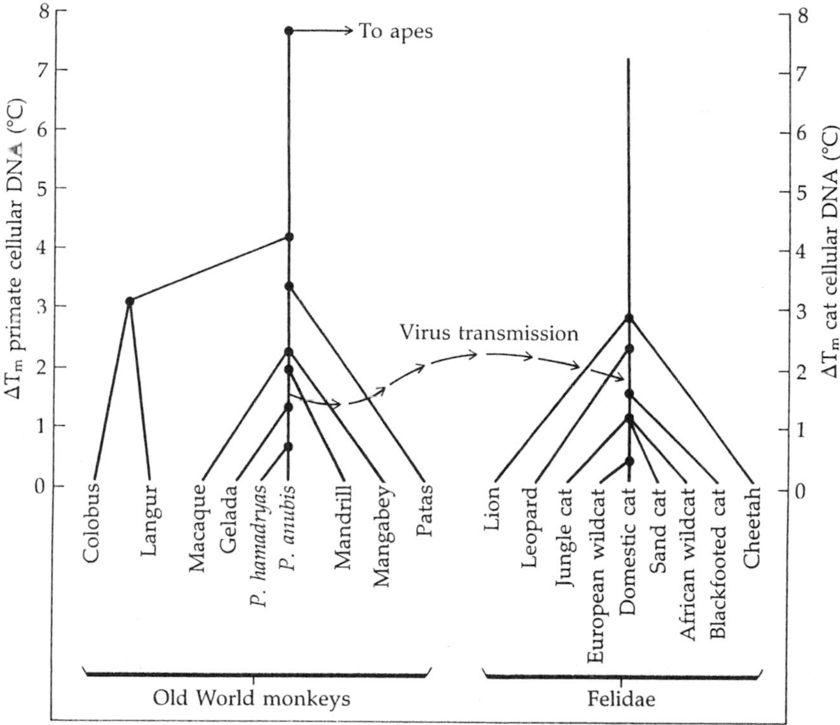

Figure 26. Domestic cats and their close relatives contain an endogenous retrovirus derived by horizontal transfer from ancestors of baboons. The time of the horizontal transfer can be determined by aligning the cat and baboon phylogenies according to the melting temperature of renatured nonrepetitive cellular DNA. In the cat phylogeny, the virus appears after the divergence of the leopard but before the divergence of the black-footed cat. In the baboon phylogeny this interval corresponds to the time between the divergence of the mandrill and the divergence of the gelada. In absolute time, the horizontal transfer appears to have occurred 8–11 million years ago. (After Benveniste 1985.)

the divergence of the black-footed cat. In terms of ΔT_m the bracketed values for the transfer are 1.6°C to 2.2°C. Assuming five million years of divergence per °C in ΔT_m (see Problem 20), the bracketed times for the transfer are 8–11 million years. In the phylogeny of baboons, the ΔT_m values demarcate the horizontal transfer as occurring after the divergence of the mandrill but prior to the divergence of the gelada.

By means of an as-yet poorly defined mechanism, retroviruses can occasionally incorporate genes of the host into their own genetic material. The best evidence on this point comes from the occurrence of **oncogenes** in certain retroviruses, which are associated with the cancerous growth of infected cells. Retroviral oncogenes, many of which are related to cellular growth factors, share unmistakeable DNA sequence similarity with homologous genes found in the host organism and related species (Bishop 1981, 1982). Retroviruses can therefore bring about the horizontal transfer of genes between species. However, it remains to be determined whether other types of genes undergo horizontal transfer, by whatever mechanisms, at a rate that is of importance in the evolutionary process.

SUMMARY

Extensive data have become available on DNA and protein differences that occur among species and polymorphisms that occur in natural populations of single species. In this way molecular biology has provided a new level of detail to studies of the evolutionary relationships among organisms and the mechanisms of evolutionary change.

Much of the molecular data can be explained by the neutral theory of molecular evolution, which holds that many nucleotide and amino acid substitutions are so nearly equivalent in their effects on fitness that their fate is largely determined by random genetic drift rather than natural selection. At one time held to be antidarwinian, and still somewhat controversial, the neutral theory has been instrumental in stimulating the development of new theories and new methods of data analysis.

Many theoretical predictions of the neutral theory have been calculated, including the probability of ultimate fixation of a neutral allele, the time to fixation, the rate of fixation, the average time between fixations, the probability of polymorphism, and the average homozygosity and heterozygosity.

For amino acid replacements in proteins, the expected proportion K of differences between two sequences that have undergone independent evolution for a period estimated as $\langle t \rangle$ years can be estimated approximately as $\langle K \rangle = -\ln(1 - \langle D \rangle)$, where $\langle D \rangle$ is the observed proportion of differences; the rate λ of amino acid substitution is estimated as $\langle \lambda \rangle = \langle K \rangle / 2\langle t \rangle$, and the units of λ are usually expressed in amino acid replacements per amino acid site per year. Rates of amino acid replacement in proteins vary over a 500-fold range, and the average rate is about 1×10^{-9} amino acid substitutions per amino acid site per year.

For nucleotide substitutions, the estimated proportion of $\langle k \rangle$ of nucleotide differences can be estimated as $\langle k \rangle = -(3/4)\ln[1 - 4\langle D \rangle/3]$, where $\langle D \rangle$ is the observed proportion of differences, and mutations between nucleotides occur at random. Rates of nucleotide substitution vary considerably among different types of genes and different positions in the codons for amino acids. In general, rates of nonsynonymous (amino-acid-changing) nucleotide substitutions parallel those of amino acid replacements and vary tremendously among different proteins. Rates of synonymous (non-amino-acid-changing) substitutions are generally more rapid and more uniform. Still faster rates of nucleotide substitution occur in intervening sequences and regions flanking genes at the 3′ (downstream) end. However, the fastest known rates of nucleotide substitution occur in pseudogenes, which appear to carry out no function in the organism. According to the neutral theory, differences in rates of nucleotide substitution or amino acid replacement within and among genes reflect the extent to which specific sequences are necessary to give correct molecular structures and are thereby constrained by natural selection.

Rates of nucleotide substitution are even faster in mammalian mitochondria than in pseudogenes, making mitochondrial DNA (mtDNA) convenient for molecular evolutionary studies of closely related organisms. Nucleotide substitutions also occur rapidly in fungal mitochondria but slowly in the chloroplast DNA (cpDNA) and mtDNA of higher plants. In terms of structural changes in DNA through addition/deletion or rearrangements, the fastest rates occur in angiosperm and fungal mtDNA, and the slowest rates occur in angiosperm cpDNA and animal mtDNA. The evolutionary significance of these differences in rate is unknown.

Average rates of nucleotide substitution or amino acid replacement demonstrate an approximate constancy over long periods of evolutionary time, known as the molecular clock. Although molecular clocks "tick" at different rates in different proteins and different types of genes, the uniformity of rates in different organisms enables molecular differences to be translated into absolute time and thus provides an important adjunct to fossil dating, and in some cases it is the only method available.

Certain tenets of the neutral theory have been misconstrued as implying that genes are not important, or not subject to natural selection or constrained by natural selection, or that neutral alleles have a selection coefficient of $s = 0$ [what is required is that $s << 1/(4N_e)$], or that the environment is not important in determining or altering the selection coefficient of an allele. In overall appraisal, the neutral theory has provided the most economical explanation so far of the great differences between rates of molecular and morphological evolution, and the great differences in rates of molecular evolution in different codon positions, different parts of genes such as intervening sequences, and pseudogenes. However, the neutral theory must still be regarded as unproven. (In a broader context, from the standpoint of scientific method, can the neutral theory ever really be proven? What kind of evidence might be considered proof?)

Gene duplication is an important process in molecular evolution because natural selection can bring about the functional divergence of duplicate genes and result ultimately in the evolution of multigene families of related genes, such as the α-globin and β-globin families. Counteracting the divergence of duplicate genes are mechanisms for concerted evolution including gene conversion and unequal crossing-over, which tend to maintain sequence homogeneity among the gene copies. For example, if the rate of gene conversion is rapid, gene copies at nonhomologous positions in the chromosome remain almost as similar as gene copies at homologous positions. Strong evidence for gene conversion has been found in the duplicate γ-globin genes in humans.

Unequal crossing-over in multigene families can result in an increase or decrease in number of genes in the family. In the α-globin duplication in humans, individuals containing a deletion of one of the copies are apparently at a selective advantage in environments in which malaria is endemic, because the frequency of the α deletion is directly related to the incidence of malaria.

Evolutionary diversification and specialization of multigene families results in multigene superfamilies of distantly related genes. One of the best characterized multigene superfamilies is that of the immunoglobulin superfamily, which includes the genes for antibodies, T-cell and other cell receptors, and certain molecules of the major histocompatibility complex.

Most eukaryotic genomes contain tandemly arranged, highly repetitive, simple-sequence or satellite DNA localized near the centromeres or telomeres, which may account for a significant fraction of the total genome. The functions, if any, of these sequences are obscure. Hypotheses include roles in chromosome structure, pairing and recombination, nuclear structure, and genome evolution. However, the functional models are strained by the great variability that occurs in the amount, composition and location of satellite DNA within and among species. It is possible that localized highly repetitive DNA represents selfish DNA that maintains itself through its ability to replicate and undergo concerted evolution.

Many eukaryotic genomes also contain highly repetitive short DNA sequences (SINEs) or long DNA sequences (LINEs), which are dispersed among the chromosomes and may account for ten percent or more of the total DNA.

The molecular structure of LINE sequences resembles that of DNA copies of processed RNA transcripts (processed pseudogenes), which result from the action of the enzyme reverse transcriptase. Dispersed highly repetitive LINE sequences may originate from normal genes that are transcribed at times in development that are particularly favorable for reverse transcription and integration into the genome. In this model, a small number of functional sequences gives rise to a much larger number of essentially functionless sequences that are free to undergo mutation and random genetic drift until cleared from the genome by deletion or other mechanisms. Studies of one family of dispersed repeated sequences in the mouse suggest that the turnover process has a half-life of approximately two million years.

Transposable elements are DNA sequences that can change their position in the genome. Transposition is frequently a replicative process resulting from the action of proteins encoded within the element itself, so transposable elements tend to increase in number in the genome unless counteracted by other effects, such as regulation, with the transposition rate decreasing as the copy number increases, or the elimination of individuals with excessive copy number through reduced fitness.

Transposable elements are ubiquitous in prokaryotes and eukaryotes. One type of transposable element in prokaryotes is the insertion sequence, typically 1000–2000 nucleotides in length, flanked by short inverted repeat sequences and containing a long open reading frame coding for the transposase and, in some cases, a protein that regulates transposition. Models that predict the distribution of numbers of insertion sequences among strains of bacteria fit the observed data quite well.

Insertion sequences can work in combination to form composite transposons, in which a DNA sequence flanked by two identical insertion sequences undergoes transposition as an intact unit. The central sequence often contains genes that are beneficial to their carriers in certain environments, such as genes for resistance to antibiotics or toxic metals. Composite transposons enable the piecewise assembly of complex transmissible plasmids containing genes for antibiotic resistance and other functions which can become widespread in bacterial populations.

Several types of transposable elements with different DNA sequence organizations occur in eukaryotes. One type of element resembles a retrovirus genome in having a central coding region with one open reading frame flanked by long direct repeat sequences. In another type of element, the central sequence is flanked by inverted repeats and resembles a typical eukaryotic gene in that it often contains intervening sequences which must be removed.

Most natural populations of *Drosophila melanogaster* contain multiple copies of the transposable element P (making them P strains), which is absent from the genome of old laboratory strains (M strains). The P element becomes mobilized in the germ line in the F_1 generation in crosses of P males with M females, and the result is a syndrome of traits termed hybrid dysgenesis, which includes hybrid sterility or reduced fertility at certain temperatures, elevated mutation rates, and chromosome breakage. Hybrid dysgenesis does not occur in the reciprocal cross. The difference in the crosses results from substances transmitted in the cytoplasm which repress transposition of the P element. Females of P strains have repressing cytoplasm (P cytotype), whereas females of M strains do not (M cytotype).

Since the P element appears to have invaded the genome of *D. melanogaster* worldwide within just the past 50 years, the reduction in fertility in hybrids is apparently more than offset by the replication and transposition of P elements in the hybrid germ line. The infection, spread and turnover of transposable elements may be a frequent process in eukaryotic genomes.

Retroviruses can undergo reverse transcription and integration into the genome. Some organisms contain, as a normal constituent of their germ line, 10–100 copies of one or more retroviral sequences, which are known as endogenous retroviruses or virogenes. Under suitable conditions the cells can be induced to release infectious retrovirus particles. Although the released viruses do not normally infect cells of their species of origin, they do infect cells of related species.

Once established in the germ line, virogenes are maintained for long periods and undergo rapid sequence evolution. A baboon Type C retrovirus, which has been retained in all Old World monkeys and which is prone to being released as infectious particles, has been found to evolve only approximately half as fast in primates of African origin as in primates of Asian origin. The proposed explanation is that Asian primates are not in contact with the African baboons that release the infectious particles, and thus the Asian virogene has not been selected for immunity to infection and is free to evolve. Interestingly, the human virogene related to the baboon Type C endogenous retrovirus has evolved at a rate characteristic of that in Asian primates.

Retroviruses can sometimes undergo horizontal transfer and integration into the germ line of species that are only remotely related phylogenetically to the species of origin. One well established example is the horizontal transfer of a retrovirus from ancestors of modern baboons to ancestors of the domestic cat, which occurred approximately 8–11 million years ago. Other examples of horizontal transfer include those from ancestors of modern primates to ancestors of carnivores, other primates, and birds; as well as transfers from ancestors of modern rodents to ancestors of carnivores, primates, and artiodactyls.

Retroviruses can incorporate host genes, such as the oncogene growth factors that contribute to cancer formation, and carry them along during horizontal transfer. However, it is not known how important or how frequent the horizontal transfer of chromosomal genes between species may be during the course of evolution.

FURTHER READINGS

Ayala, F. J. (ed.). 1976. *Molecular Evolution*. Sinauer Associates, Sunderland, Massachusetts.

Clarke, B. C., A. Robertson and A. J. Jeffreys. 1986. *The Evolution of DNA Sequences*. The Royal Society Press, London.

Dover, G. A. and R. B. Flavell (eds.). 1982. *Genome Evolution*. Academic Press, New York.

Dykhuizen, D. E. and D. L. Hartl. 1983. Selection in chemostats. *Microbiol. Rev.* 47: 150–168.

Engels, W. R. 1983. The P family of transposable elements in *Drosophila*. *Annu. Rev. Genet.* 17: 315–344.

Evered, D. and G. M. Collins (eds.). 1984. *Origins and Development of Adaptations*. Pitman, London.

Gershowitz, H., D. L. Rucknagel and R. E. Tashian (eds.). 1986. *Evolutionary Perspectives and the New Genetics*. Alan R. Liss, New York.

Gillespie, J. H. 1986. Rates of molecular evolution. *Annu. Rev. Ecol. Syst.* 17: 637–665.

Goodman, M. (ed.). 1982. *Macromolecular Sequences in Systematic and Evolutionary Biology*. Plenum Press, New York.

Gutfreund, H. 1981. *Biochemical Evolution*. Cambridge University Press, Cambridge.

Hall, B. G. 1982. Evolution on a petri dish: The evolved β-galactosidase system as a model for studying acquisitive evolution in the laboratory. *Evol. Biol.* 15: 85–150.

Hall, J. G. and R. K. Koehn. 1983. Evolution of enzyme catalytic efficiency and adaptive inference from steady-state kinetic data. *Evol. Biol.* 16: 53–96.

Kimura, M. 1982. *Molecular Evolution, Protein Polymorphism and the Neutral Theory*. Springer-Verlag, New York.

Kimura, M. 1983. *The Neutral Theory of Molecular Evolution*. Cambridge University Press, Cambridge, England.

Kimura, M. and T. Ohta. 1971. *Theoretical Aspects of Population Genetics*. Princeton University Press, Princeton, New Jersey.

Krieber, M. and M. R. Rose. 1986. Molecular aspects of the species barrier. *Annu. Rev. Ecol. Syst.* 17: 465–485..

MacIntyre, R. J. 1982. Regulatory genes and adaptation: Past, present, and future. *Evol. Biol.* 15: 247–286.

MacIntyre, R. J. (ed.). 1985. *Molecular Evolutionary Genetics*. Plenum Press, New York.

Maeda, N. and O. Smithies. 1986. The evolution of multigene families: Human haptoglobin genes. *Annu. Rev. Genet.* 20: 81–108.

McDonald, J. F. 1983. The molecular basis of adaptations: A critical review of relevant ideas and observations. *Annu. Rev. Ecol. Syst.* 14: 77–102.

Milkman, R. (ed.). 1982. *Perspectives on Evolution*. Sinauer Associates, Sunderland, Massachusetts.

Mortlock, R. P. 1982. Regulatory mutations and the development of new metabolic pathways by bacteria. *Evol. Biol.* 14: 205–268.

Nei, M. 1975. *Molecular Population Genetics and Evolution*. American Elsevier, New York.

Nei, M. 1987. *Molecular Evolutionary Genetics*. Columbia University Press, New York.

Nei, M. and D. Graur. 1984. Extent of protein polymorphism and the neutral mutation theory. *Evol. Biol.* 17: 73–118.

Nei, M. and R. K. Koehn (eds.). 1983. *Evolution of Genes and Proteins*. Sinauer Associates, Sunderland, Massachusetts.

Ohta, T. and K. Aoki (eds.). 1985. *Population Genetics and Molecular Evolution*. Springer-Verlag, New York.

Riley, M. 1985. Discontinuous processes in the evolution of the bacterial genome. *Evol. Biol.* 19: 1–36.

Sibley, C. G. and J. E. Ahlquist. 1986. Reconstructing bird phylogeny by comparing DNAs. *Sci. Amer.* 254: 82–92 (February 1986).

Thorpe, J. P. 1982. The molecular clock hypothesis: Biochemical evolution, genetic differentiation and systematics. *Annu. Rev. Ecol. Syst.* 13: 139–168.

QUANTITATIVE GENETICS 4

No study of population genetics is sufficient without consideration of traits influenced by multiple genetic and environmental factors. Previous chapters have concentrated on genetic variation affecting allozymes, blood groups, and visible genetic abnormalities, which are simple Mendelian traits in that each is determined by a single gene. For example, the human glandular disease cystic fibrosis results from a single homozygous recessive mutation, and the ABO blood groups are controlled by the alleles of a single gene. In this chapter, we consider more complex traits influenced by alleles of multiple genes, which are called **polygenic traits**. Most polygenic traits are influenced by the environment to varying degrees, and they are often called **multifactorial traits** to emphasize their determination by multiple genetic and environmental factors. For example, variation in human weight is partly due to genetic differences among individuals and partly due to environmental factors such as exercise and level of nutrition.

TYPES OF QUANTITATIVE TRAITS

Multifactorial traits may be considered as resulting from the combined effects of many quantities, some genetic in origin and some environmental, and for this reason they are often called **quantitative traits**. The study of quantitative traits constitutes *quantitative genetics*.

Three types of quantitative traits may be distinguished:

1. Traits for which there is a continuum of possible phenotypes. These are called **continuous traits**, and examples include height, weight, milk yield, and growth rate. The distinguishing feature of continuous traits is that the phenotype can take on any one of a continuous range of values from minimum to maximum. In theory, there are infinitely many possible phenotypes, among which discrimination is limited only by the precision of the instrument used for measurement. In the case of height, for example, given any two individuals, no matter how similar in height, there may always exist another individual whose height is intermediate between them, provided that the trait is measured with sufficient accuracy. However, in practice, similar phenotypes are often grouped together for purposes of analysis.

2. Traits for which the phenotype is expressed in discrete, integral classes. These are called **meristic traits**, and examples include number of offspring or litter size, number of ears on a stalk of corn, number of flowers on a petal, and number of bristles on a fruit fly. The distinguishing feature of meristic traits is that the phenotype of an individual is given by a simple integer which equals the number of elements of the trait that the individual displays. For example, a popular meristic trait used in experimental studies of quantitative genetics in *Drosophila* is the number of bristles which occur on the abdominal sternites, typically the sixth in females and the fifth in males (Figure 1). Normally there are 14 to 24 bristles per sternite. A male with 19 bristles on the fifth abdominal sternite therefore has a phenotype of 19. When the number of possible phenotypes of a meristic trait is large, as it is with abdominal bristle number, then the line between continuous traits and meristic traits becomes indistinct because each possible value of the meristic trait can be associated with a small range of values for the quantitative trait. To be concrete, for a continuous trait which ranges in value from 14 to 24, the phenotypic measurements may be rounded to the nearest integer (for example, individuals with phenotypes in the range 18.5 to 19.5 may be grouped together and all considered to have a phenotype of 19). The grouping results in a distribution of phenotypes like that of a meristic trait.

3. The third category of quantitative traits consists of discrete traits, which are either present or absent in any one individual. In these cases, the multiple genetic and environmental factors combine to determine an

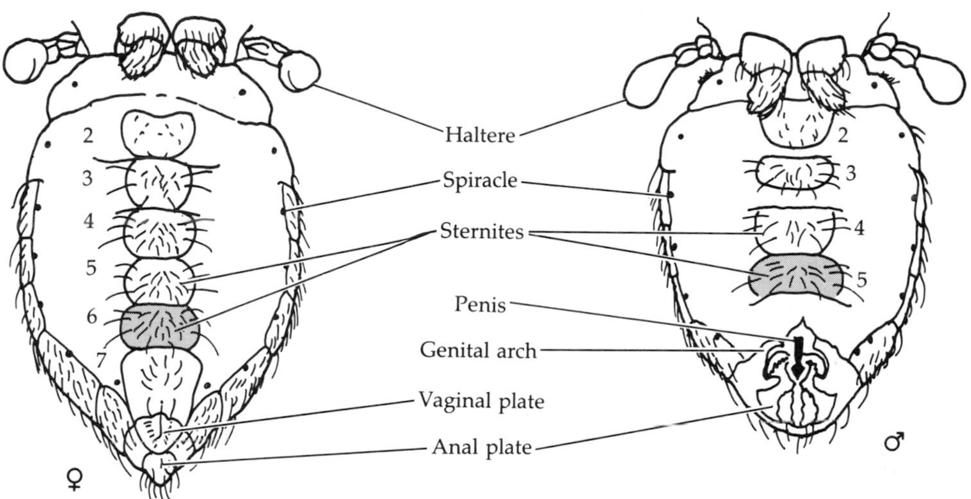

Figure 1. Ventral abdomen of females and males of *Drosophila melanogaster*, indicating the sternites. The number of sternital bristles (also called abdominal bristles) provides a convenient trait for experimental studies in quantitative genetics. The enumerated bristles are usually those on the fifth or sixth (shaded) sternite in females (or their sum), and the fourth or fifth (shaded) sternite in males (or their sum).

underlying risk or *liability* toward the trait. Liability values are not directly observable. However, individuals which actually express the trait are assumed to be ones whose liability values are greater than some *threshold* or triggering level. Traits of this type are known as **threshold traits**, and examples in human genetics include diabetes and schizophrenia. With threshold traits, studies of affected individuals and their relatives permit inferences to be made about underlying values of liability. These methods are discussed later in this chapter.

Quantitative traits are of utmost importance to plant and animal breeders, since agriculturally important characteristics such as yield of grain, egg production, milk production, efficiency of food utilization by domesticated animals, and meat quality are all quantitative traits. Even as modern methods of genetic engineering are applied to animal and plant improvement, quantitative genetics plays a continuing important role because commercially desirable traits result from complex interactions among many genes. In addition to being essential ingredients in plant and animal improvement programs, the principles of quantitative genetics, appropriately modified and interpreted, can be applied to the analysis of quantitative traits in humans and natural populations of plants and animals.

THE NORMAL DISTRIBUTION

The normal distribution plays a central role in quantitative genetics, as it does in much of classical statistics, and it will be introduced here using abdominal bristle number in *Drosophila* as an example.

Table 1 gives the distribution of number of abdominal bristles observed among 500 males in a population of *Drosophila melanogaster*. The measurements of phenotypes, which in this example are obtained by direct counts of the bristles, are called the **phenotypic values**. Phenotypic values of bristle number in this sample range from 13 to 25. The most frequent phenotypic value is 19.

The individual animals in Table 1 are of no particular interest in themselves, but they are of great interest insofar as they constitute a representative sample of the population from which they came. Inferences about the distribution of phenotypic values in the entire population are derived from the phenotypic values in the sample. Usually, the population parameters of paramount interest are the **mean**, or average, phenotypic value of the trait, and the variance. The **variance** is a measure of dispersion of the phenotypic values around the mean, and it is defined as the average of the squared differences between the phenotypic values and the mean. Since the mean of a population is typically denoted μ and the variance as σ^2, we should, to be consistent with the convention established in this book, designate the estimates of μ and σ^2 as $\langle\mu\rangle$ and $\langle\sigma^2\rangle$, respectively. However, you should note that estimates of μ and σ^2 are often designated $\bar{x}$ and s^2, respectively.

In order to cast Table 1 into the symbols usually used for the distribution

Table 1. Abdominal bristle number.[a]

Number of bristles	Number of animals	Number of bristles	Number of animals
13	1	20	92
14	5	21	48
15	26	22	29
16	44	23	11
17	66	24	3
18	78	25	2
19	95	Total:	500

(Based on data provided by T. Mackay.)

[a] Number of bristles on the fifth sternite in a sample of 500 males from an experimental population of *Drosophila melanogaster.*

of phenotypic values, let n_i be the number of males which have phenotypic value x_i. In particular,

$$\begin{aligned} x_1 &= 13 \text{ and } n_1 = 1 \\ x_2 &= 14 \text{ and } n_2 = 5 \\ x_3 &= 15 \text{ and } n_3 = 26 \\ &\cdots\cdots\cdots\cdots \\ x_{13} &= 25 \text{ and } n_{13} = 2 \end{aligned}$$

Estimates of the population parameters μ and σ^2 for the mean and variance in bristle number, based on the sample values, are as follows, where $n = \Sigma n_i$ is the total number of individuals in the sample:

$$\langle\mu\rangle = \bar{x} = \Sigma n_i x_i / n \qquad \textbf{(4.1)}$$

$$\langle\sigma^2\rangle = s^2 = \Sigma n_i(x_i - \bar{x})^2/(n - 1) \qquad \textbf{(4.2)}$$

Summation is over all possible values of i. For the data in Table 1, $\langle\mu\rangle = \bar{x} = 9369/500 = 18.7$ bristles, and $\langle\sigma^2\rangle = s^2 = 4.3$.

An alternative to Equation 4.2 is often easier to use in computation, namely

$$\langle\sigma^2\rangle = s^2 = (\Sigma n_i x_i^2 - n\bar{x}^2)/(n - 1) \qquad \textbf{(4.3)}$$

For the data in Table 1, $\Sigma n_i x_i^2 = 177{,}705$ and $n\bar{x}^2 = 175{,}556$.

Use of $n - 1$ in the denominator of Equations 4.2 and 4.3 is a correction for sampling error, but when n is reasonably large, the correction is minor. When n is used in Equation 4.3 instead of $n - 1$, it can be seen that $\langle\sigma^2\rangle$ equals the mean of the squares of the phenotypic values minus the square of the mean phenotypic value. This formulation of the variance formed the basis of the calculation in Figure 12 in Chapter 2.

Another term frequently encountered in quantitative genetics is the **standard deviation**, which is simply the square root of the variance. As the variance is symbolized σ^2, the standard deviation is symbolized σ. The value of σ is estimated as the square root of the estimated variance in Equations 4.2 or 4.3. In the bristle-number example, $\langle\sigma\rangle = s = \sqrt{4.3} = 2.1$.

PROBLEM 1

The flour beetle *Tribolium castaneum* is a popular organism for experimental studies in quantitative genetics because it is small, has a short generation time with many offspring, and can easily be reared on dry medium in small culture vessels. The pupal stage of the life cycle, during which larvae undergo metamorphosis into adults, is a convenient stage for detailed studies because pupae are immobile and can be measured, weighed, and otherwise manipulated. Shown below are the weights of 32 male pupae. Using the values of n_i and x_i provided, estimate the mean, variance, and standard deviation of pupa weight in the population from which these males were sampled. Treat the data as if each animal within a particular range had the phenotypic value of the midpoint of the range.

Pupa weight (range, in micrograms)	Number n_i of males	Midpoint of range (x_i, in micrograms)
1800–1900	1	1850
1900–2000	1	1950
2000–2100	5	2050
2100–2200	7	2150
2200–2300	5	2250
2300–2400	4	2350
2400–2500	8	2450
2500–2600	1	2550

ANSWER From Equation 4.1, $\langle\mu\rangle = 71{,}900/32 = 2246.88$. From Equation 4.2, $\langle\sigma^2\rangle = 969{,}687.52/31 = 31{,}280.24$. Using Equation 4.3, $\Sigma n_i x_i^2 = 162{,}520{,}000$ and $n\bar{x}^2 = 161{,}550{,}312.5$, hence $\langle\sigma^2\rangle = 31{,}280.24$. The extra digits have been carried along to avoid round-off error. The estimated standard deviation is $\langle\sigma\rangle = \sqrt{(31{,}280.24)} = 176.86$. These data are presented graphically later in this chapter in Figure 16. (Data from F. D. Enfield.)

Use of only the mean and variance to summarize a quantitative trait in a large population might seem like a gross oversimplification, but in many cases these two numbers are often adequate to specify the entire distribution. Indeed, in many cases, quantitative traits conform to the familiar, bell-shaped normal distribution, which is illustrated for *Drosophila* bristle number in Figure 2. The data are those in Table 1, plotted as a histogram, the overall shape of which is closely approximated by the smooth normal curve. In a **normal distribution**, the height of the curve y for any phenotypic value x is given by

$$y = \frac{1}{\sqrt{2\pi}\sigma} \exp[-(x - \mu)^2/2\sigma^2] \qquad \textbf{(4.4)}$$

in which μ represents the mean of the population, σ^2 the variance, and exp[] equals $e^{[]}$. As usual, $\pi = 3.14159$ and $e = 2.71828$ are constants. The theoretical range of x is minus infinity to plus infinity. The curve defined by Equation 4.4 is called the *normal density* in statistics, but in population genetics the term normal distribution is often used as a synonym. The true mean μ and variance σ^2 of the population are unknown, but if the sample of individuals from the population is sufficiently large and representative, then estimates based on Equations 4.1 and 4.2 are expected to be close to μ and σ^2. The smooth curve in Figure 2 is that of a normal distribution with $\mu = 18.7$ and $\sigma = 2.1$.

If a trait is normally distributed, and reliable estimates of the mean and variance are known, then certain probability statements about the distribution of phenotypes in the population can immediately be made. These follow

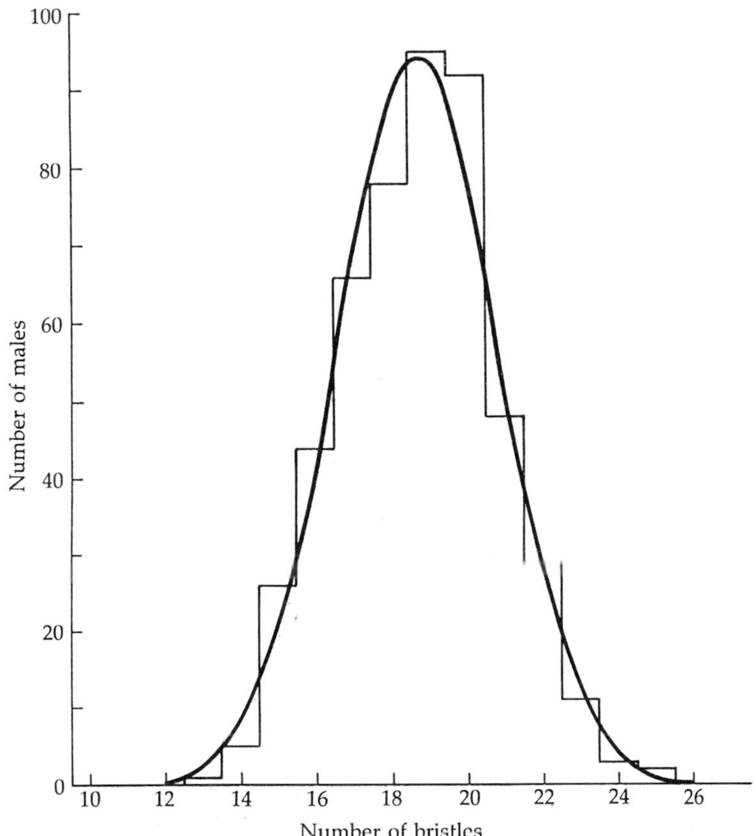

Figure 2. Number of bristles on the fifth abdominal sternite in males of a strain of *Drosophila melanogaster*. The smooth curve is that of a normal distribution with mean 18.7 and standard deviation 2.1. (Data from T. Mackay.)

from the fact that the area under the normal curve in Equation 4.4 between any two x values equals the proportion of individuals whose phenotypic values lie within that range. For example, in a normal distribution, 31.7 percent of the individuals have phenotypic values which deviate from the mean by one or more standard deviations, and 4.6 percent have phenotypic values which deviate by two or more standard deviations. The probabilities associated with deviations of various magnitudes are summarized graphically in Figure 3. As an example of the use of Figure 3, consider the distribution of weight of *Tribolium* pupae summarized in Problem 1, for which $\langle\mu\rangle =$ 2246.88 μg and $\langle\sigma\rangle =$ 176.86 μg. Assuming a normal distribution of pupa weight, what proportion of pupae are expected to weigh more than 1.5 standard deviations above the mean? Figure 3 gives the probability of a deviation as large or larger than 1.5σ as 0.134, but this proportion includes

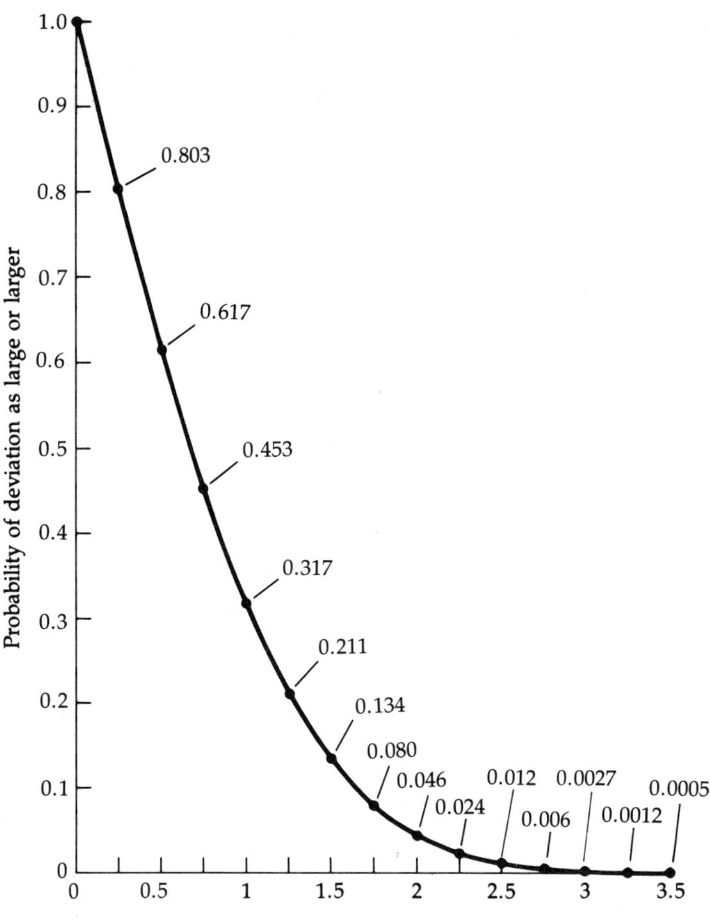

both negative and positive deviations from the mean. Because the normal distribution is symmetrical, half the deviations are negative and half are positive, thus the probability of a positive deviation as great as or greater than 1.5σ equals $0.134/2 = 0.067 = 6.7$ percent. That is, 6.7 percent of beetles are expected to have pupa weights greater than $2246.88 + (1.5)(176.86) = 2512.17$ μg. Similarly, 6.7 percent of beetles are expected to have pupa weights smaller than $2246.88 - (1.5)(176.86) = 1981.59$ μg.

The proportion of individuals with phenotypic values that deviate from the mean by *less* than a specified value can be obtained from Figure 3 by subtracting the tabulated values from 1.0. For example, the proportion of individuals with phenotypic values which differ from the mean by less than one standard deviation equals $1 - 0.317 = 0.683 = 68.3$ percent. Three key proportions of this type are commonly quoted in regard to the normal distribution:

◀ **Figure 3. The x-axis gives the deviation from the mean in multiples of the standard deviation for a hypothetical observation from a normal distribution. The y-axis gives the probability of actually obtaining a deviation as large or larger. For the normal distribution, half the observations are smaller than the mean and half are larger. For example, the overall probability of a deviation of two standard deviations or more is 0.046: the probability that a particular observation will be at least two standard deviations below the mean is therefore 0.023, and the probability that it will be at least two standard deviations above the mean is 0.023. The exact probabilities corresponding to deviations of 0.25, 0.50, 0.75, 1.0, . . . are noted on the graph. For deviations x between 0 and 2, the curve is approximated by Prob(x) = $1 - [1 - \exp(-2x^2/\pi)]^{1/2}$. For x greater than 2, the curve is approximated by Prob(x) = $(0.7979/x)\exp(-x^2/2)$.**

1. Approximately 68 percent of the population have phenotypic values within *one* standard deviation from the mean.
2. Approximately 95 percent of the population have phenotypic values within *two* standard deviations from the mean.
3. Approximately 99.7 percent of the population have phenotypic values within *three* standard deviations from the mean.

Applying these to the *Tribolium* data, and assuming μ = 2246.88 μg and σ = 176.86 μg, we can say that approximately 68 percent of the population have pupa weights in the range $\mu \pm \sigma$ (which is 2070.0 to 2423.8 μg), approximately 95 percent have weights in the range $\mu \pm 2\sigma$ (1893.2 to 2600.6 μg), and approximately 99.7 have weights in the range $\mu \pm 3\sigma$ (1716.3 to 2777.5 μg).

The frequent occurrence of the normal distribution in practice results in part from the fact that the summation of many random, independent quantities conforms to the normal distribution. Many real variables are determined by such sums. For example, the multiple genetic and environmental factors which determine quantitative traits may be approximately additive in their effects. The prevalence of the normal distribution, and its convenient mathematical characteristics, make the distribution very important in quantitative genetics. As illustrated in Figure 2, the normal distribution is of value not only when applied to continuous traits but also when applied to meristic traits in which the number of possible phenotypes is relatively large.

PROBLEM 2

Key aspects of the theory of quantitative genetics have been verified in *Drosophila melanogaster* in a series of classic experiments (Clayton, Morris and Robertson 1953, 1956; Clayton, Knight, Morris and Robertson 1956; Clayton and Robertson 1956). The trait of interest was the sum of the number of bristles on the fourth and fifth abdominal sternites. The original or *base* population was a sample from a large population cage that had been established some five years earlier from African flies. Among females in the base population, the number of bristles was distributed normally with an estimated mean of approximately 39 and estimated variance of 12.53. What fraction of females were expected to have bristle scores

____ Greater than 44?	____ Less than 30?
____ Less than 34?	____ Between 36 and 42?
____ Greater than 48?	

(Hint: To avoid the necessity of interpolating in Figure 3, the following numbers are provided: the probabilities associated with deviations of 0.85σ, 1.41σ, and 2.54σ are 0.395, 0.159, and 0.011, respectively.)

ANSWER Put $\mu = 39$ and $\sigma = (12.53)^{1/2} = 3.54$. A bristle number of 44 is a positive deviation from the mean of $(44 - 39)/3.54 = 1.41$ standard deviations (σ). The hint gives 0.159 as the probability of a deviation at least as large, which includes both positive and negative deviations; thus the probability of a positive deviation at least as large as 1.41σ is $0.159/2 = 0.080$, so 8 percent of females have 44 or more bristles. A bristle score of 34 is a negative deviation of $(39 - 34)/3.54 = 1.41\sigma$, so 8 percent of females have 34 or fewer bristles. Similarly, 0.55 percent of females have 48 or more bristles, and 0.55 percent have 30 or fewer. Finally, 36 bristles is a negative deviation of 0.85σ, and 42 bristles is a positive deviation of 0.85σ. The hint says that the probability of a deviation at least as large as 0.85σ is 0.395, so the probability of a smaller deviation is $1 - 0.394 = 0.60$. Thus, 60 percent of females are expected to have between 36 and 42 bristles.

Armed with the normal distribution as an adequate approximation of the phenotypic distribution of many continuous or meristic traits, we can approach the issue of how a population is expected to change genetically and phenotypically when certain phenotypes are favored by selection.

ARTIFICIAL SELECTION

The deliberate choice of a select group of individuals to be used for breeding constitutes **artificial selection**. The most common type of artificial selection is **directional selection**, in which phenotypically superior animals or plants are chosen for breeding (see Figure 15 in Chapter 2). Although artificial selection has been practiced successfully for thousands of years (for example, in the body size of domesticated dogs), only during this century have the genetic principles underlying its successes become clear. Moreover, application of genetic principles permits prediction of the rapidity and amount by which a population can be altered through artificial selection in any particular generation or small number of generations.

Artificial selection in outcrossing, genetically heterogeneous populations is virtually always successful in the sense that the mean phenotype of the population changes over generations in the direction of selection, provided the population has not previously been subjected to long-term artificial selection for the trait in question. In experimental animals, the mean of almost any quantitative trait can be altered in whatever direction desired by artificial selection. For example, in *Drosophila*, body size, wing size, bristle number, growth rate, egg production, insecticide resistance, and many other traits can be increased or decreased by selection. In domesticated animals and plants, birth weight, growth rate, milk production, egg production, grain yield, and countless other traits respond to selection. Figure 4 shows the results of a long-term selection program involving oil content in corn.

The general success of artificial selection in outcrossing species indicates that a wealth of genetic variation affecting quantitative traits exists. On the other hand, the mean phenotype of a population cannot usually be changed through artificial selection in genetically uniform populations, for the reason that genetic variation is essential to progress under artificial selection. When practiced within inbred and essentially homozygous lines, artificial selection

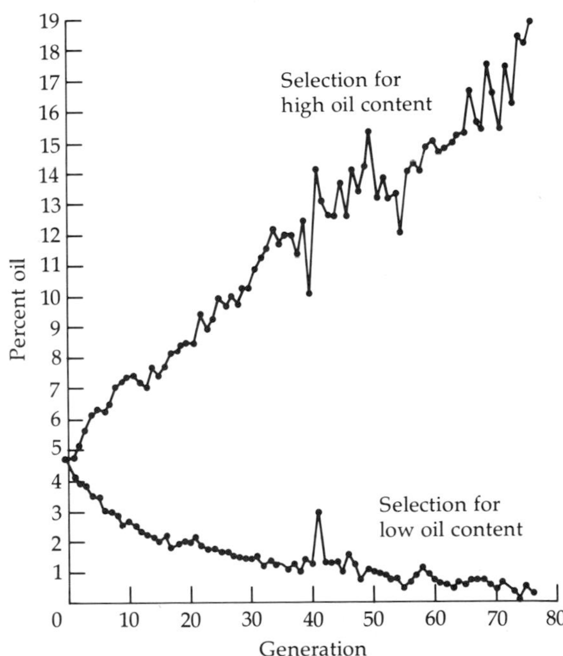

Figure 4. Results of a famous long-term experiment selecting for high and low oil content in corn seeds. Begun in 1896, the experiment has the longest duration of any on record and still continues at the University of Illinois. Note the steady, linear rise in oil content shown by the upper curve. The lower curve started on a roughly linear path and continued so for about ten generations, but then the response tapered off, presumably because 0 percent oil is an absolute lower limit for the trait. (After Dudley 1977.)

consistently results in failure—provided the inbred lines are truly homozygous. In genetically homozygous populations, the only source of genetic variation comes from new mutations.

Prediction Equation for Individual Selection

When individuals are selected for breeding based solely on their own individual phenotypic values, the type of artificial selection is called **individual selection**. Figure 5 illustrates a variety of individual selection called **truncation selection**. The curve in panel (a) represents the normal distribution of a quantitative trait in a population, and the shaded part of the distribution to the right of the phenotypic value denoted T indicates those individuals selected for breeding. The value T is called the **truncation point**. The mean phenotype in the entire population is denoted μ, and that of the selected parents is denoted μ_S. When the selected parents are mated at random, their

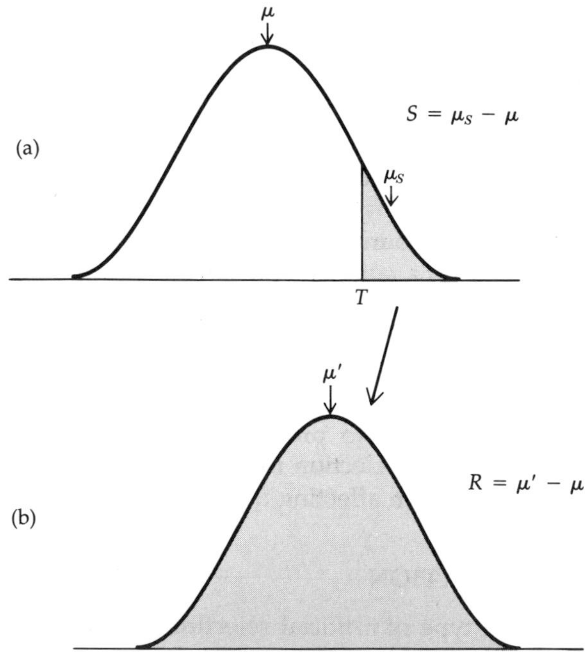

Figure 5. Diagram of truncation selection. (a) Distribution of phenotypes in parental population, mean μ. Individuals with phenotypes above the truncation point (T) are saved for breeding the next generation. The selected parents are denoted by the shading and their mean phenotype by μ_S. (b) Distribution of phenotypes in offspring generation derived from the selected parents. The mean phenotype is denoted μ'. Note that μ' is greater than μ but less than μ_S. The quantity S is called the selection differential, and R is called the response to selection.

offspring have the phenotypic distribution shown in panel (b), where the mean phenotype is denoted μ'.

An example of truncation selection for seed weight in edible beans is shown in Figure 6. In this example, T = 650 mg, μ = 403.5 mg, μ_S = 691.7 mg, and μ' = 609.1 mg. In this case—as is typical of truncation selection—the offspring mean μ' is greater than μ but less than μ_S. The reason μ' is greater than μ is that some of the selected parents have favorable genotypes and therefore pass favorable genes on to their offspring. At the same time, μ' is generally less than μ_S for two reasons:

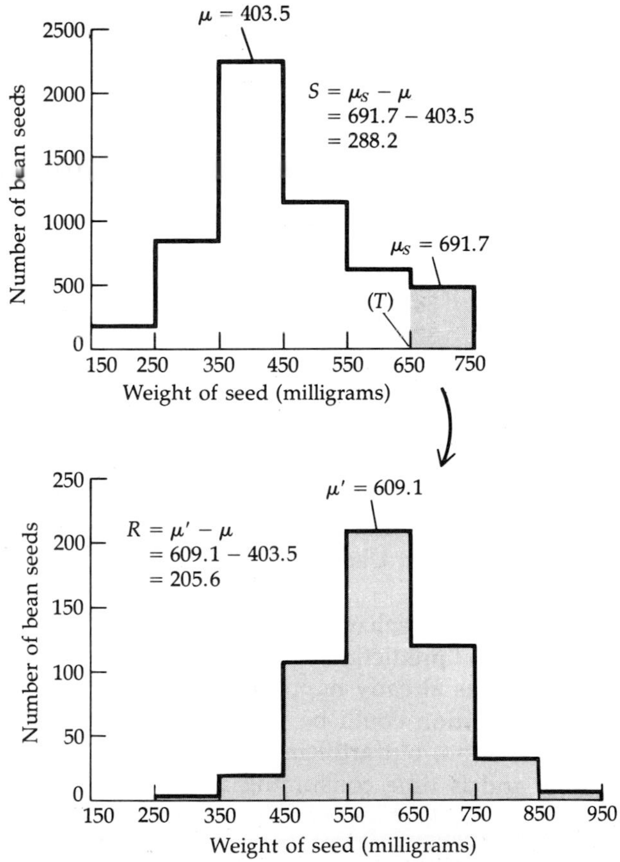

Figure 6. Truncation selection experiment for seed weight in edible beans of the genus *Phaseolus*, laid out as in Figure 5. The truncation point (T) is 650 mg. The selection differential S is the difference in means between the selected parents and the whole population. The response R is the difference in means between the progeny generation and the entire population in the previous generation. The quantity R/S is the realized heritability. (Data from Johannsen 1903.)

1. Because some of the selected parents do not have favorable genotypes. Rather, their exceptional phenotypes result from chance exposure to exceptionally favorable environments.
2. Because alleles, not genotypes, are transmitted to the offspring, and exceptionally favorable genotypes are disrupted by Mendelian segregation and recombination.

The difference in mean phenotype between the selected parents and the entire parental population is called the **selection differential** and is designated S. In symbols,

$$S = \mu_S - \mu \tag{4.5}$$

The difference in mean phenotype between the progeny generation and the previous generation is called the **response** to selection and is designated R. Symbolically,

$$R = \mu' - \mu \tag{4.6}$$

In Figure 6, $\langle S \rangle = 288.2$ mg and $\langle R \rangle = 205.6$ mg.

In quantitative genetics, any equation that defines the relationship between the selection differential S and the response to selection R is known as a **prediction equation**. Since selection can be applied to a population in many different ways (others will be discussed later in this chapter), there are many different prediction equations corresponding to the different modes of selection. For truncation selection, which is the type of selection illustrated in Figure 5, the prediction equation reads

$$R = h^2 S \tag{4.7}$$

where h^2 is a quantity called the **heritability** of the trait. (For historical reasons, heritability is symbolized as h^2, not as h.) Heritability can be understood at several different levels. At one level, heritability can be interpreted as a mere description of what happens when artificial selection is practiced. In Figure 6, for example, $\langle S \rangle = 288.2$ and $\langle R \rangle = 205.6$, so $\langle h^2 \rangle = \langle R \rangle / \langle S \rangle = 205.6/288.2 = 71.3$ percent. When estimated like this from an observed result, h^2 is called the **realized heritability**, and it simply summarizes the observed result.

PROBLEM 3 Below are data on the number i of sternital bristles in samples from two consecutive generations G1 and G2 of an experiment in directional selection for increased bristle number. In the G1 generation, individuals with 22 or more bristles (enclosed in brackets) were mated together at random to form the G2 generation. Estimate the realized heritability of the number of sternital bristle in this experiment. (In order to make the sexes comparable, the value of two has been added to the bristle number in males. Data provided by T. Mackay.)

i	G1	G2	*i*	G1	G2	*i*	G1	G2
15	0	2	20	20	13	25	[1]	3
16	21	4	21	12	14	26	0	2
17	5	7	22	[13]	12	27	0	0
18	18	16	23	[3]	6	28	0	2
19	17	17	24	[5]	3			

ANSWER Estimates of the means are $\langle\mu\rangle = 2220/115 = 19.3$, $\langle\mu_S\rangle = 500/22 = 22.7$, $\langle\mu'\rangle = 2035/101 = 20.1$. Then $\langle S\rangle = 22.7 - 19.3 = 3.4$ (Equation 4.5), and $\langle R\rangle = 20.1 - 19.3 = 0.8$ (Equation 4.6). Realized heritability estimate from Equation 4.7 is $\langle h^2\rangle = 0.8/3.4 = 23.5$ percent.

The data in Problem 3 are taken from experiments by Mackay (1984) that demonstrate the potential significance of new mutations in quantitative genetics. The base population was formed by a cross which mobilizes the transposable element P and results in new insertions in the germ line and a syndrome of partial infertility and other reproductive abnormalities known as *hybrid dysgenesis* (discussed in Chapter 3). A genetically identical base population was formed by the reciprocal cross, in which the P element is not mobilized and hybrid dysgenesis does not occur. In the dysgenic cross, the realized heritability in abdominal bristle number was increased by 40 percent as compared with the nondysgenic control. More strikingly, the phenotypic variance of bristle number in the selected dysgenic lines increased by a factor of three over the course of eight generations. These results demonstrate that the genetic variation affecting quantitative traits may even include insertions of transposable elements.

PROBLEM 4

From the population of *Drosophila* discussed in Problem 2, Clayton et al. (1956b) established five initial populations and practiced directional selection for an increased number of sternital bristles. The data shown below are the results after one generation of selection. The symbols μ, μ_S, μ', S, and R are estimates obtained as in Figure 5, and the means are the averages of males and females taken together. Complete the columns headed S and R, and estimate the realized heritability in each case. What is the overall mean h^2? What is the range of h^2 (i.e., the minimum and maximum values)? Observe that the minimum of h^2 is about 26 percent smaller than the mean and that the maximum is about 30 percent larger than the mean. This sort of variation in realized heritability from experiment to experiment is not unusual, so caution must be exercised in interpreting the value obtained in any one experiment.

Population	$\langle\mu\rangle$	$\langle\mu_S\rangle$	$\langle\mu'\rangle$	$\langle S\rangle$	$\langle R\rangle$	Realized $\langle h^2\rangle$
H1	35.3	40.0	37.9			
H2	35.3	40.0	37.2			
H3	33.5	38.2	35.8			
H4	33.0	37.7	36.3			
H5	34.0	38.7	36.7			

ANSWER $\langle S\rangle$ = 4.7 in all cases. $\langle R\rangle$ (top to bottom) = 2.6, 1.9, 2.3, 3.3, and 2.7. Realized $\langle h^2\rangle$ (top to bottom) = 0.55, 0.40, 0.49, 0.70, and 0.57. Mean of realized $\langle h^2\rangle$ = 0.54; range, 0.40 to 0.70.

When h^2 is interpreted as realized heritability, then Equation 4.7 is hardly a "prediction equation" inasmuch as it merely describes what has already happened in one generation of selection. Of course, the equation could be used to predict the result of the next generation of selection, but artificial selection is impossible in many natural populations and is time consuming and expensive in many domesticated plants and animals. It would therefore be useful if one could estimate heritability without actually performing any artificial selection. If the heritability h^2 could be estimated in such a manner, then Equation 4.7 would be a true prediction equation in the sense that the response R could be predicted for any selection differential S, based on the estimated value of h^2. Such an estimate of h^2 is indeed possible, but it involves an understanding of heritability at a level that includes the underlying genetic basis of quantitative traits.

An understanding of the genetics behind Equation 4.7 requires three items: (1) a concept of how alternative alleles of a gene affect a quantitative trait; (2) a determination of how selection changes the allele frequencies; and (3) a calculation of how much the mean of the trait increases as a result of the change in allele frequency. Some detail is required to establish these three items, but the detail is necessary in order to understand the genetic meaning of heritability. Before proceeding, it will be well to illustrate the nature of the arguments (and one of the important conclusions) with the following problems.

PROBLEM 5 Nilsson-Ehle (1909) was the first to show that a trait with a nearly continuous distribution of phenotypes could result from the joint effects of several genes. The trait of interest is the intensity of red pigment in the glume of wheat *Triticum vulgare*, which Nilsson-Ehle found to result from three unlinked genes, each with two alleles. The situation is exceptionally simple for a quantitative trait because the environment has a negligible effect on phenotype, because the alleles of each gene are additive (i.e., heterozygotes have a phenotype that is exactly intermediate between homozygotes), and because the genetic effects are also additive across genes (i.e., the total genetic effect of any three-gene genotype is just the sum of the separate effects of each gene). To simplify matters, consider just two of the genes, and let the genes and their alleles be denoted (*A*, *a*) and (*B*, *b*). With additivity within and across genes, we may assume that the genotype *aa bb* has a color score of 0 (white) and that each *A* or *B* allele in the genotype contributes one unit of red pigment. The data below show the nine possible two-gene genotypes, their frequencies with random mating when the allele frequencies of *A* and *B* are both 1/2, and the color score of each genotype assuming additivity. (The column headed "Phenotype with dominance" is discussed in Problem 6.) The mean color score of the population is 2. Indeed, when the allele frequencies of *A* and *B* are both *p*, then the mean of a population with random mating can be shown to equal 4*p*. Suppose that the two lowest phenotypic classes (i.e., 0 and 1) are selected as parents of the next generation. Calculate μ_S, μ', *S*, *R*, and $h^2 = R/S$. (Hint: Calculate the allele frequency of *A* and *B* among selected parents, then use the *mean* = 4*p* formula to calculate the mean of the offspring with random mating.)

Genotype	Frequency	Phenotype with additivity	Phenotype with dominance
AA BB	1/16	4	2
AA Bb	2/16	3	2
AA bb	1/16	2	1
Aa BB	2/16	3	2
Aa Bb	4/16	2	2
Aa bb	2/16	1	1
aa BB	1/16	2	1
aa Bb	2/16	1	1
aa bb	1/16	0	0

ANSWER $\mu = 4(1/2) = 2$ is given. Selected parents consist of genotypes *Aa bb*, *aa Bb*, and *aa bb* with respective frequencies 2/5, 2/5, and 1/5. Mean of parents = $\mu_S = (2/5)(1) + (2/5)(1) + (1/5)(0) = 4/5$. Allele frequency of *A* among parents = $(1/2)(2/5) = 1/5$, which also equals the allele frequency of *B*. Mean among offspring is therefore $\mu' = 4(1/5) = 4/5$. Then $S = (4/5) - 2 = -6/5$ and $R = (4/5) - 2 = -6/5$, so $h^2 = R/S = 1.0$. As you will see in Problem 6, the high heritability is due to the additivity within and across genes and not merely to the fact that environmental effects are negligible.

PROBLEM 6

The column headed "Phenotype with dominance" in Problem 5 refers to a hypothetical situation in which the *A* and *B* alleles are dominant but additivity across genes remains. Thus, genotypes *AA*, *Aa*, *BB*, and *Bb* each add one unit of red pigment to the phenotype. In this case, it can be shown that the mean of a random-mating population with allele frequencies of *A* and *B* both equal to *p* is given by $2p(1 + q)$, where $q = 1 - p$. Suppose that the two lowest phenotypic classes (i.e., 0 and 1) are selected as parents of the next generation. Calculate μ, μ_S, μ', *S*, *R*, and h^2.

ANSWER In the original population, $p = q = 1/2$, so $\mu = 2(1/2)[1 + (1/2)] = 3/2$. Selected parents consist of genotypes *AA bb*, *Aa bb*, *aa BB*, *aa Bb*, and *aa bb* with respective frequencies 1/7, 2/7, 1/7, 2/7, and 1/7. Mean of parents = $\mu_S = (1/7)(1) +$

(2/7)(1) + (1/7)(1) + (2/7)(1) + (1/7)(0) = 6/7. Allele frequency of *A* among parents is p = (1/7) + (1/2)(2/7) = 2/7, which is also the allele frequency of *B*. Mean of offspring is therefore μ' = 2(2/7)[1 + (5/7)] = 48/49. Thus, S = (6/7) − (3/2) = −9/14 and R = (48/49) − (3/2) = −51/98, so h^2 = 51/63 = 0.81.

Although environmental effects on the trait in Problem 6 are still negligible, the heritability has become less than 1.0. This perhaps surprising result occurs because certain genetic effects (such as those resulting from dominance or, in other examples, nonadditivity across genes) are not useful in changing a population by means of the type of individual selection discussed here.

Genetic Basis of Quantitative Traits

The type of reasoning exemplified in Problems 5 and 6 will now be generalized to traits influenced by many genes and also influenced by effects of environment. Figure 7 shows the normal density of the trait in question in a hypothetical randomly mating population. The mean is denoted μ and the variance σ^2. In truncation selection, all individuals with phenotypes above the truncation point T are saved for breeding, and the shaded area B of the density represents the proportion of the population selected. (The total area under any normal density equals 1.) The height of the normal density at the point T is denoted Z, and, as before, the mean phenotype among the selected

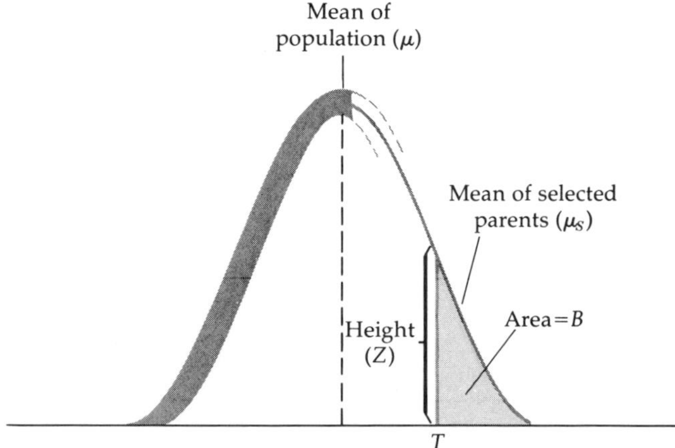

Figure 7. Normal distribution of a quantitative trait in a hypothetical population, showing some important symbols used in quantitative genetics. Here μ is the mean of the population, T the truncation point, Z the height (ordinate) of the normal density at the point T, B (shaded) the area under the normal curve to the right of T, and μ_S the mean among selected parents.

individuals is called μ_S. One of the special properties of the normal density to be used below is that

$$(\mu_S - \mu)/\sigma^2 = Z/B \qquad \textbf{(4.8)}$$

To determine the amount of increase in mean phenotype of a population resulting from one generation of truncation selection, we first imagine a gene that affects the trait in question and that has alleles A and A' with respective allele frequencies p and q. Because of random mating, genotypes AA, AA', and $A'A'$ are present in the population with frequencies p^2, $2pq$, and q^2, respectively, but the individual genotypes cannot be identified through their phenotypic values because of the variation in phenotype caused by environmental factors and genetic differences in other genes. If the genotypes could be identified, their individual distributions of phenotypic value might appear as shown in Figure 8. Each distribution is normal and has the same variance, but the means are very slightly different. The mean phenotypes of AA, AA', and $A'A'$ genotypes are denoted $\mu^* + a$, $\mu^* + d$, and $\mu^* - a$, respectively. The symbols a and d serve as convenient representations of the effects of the alleles in question on the quantitative trait. The difference between means of homozygotes is $(\mu^* + a) - (\mu^* - a) = 2a$, and d/a serves as a measure of dominance. The relationship $d = a$ means that A is dominant, $d = 0$ implies additivity (i.e., heterozygotes exactly intermediate in phenotype between the homozygotes), and $d = -a$ means that A' is dominant. (Use of a and d in

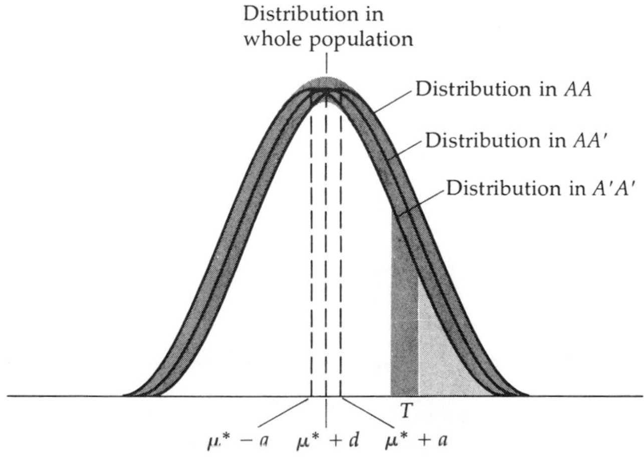

Figure 8. Same distribution as in Figure 7, showing the slightly different distributions of phenotypic value among the three genotypes (AA, AA', and $A'A'$) for a gene with two alleles that contributes to the quantitative trait. The means of the distributions of AA, AA', and $A'A'$ are symbolized $\mu^* + a$, $\mu^* + d$, and $\mu^* - a$, respectively.

Table 2. Calculation of μ^*, a, and d for alleles at a locus affecting coat coloration in guinea pigs.[a,c]

Genotype	Amount of black coloration[b]		
$c^r c^r$ (AA)	1.202	=	$\mu^* + a = 1.075 + .127$
$c^r c^d$ (AA')	1.059	=	$\mu^* + d = 1.075 - .016$
$c^d c^d$ $(A'A')$	0.948	=	$\mu^* - a = 1.075 - .127$

$$\mu^* = (1.202 + .948)/2 = 1.075$$
$$a = 1.202 - 1.075 = .127$$
$$d = 1.059 - 1.075 = -.016$$

[a] The calculations to be carried out first are those beneath the data; then the right-hand column is completed.
[b] Here the amount of black coloration is measured as arcsin $\sqrt{x}$, where x is the percentage of black coloration on the animal. For $c^r c^r$, $c^r c^d$, and $c^d c^d$ genotypes, the corresponding x values are 87 percent, 76 percent, and 66 percent, respectively.
[c] Data from Wright (1968).

this manner simplifies some of the subsequent formulas.) Calculation of a and d for an actual example involving two alleles that affect coat coloration in guinea pigs is illustrated in Table 2. In this case, $\langle a \rangle = 0.127$, $\langle d \rangle = -0.016$ (the negative sign on d means that the c^d allele is partially dominant), and $\langle d/a \rangle = -0.126$. In Figure 8, the mean phenotype in the entire population is

$$\mu = p^2(\mu^* + a) + 2pq(\mu^* + d) + q^2(\mu^* - a) \qquad \textbf{(4.9)}$$

PROBLEM 7 Using the data in Table 2, calculate the average coat color score of a random-mating population of guinea pigs in which the allele frequency of c^r is $p = 0.2$ and that of c^d is $q = 0.8$.

ANSWER Working from Equation 4.9, $\mu = (0.2)^2(1.202) + 2(0.2)(0.8)(1.059) + (0.8)^2(0.948) = 0.99368$.

PROBLEM 8 Values of μ^*, a, and d depend on the scale of measurement of the trait. In Table 2, for example, different results are obtained when phenotypic values are measured in terms of x (the percentage of black coloration) rather than in terms of arcsin $\sqrt{x}$. The appropriate scale of measurement for a trait is a scale in which the distribution of phenotypes is approximately normal, because normality is one of the underlying assumptions of the theory. Very often, data in percentages (x) are not distributed in a normal distribution, but become nearly normal when expressed as arcsin $\sqrt{x}$. [Another "normalizing transformation" useful in many other examples is ln(x).] Wright (1968) provides the following data pertaining to the c^r and c^a (albino) alleles in guinea pigs. Calculate μ^*, a, d, and the degree of dominance (d/a) for the proportions of black (x) and for the transformed proportions arcsin $\sqrt{x}$.

Genotype	Proportion of black coloration (x)	arcsin $\sqrt{x}$
c^rc^r	0.87	1.202
c^rc^a	0.44	0.725
c^ac^a	0	0

ANSWER For the proportions, $\langle\mu^*\rangle = 0.435$, $\langle a\rangle = 0.435$, $\langle d\rangle = 0.005$, $\langle d/a\rangle = 0.011$. For the transformed scale, $\langle\mu^*\rangle = 0.601$, $\langle a\rangle = 0.601$, $\langle d\rangle = 0.124$, $\langle d/a\rangle = 0.206$. The alleles are nearly additive in terms of proportions, but there is substantial dominance of the c^r allele on the transformed scale.

Effects of the scale of measurement are known as **scaling effects**. Since estimates of the additive and dominance values of alleles depend on scaling, so does the heritability.

In the next section we calculate the change in allele frequency which occurs as a result of directional selection.

Change in Gene Frequency

Suppose for the moment that we were practicing artificial selection for increased amount of black coat coloration in the guinea pigs in Table 2. Selection for black coat coloration in a population containing both the c^r (i.e., A) and c^d (i.e., A') alleles would be successful in increasing the allele frequency of A, and the average amount of black coloration among individuals of the next generation would increase. Therefore, in order to calculate the expected increase in black coloration in one generation of selection, we must first calculate the corresponding change in the allele frequency of A. An equation for change in allele frequency with natural selection was derived in Chapter 2 (Equation 2.26), which remains valid for artificial selection if we agree to interpret the "fitness" of an individual as the probability that the individual is included among the group selected as parents of the next generation. With this interpretation of fitness, differences in fitness (i.e., reproductive success) of AA, AA', and $A'A'$ genotypes correspond to the differences in area to the right of the truncation point in Figure 8, because only those individuals in the shaded area are allowed to reproduce. The differences in area are easy to calculate if you shift or slide each curve horizontally until its mean coincides with μ^*. The $A'A'$ curve must slide a units to the right, and the AA' and AA curves must slide d and a units to the left, respectively. This shifting brings the distributions into coincidence, but it slides the truncation points slightly out of register, as shown in Figure 9. The difference in "fitness" between AA and AA', denoted $w_{11} - w_{12}$ (as in Chapter 2), is equal to the small area indicated in Figure 9, as is the difference in fitness between AA' and $A'A'$, denoted $w_{12} - w_{22}$. The areas corresponding to $w_{11} - w_{12}$ and $w_{12} - w_{22}$ are approximately rectangles, and the area of a rectangle is the product of the base and the height. Therefore, since Z represents the height of the normal distribution at the point T, we can make the following approximations:

$$\begin{aligned} w_{11} - w_{12} &= Z[(T - d) - (T - a)] = Z(a - d) \\ w_{12} - w_{22} &= Z[(T + a) - (T - d)] = Z(a + d) \end{aligned} \quad \textbf{(4.10)}$$

The average fitness $\overline{w}$ of the entire population simply equals B, because B is the proportion of the population saved for breeding. From Equation 2.26 in Chapter 2 we know that $\Delta p = pq[p(w_{11} - w_{12}) + q(w_{12} - w_{22})]/\overline{w}$, where Δp is the change in frequency of the allele A in one generation of selection. Substituting from Equation 4.10 and using $\overline{w} = B$ leads to

$$\Delta p = pq[pZ(a - d) + qZ(a + d)]/B$$

or, since $p + q = 1$,

$$\Delta p = (Z/B)pq[a + (q - p)d] \quad \textbf{(4.11)}$$

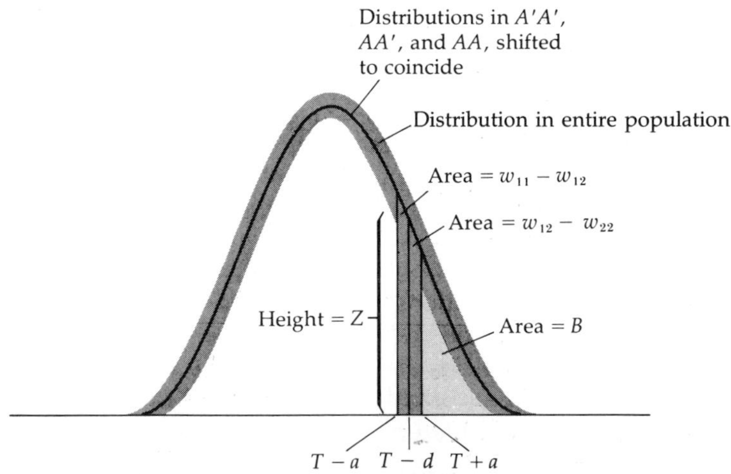

Figure 9. Same distribution as in Figures 7 and 8, but with the distributions of AA, AA', and $A'A'$ shifted laterally to coincide. Shifting the distributions slides the truncation points slightly out of register, so the truncation points for AA, AA', and $A'A'$ become $T - a$, $T - d$, and $T + a$, respectively. The small area that is denoted $w_{11} - w_{12}$ is the difference between the proportions of AA and AA' genotypes that are included among the selected parents, and the area $w_{12} - w_{22}$ is the difference in the proportion of AA' and $A'A'$ genotypes included among the selected parents.

An equation corresponding to Equation 4.11 could be obtained for any gene affecting the trait, but the values of p, a, and d would differ for each gene. The quantity in square brackets in Equation 4.11 is called the **average excess**. A generalization which includes nonrandom mating is found in Falconer (1985).

PROBLEM 9

Consider the hypothetical guinea pig population in Problem 7, in which $a = 0.127$, $d = -0.016$, $p = 0.2$, and $q = 0.8$. Suppose, for purposes of illustration, that the only genetic effects on coat color in this population result from the c^r and c^d alleles. Suppose further that there are such environmental effects on coat color that the overall distribution of coat color in the population is normal with variance $\sigma^2 = 0.0169$. Suppose finally that truncation selection is practiced in such a way that the darkest 40 percent of the animals are used for breeding. Thus, $B = 0.40$, and for this value of B, the quantity Z/B turns out to be 7.46. (Calculation of Z/B for a given value of B is discussed in a few pages.) Use Equation 4.11 to calculate the allele frequencies of c^r (p') and c^d (q') in the next generation. Then use these allele frequencies to calculate the mean phenotype in the population.

ANSWER $\Delta p = (7.46)(0.2)(0.8)[0.127 + (0.8 - 0.2)(-0.016)] = 0.1401286$, or about 0.14. Thus, $p' = p + \Delta p = 0.2 + 0.14 = 0.34$, and $q' = 1 - p' = 0.66$. Mean phenotype in the progeny generation is thus $\mu' = (0.34)^2(1.202) + 2(0.34)(0.66)(1.059) + (0.66)^2(0.948) = 1.0272$.

Change in Mean Phenotype

Equation 4.11 provides an expression for Δp which can be used in the manner illustrated in Problem 9 in order to calculate the mean phenotypic value of coat color after one generation of selection. In the next generation, the allele frequencies of A and A' are $p + \Delta p$ and $q - \Delta p$, respectively. With random mating, the mean phenotype in this generation is given by Equation 4.9 as

$$\begin{aligned}\mu' = &(p + \Delta p)^2(\mu^* + a)\\ &+ 2(p + \Delta p)(q - \Delta p)(\mu^* + d)\\ &+ (q - \Delta p)^2(\mu^* - a).\end{aligned}$$

When the right-hand side of this expression is multiplied out and terms in $(\Delta p)^2$ are ignored because Δp is usually very small, then μ' is found to approximate

$$\mu' = \mu + 2[a + (q - p)d]\Delta p \qquad \textbf{(4.12)}$$

The approximation in Equation 4.12 is rather good even for the relatively large value of Δp in the guinea pig example in Problem 9. In that example, the exact value of μ' was calculated as 1.0272, whereas the approximation in 4.12 yields $\mu' = 1.0266$.

Equation 4.12 warrants a little more development since it yields the prediction equation $R = h^2S$ (Equation 4.7) and also provides an expression for h^2 in terms of the parameters a, d, and p that can be interpreted genetically. First, rewrite Equation 4.12 as

$$\mu' - \mu = 2[a + (q - p)d]\Delta p$$

Then substitute for Δp from Equation 4.11, which yields

$$\mu' - \mu = (Z/B)2pq[a + (q - p)d]^2$$

Now use the expression for Z/B given in Equation 4.8 to obtain

$$\mu' - \mu = (\mu_S - \mu)2pq[a + (q - p)d]^2/\sigma^2$$

Finally, substitute from Equations 4.5 and 4.6 for the selection differential S and the response R, which yields

$$R = (S)2pq[a + (q - p)d]^2/\sigma^2$$

However, $R = h^2S$ also (Equation 4.7), so

$$h^2 = 2pq[a + (q - p)d]^2/\sigma^2 \qquad \textbf{(4.13)}$$

Equation 4.13 for h^2 is the one we were after, as it defines the heritability in terms of p, q, a, and d—each of which has a genetic meaning.

PROBLEM 10 For the guinea pig population in Problem 9, calculate the heritability using Equation 4.13. Then calculate S from Equations 4.5 and 4.8, and R from Equation 4.7. Use the calculated values of R and S along with the value of μ in Problem 7 to obtain μ_S and μ'.

ANSWER Equation 4.13 implies that $h^2 = 2(0.2)(0.8)[0.127 + (0.8 - 0.2)(-0.016)]^2/(0.0169) = 0.26098$. From Equations 4.5 and 4.8, $S = (0.0169)(7.46) = 0.12607$. Applying Equation 4.7, $R = (0.26098)(0.12607) = 0.03290$. From Problem 7, $\mu = 0.99368$, so $\mu_S = 0.99368 + 0.12607 = 1.120$ and $\mu' = 0.99368 + 0.03290 = 1.0266$.

Equation 4.13 is valid when a single gene affects the trait in question. However, when many genes affect the trait, the right-hand side of the equation must be replaced by a summation of such terms, one for each gene. That is, for many genes, $R = h^2S$ where

$$h^2 = \frac{\Sigma 2pq[a + (q - p)d]^2}{\sigma^2} \qquad \textbf{(4.14)}$$

in which the summation is over all genes that affect the trait. (However, each gene may have different values of a, d, p, and q.) As will be discussed in further detail later, the quantity

$$\sigma_a^2 = \Sigma 2pq[a + (q - p)d]^2 \tag{4.15}$$

is called the **additive genetic variance** of the trait. Although the individual components in the additive genetic variance are difficult to identify except in contrived examples like the one involving guinea pigs, the collective effects (represented by the summation) can be estimated.

Number of Genes Affecting Quantitative Traits

The number of genes which contribute to quantitative traits is not always large. Problem 5 gave an example of seed color in wheat in which the number of genes was three. When the number of genes is relatively small, the number can often be estimated from the means and variances observed in different strains and their hybrids and backcrosses. Figure 10 gives an example with two alleles of each of three genes, which, as in the wheat example, are additive in their effects. The solid dots represent alleles that contribute one unit to phenotype and the open circles represent alleles that contribute nothing. The parental strains are homozygous for all unfavorable or all favorable alleles, and they differ by six units in phenotypic value, and the variance in phenotypic value in the F_2 generation equals 3/2 units^2. For two additive alleles of each of n unlinked genes, the difference D in phenotypic value between parental inbred lines is $2n$ units, and the variance σ_g^2 in the F_2 generation is $n/2$ units^2.

In order to obtain an estimate of the number of genes which is independent of the units of measurement, a ratio is needed to make the units cancel. One possibility is

$$\langle n \rangle = D^2/8\sigma_g^2 \tag{4.16}$$

With n unlinked, additive genes, $\langle n \rangle = (2n)^2/[8(n/2)] = n$, as it should. Equation 4.16 is based on the assumptions of complete additivity, equal effects of all genes, and no linkage. When the assumptions are violated, application of the equation usually results in estimates of gene number which are smaller than the actual numbers. For this reason, the quantity estimated in Equation 4.16 is called the **effective number of genes** because it defines a lower limit to the actual number.

The variance in Equation 4.16 is called the *genetic variance*, which is the variance in phenotypic value which results from genetic differences among individuals. When the environment contributes an amount σ_e^2 to the phen-

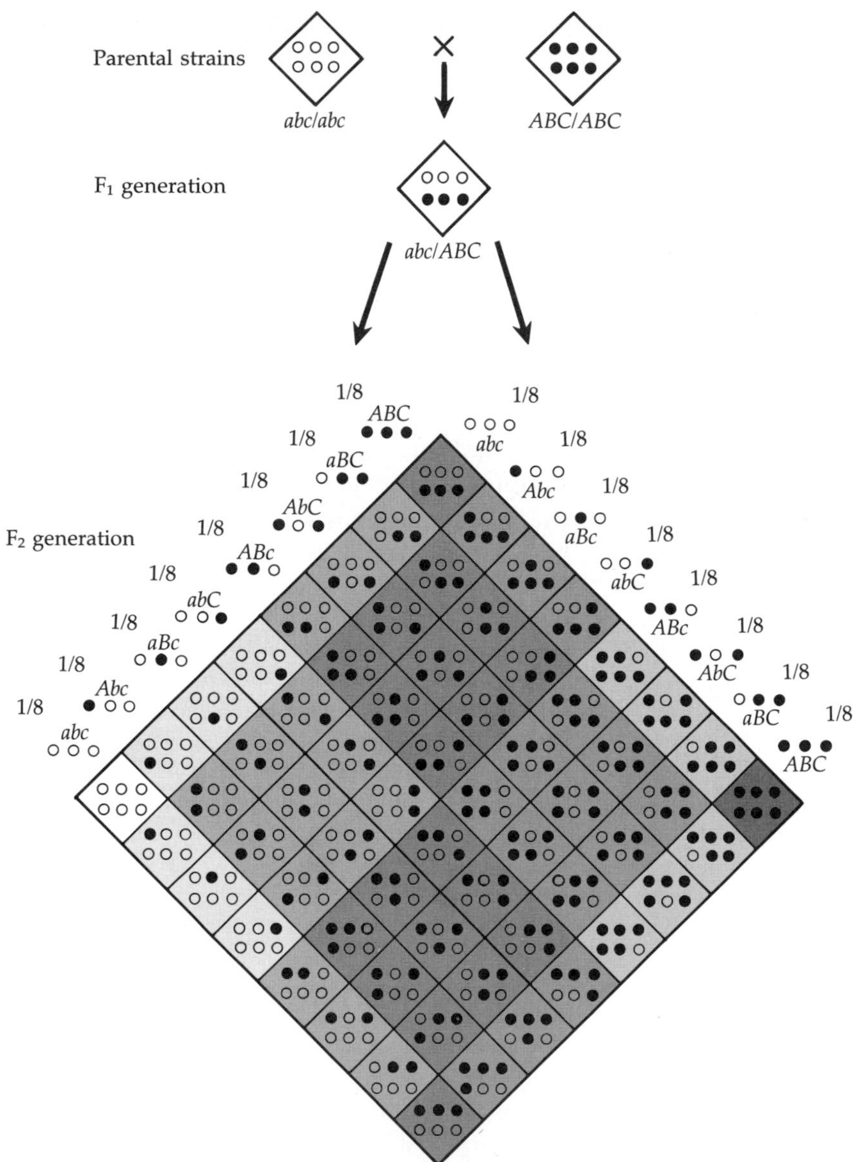

otypic variance, then the variance within the parental inbred lines, or within the F_1 population, equals σ_e^2. This is because the populations are genetically uniform, and the only source of variation in phenotype results from the environment. However, within the F_2 generation, variation in phenotype results in part from genetic variation and in part from environmental variation, and the total variance in phenotypic value equals the summation σ_g^2 +

Figure 10. Segregation of three unlinked, additive genes affecting a quantitative trait. Each upper-case allele in a genotype contributes one unit to the phenotypic value. In the F_2 generation, the mean phenotypic value equals 3.0 and the variance is 1.5.

σ_e^2. Therefore, subtraction of the F_1 variance from the F_2 variance gives an estimate of σ_g^2 because

$$\sigma_g^2 = [\sigma_g^2 + \sigma_e^2] - \sigma_e^2 \qquad \textbf{(4.17)}$$

Further discussion of the genetic variance occurs later in this chapter, and Lande (1981) gives several alternative methods of estimating the genetic variance using data from inbreds, hybrids and backcrosses.

The scale in which phenotype is measured is important in using Equation 4.16 because the genes and alleles are assumed to be additive. In the ideal additive case, a plot of the means and variances of inbreds, hybrids and backcrosses forms a triangle of the type shown in Figure 11(a). In actual cases, the phenotypes must be measured using a scale which yields an approximate triangle. With fruit in tomatoes, for example, an obvious scale of weight is in grams, but the scale giving the triangle in Figure 11(b) is x = log[(weight in grams) − 0.153] (Lande 1981). Using this scale, D = 1.552 and σ_g^2 = 0.0426, from which Equation 4.16 implies $\langle n \rangle = (1.552)^2/[8(0.0426)] = 7$ as the effective number of genes.

PROBLEM 11

In analyzing data on oil content in maize, Lande (1981) found that values of log[(percent oil in kernels) + 1.87] approximated a triangular form like those in Figure 11. Using this scale of measurement, the means of two inbred lines were 0.513 and 1.122, respectively, and the phenotypic variances of the inbred lines, the F_1 generation, and the F_2 generation were 0.00142, 0.00053, 0.00030, and 0.00303, respectively. Use Equations 4.16 and 4.17 to estimate the effective number of genes. Estimate the environmental variance in two different ways: (1) as the mean of the variance in phenotypic value in the parental inbred lines; and (2) as equal to the variance in phenotypic value in the F_1 generation.

ANSWER Using the first estimate, $\langle \sigma_e^2 \rangle = (0.00142 + 0.00053)/2 = 0.00098$, $\langle \sigma_g^2 \rangle = 0.00303 - 0.00098 = 0.00205$, $D = 0.513 - 1.122 = -0.609$, and $\langle n \rangle = (-0.609)^2/[8(0.00205)] = 22.6$. Using the second estimate, $\langle \sigma_g^2 \rangle = 0.00303 - 0.00030 = 0.00273$, and $\langle n \rangle = 17.0$.

The effective number of genes for oil content in maize kernels calculated in Problem 11 bracket those obtained by Lande (1981) using other estimates of

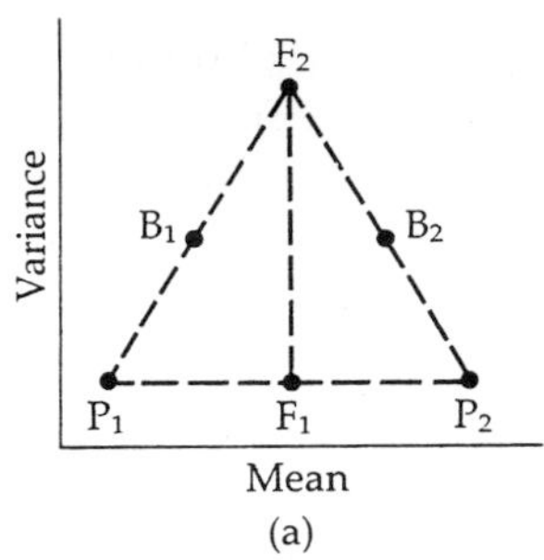

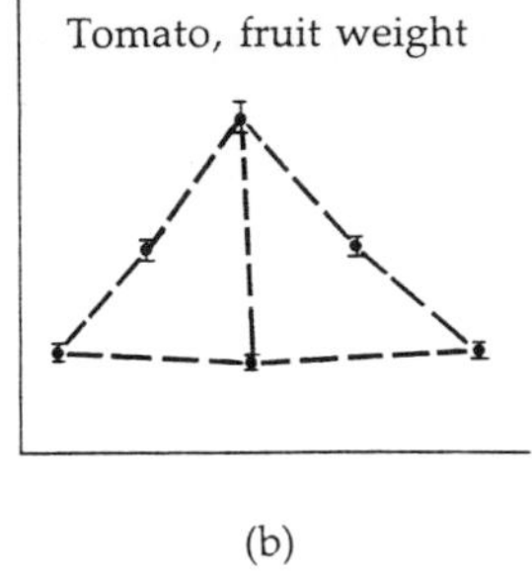

Figure 11. (a) Expected triangular relation between means and variances of phenotypic value among inbred parents (P), backcross progeny (B) and hybrids (F) for an ideal quantitative trait determined by unlinked and completely additive genes. (b) Observed relation for fruit weight in tomato, in which the index of fruit weight is ***phenotypic value*** **= log[(*****weight in grams*****) − 0.153]. (After Lande 1981.)**

the environmental variance. A summary of the estimated number of genes affecting quantitative traits is given in Table 3. With the notable exception of oil content in maize kernels, the effective number of genes is less than ten. These examples demonstrate that genetic variation in quantitative traits can result from the segregation of a relatively small number of genes. However, the effective number of genes represents the minimum number, and the actual number of genes is likely to be larger. In some cases the estimated number of genes is very large. For example, at least 40 genes contribute antigens that are important in the rejection of skin transplants in mice, and at least 150 genes contribute to body weight in the mouse and *Tribolium*. It should be emphasized that the number of genes which affect quantitative traits may differ according to the precise definition of *affect*. For example, the

Table 3. Minimum number of genes affecting quantitative traits.

Organism	Trait	Minimum number of genes
Goldenrod	Date of stamen maturation	6–7
Tomato	Fruit weight	7–11
Maize	Percent oil in kernels	17–22
Hawaiian *Drosophila*	Head size	6–9
Fish	Eye diameter	5–7
Laboratory rat	Coat coloration	5–8
Human	Skin color	4–6

(Data from Lande 1981.)

overall long-term selection response illustrated in Figure 4 may result from a large number of genes, but most of the response that occurs at any one time may result from changes in allele frequency in only a few of them.

Intensity of Selection

The prediction equation $R = h^2S$ is often written in another, completely equivalent way as

$$R = i\sigma h^2 \tag{4.18}$$

where, as before, σ is the phenotypic standard deviation of the trait under selection. The quantity i is called the **intensity of selection**, and it proves to be a useful way to compare different selection programs. Comparing Equations 4.7 and 4.18, we see that $i\sigma = S$. However, Equations 4.5 and 4.8 say that $S = \sigma^2 Z/B$. Consequently,

$$i = S/\sigma = \sigma Z/B \tag{4.19}$$

is the definition of i in terms of the symbols in Figure 7. In biological terms, i equals the selection differential S expressed in units of the standard deviation σ in phenotypic value. Judging from Equation 4.19, i would seem to depend on σ, Z, and B. However, i depends only on B—the proportion of the population saved for breeding—which is the feature that makes i useful in comparing otherwise very different breeding programs. For the hypothetical guinea pig population in Problem 9, for example, $Z/B = 7.46$ and $\sigma^2 = 0.0169$. Thus, $\sigma = (0.0169)^{1/2} = 0.13$ and $i = (0.13)(7.46) = 0.97$.

For any value of B, the corresponding value of i can be calculated from tables pertaining to the normal distribution that are published in many mathematical or statistical handbooks. A graphical representation of the relationship between i and B is shown as the gray curve in Figure 12. The other curves (dashed and dotted lines) are approximations for various parts of the actual curve (Simmonds 1977). The appropriate range of values for each approximation is given in the legend of Figure 12, and, for most purposes, the approximations are sufficiently accurate. For any specified value of B, the coresponding value of i can therefore be obtained from Figure 12, and Z/B can be calculated as $Z/B = i/\sigma$.

APPLICATIONS TO NATURAL POPULATIONS

At this point, it is reasonable to wonder what relevance the principles of quantitative genetics may have for natural populations. Up to now, focus has been mainly on the principles as they apply in plant or animal breeding.

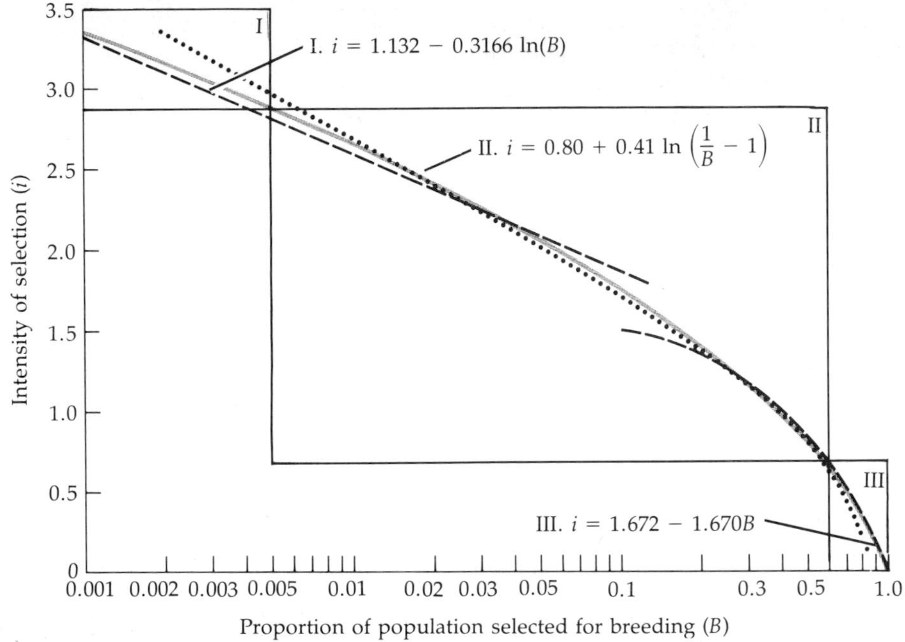

Figure 12. The gray curve is the relationship between the intensity of selection (i) and the proportion of the population saved for breeding (B). Note that the scale of B is logarithmic. The dashed and dotted lines are approximations to the actual curve devised by Simmonds (1977). As elsewhere in this book, the symbol ln means logarithm to the base e. The formulas can be converted into logarithms to the base 10 (log) by means of the relationship $\ln(x) = 2.30258[\log(x)]$. Approximation I is suitable for the range $B = 0.001$ to 0.005 or $i = 3.4$ to 2.9; approximation II for the range $B = 0.005$ to 0.6 or $i = 2.9$ to 0.65; and approximation III for the range $B = 0.6$ to 1.0 or $i = 0.65$ to 0. These approximations are sufficiently accurate for most purposes.

However, their relevance to natural populations becomes immediately apparent when one realizes that *fitness is a quantitative trait*.

Fundamental Theorem of Natural Selection

Although the mode of natural selection is not usually truncation selection as outlined in Figure 5, there are sufficient similarities between natural selection and truncation selection that it is useful to think about natural selection in the framework of truncation selection (Kimura and Crow 1978). This approach to natural selection is illustrated in Problems 12–14. As might be guessed from the earlier discussion of heritability, the additive genetic variance in fitness is an important quantity in evolution. Accordingly, let σ_a^2 in Equation 4.15 represent the additive genetic variance *in fitness*, and let σ^2

represent the total phenotypic variance in fitness. An important theoretical relation, due to Fisher (1930), is that the increase in average fitness in one generation of natural selection in a population with nonoverlapping generations equals the additive genetic variance in fitness divided by the average fitness. This relationship is called the **fundamental theorem of natural selection**. Because the increase in fitness in one generation of natural selection is just the response *R*, Fisher's theorem can be written as

$$R_w = \sigma_a^2 / \overline{w} \tag{4.20}$$

where R_w is the increase in mean fitness ($\Delta\overline{w}$) and $\overline{w}$ is the average fitness of the parental generation, which corresponds to μ in Figure 5.

Since σ_a^2 is a variance, it must always be positive (or possibly zero), and $\overline{w}$ must also be positive (indeed, since the fitnesses are relative, they can always be scaled to make $\overline{w}$ equal one). Therefore, the implication of Equation 4.20 is that the average fitness of a population must always increase. While this is true in the simplest kinds of selection models, it is not true in general (for some simple examples, see Karlin and McGregor 1974; Hartl 1975). Perhaps the most accurate summary of the fundamental theorem is that of Wright (1931, quoted in Provine 1986, p. 272): "The rate of increase in fitness of any population at any time is equal to its genetic variance in fitness at that time [with $\overline{w}$ scaled to 1], except as affected by mutation, migration, change of environment and the effects of random sampling."

PROBLEM 12

Use Fisher's theorem (Equation 4.20) to show that, for fitness, the selection differential *S* equals $\sigma^2/\overline{w}$ and the intensity of selection *i* equals $\sigma/\overline{w}$, where σ^2 is the total phenotypic variance in fitness.

ANSWER Substitute from Equation 4.7 into Equation 4.20 to obtain $h^2S = \sigma_a^2/\overline{w}$. Equations 4.13 and 4.15 imply that $h^2 = \sigma_a^2/\sigma^2$. Therefore, $S = \sigma^2/\overline{w}$. Substituting this result into Equation 4.19 yields $i = \sigma/\overline{w}$.

Two comments about the relations in Problem 12 are in order. First, the formulations make the assumption that fitness is normally distributed. This is a doubtful assumption, and in many cases it might be more appropriate if the scale of measurement were the logarithm of fitness rather than fitness itself. Consequently, the relations in Problem 12 should best be considered as rough approximations. The second comment pertains to the finding $i = \sigma/\overline{w}$. Crow (1966) calls the quantity i^2, which equals $\sigma^2/\overline{w}^2$, the **index of opportunity for selection** and explains its significance as follows

> [It] tells how much potential genetic selection is inherent in the pattern of births and deaths. . . . Of course these differences may not be her-

itable—hence I speak of opportunity for selection. If [fitness were] completely heritable, the index tells its rate of increase. Otherwise, and this is the situation in practice, it provides only an upper limit.

PROBLEM 13 Crow and Morton (1955) have compiled data on a number of fitness-related traits in a variety of organisms. Below are the values of $\sigma^2/\overline{w}^2$, adjusted to what the values would be in a population of constant size. Estimate the intensity of selection (i) in each case. Then, using approximation II in Figure 12 [i.e, $i = 0.8 + 0.41 \ln(1/B - 1)$] estimate the corresponding value of B in each case. The value of B is the proportion of the population that would need to be selected in order to produce the same intensity of selection for an ordinary quantitative trait as that observed for the fitness-related traits. Because the distribution of fitness is very likely nonnormal, the approximation is a crude one, but interesting for comparative purposes. Bear in mind that the intensity of selection is only one aspect of the situation. Despite the considerable selection intensities for these fitness-related traits, such traits tend to have rather low heritabilities, so the response to natural selection is nevertheless small in any one generation.

	$\langle\sigma^2/\overline{w}^2\rangle$ (adjusted)	$\langle i\rangle$	$\langle B\rangle$ (approx. II)
Drosophila females (lifetime egg production)	0.68		
Drosophila females (adult progeny)	0.91		
Drosophila males	1.59		
Lymnaea (a snail)	0.84		
U.S. women (born in 1866)	0.82		
British women	0.94		

ANSWER $\langle i\rangle = \sigma/\overline{w} = [\sigma^2/\overline{w}^2]^{1/2}$ = 0.82, 0.95, 1.26, 0.92, 0.91, and 0.97, respectively (from top to bottom). Then, from approximation II, $\langle B\rangle = \{\exp[(i - 0.8)/(0.41)] + 1\}^{-1}$ and (from top to bottom) B = 0.49, 0.41, 0.25, 0.43, 0.43, and 0.40, respectively.

Effects of Gene Flow

Natural selection for quantitative traits is often impeded by migration into the selected population. A classic example is found in the evolution of heavy metal tolerance. **Heavy metal tolerance** refers to the resistance which certain plants, particularly grasses, have evolved toward high soil concentrations of such metals as copper, zinc, or lead in the 50 to 100 years since the abandonment of mine sites contaminated by ore tailings and seepage (Jain and Bradshaw 1966; Antonovics et al. 1971). Heavy metal tolerance is a quantitative trait with a high heritability, but evolution of tolerance is hindered by genetic migration due to windblown pollen from areas upwind of the mine.

Despite the high heritability and intensity of selection, the effects of selection are counterbalanced by gene flow.

PROBLEM 14

McNeilly (1968) has studied copper tolerance in the grass *Agrostis tenuis* at an abandoned mine site in Wales so situated that the wind always blows from the same direction. The data below show the mean index of copper tolerance among seedlings (μ) and among surviving adult plants (μ_S) at six positions along a transect running perpendicular to the wind. The soil concentration of copper decreases from 2700 ppm at site 1 to 156 ppm (nontoxic) at site 6, and the phenotypic variance in copper tolerance is about 243.4. Estimate S, i, and B at each site, using the approximations in Figure 12 as appropriate.

	$\langle\mu\rangle$	$\langle\mu_S\rangle$	$\langle S\rangle$	$\langle i\rangle$	$\langle B\rangle$
1. Center of mine	37	60			
2. On mine	39	58			
3. Near mine edge	34	54			
4. On mine edge	29	45			
5. Just off mine	20	27			
6. Off mine (18 m from edge)	15	16			

ANSWER Use $S = \mu_S - \mu$; for sites 1-6, $\langle S\rangle$ = 23, 19, 20, 16, 7, and 1, respectively. Then $i = S/\sigma$ with $\langle\sigma\rangle = (243.4)^{1/2} = 15.6$ given; for sites 1–6, $\langle i\rangle$ = 1.47, 1.22, 1.28, 1.03, 0.45, and 0.06, respectively. For sites 1–4, it is better to use approximation II ($B = \{\exp[(i - 0.8)/(0.41)] + 1\}^{-1}$); answers are $\langle B\rangle$ = 0.16, 0.26, 0.24, and 0.36, respectively. For sites 5–6, use approximation III [$B = (1.672 - i)/1.670$]; answers are $\langle B\rangle$ = 0.73 and 0.97, respectively.

Stabilizing Selection

When selection eliminates phenotypically extreme individuals and preserves those near the population mean, the type of selection is **stabilizing selection** (Figure 15b in Chapter 2). Stabilizing selection is generally thought to be the most common form of natural selection at the phenotypic level (Wright 1977). An example of stabilizing selection in humans affects birth weight: babies that are too small or too large have reduced survival relative to those of intermediate birth weight.

The change in frequency of a gene which affects a quantitative trait subject to stabilizing selection can be approximated using a geometric method similar to that illustrated in Figures 8 and 9 for directional selection. The situation with stabilizing selection is shown in Figure 13, in which the solid distribution represents the entire population and the dashed curve the distribution among individuals of a genotype in which the mean deviates from that of the entire population by an amount a. With stabilizing truncation selection there are

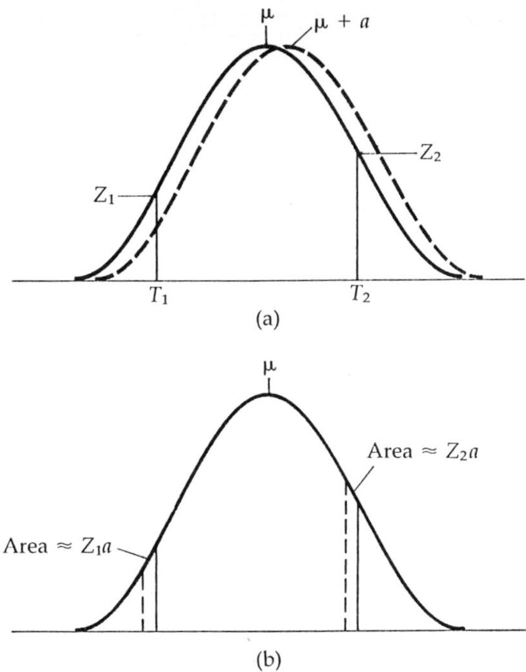

Figure 13. Stabilizing selection modeled as double truncation selection. (a) Distributions of phenotypic value of a genotype with mean $\mu + a$ (dashed distribution) and that of the entire population, mean μ. The truncation points are denoted T_1 and T_2. (b) Shifting the dashed distribution brings the means into coincidence but shifts the truncation points out of register. Approximations for change in gene frequency can be derived from the geometry of the distributions, as in Figures 8 and 9.

two truncation points, denoted T_1 and T_2. As indicated in Figure 13(a), the genotype with the dashed distribution loses more individuals on the right-hand side, and fewer on the left-hand side, relative to the entire population. The difference in fitness can be approximated as shown in Figure 13(b), which results from sliding the dashed distribution in (a) to the left by a units. The area approximated by Z_1a is the difference on the left favoring the dashed distribution, and the area approximated by Z_2a is the area on the right discriminating against the dashed distribution. The relative fitness of the genotype with the dashed distribution, relative to the entire population, is proportional to the difference in the areas, which equals $(Z_1 - Z_2)a$.

With three genotypes, AA, AA' and $A'A'$ with means $\mu^* + a$, $\mu^* + d$, and $\mu^* - a$, respectively, the differences in fitness with stabilizing selection are given approximately by

$$w_{11} - w_{12} = (Z_1 - Z_2)(a - d)$$
$$w_{12} - w_{22} = (Z_1 - Z_2)(a + d)$$

which is analogous to Equation 4.10 for directional selection. The analogue of Equation 4.11 is therefore

$$\Delta p = \left(\frac{Z_1 - Z_2}{B}\right) pq[a + (q - p)d] \quad \textbf{(4.21)}$$

When the distribution of phenotypes is normal, it can be demonstrated that $(Z_1 - Z_2)/B = (\mu_S - \mu)/\sigma^2$, which is analogous to Equation 4.8. In the context of stabilizing selection, the mean μ_S of the individuals which survive and reproduce is often denoted μ_o representing the *optimum* phenotype. Substituting into Equation 4.21 yields

$$\Delta p = \left(\frac{\mu_o - \mu}{\sigma}\right) pq \left(\frac{a + (q - p)d}{\sigma}\right) \quad \textbf{(4.22)}$$

In words, the change in gene frequency is determined by the product of the selection differential in units of the phenotypic standard deviation [$i = (\mu_o - \mu)/\sigma$] and the average excess in units of the phenotypic standard deviation [$a + (q - p)d]/\sigma$. This is a very general result for weak selection of all kinds (Milkman 1978).

Equation 4.22 demonstrates that the intensity of selection on genes affecting a quantitative trait is proportional to the deviation between the population mean μ and the optimum phenotype μ_o. A very important result emerges when $\mu_o = \mu$, namely that $\Delta p = 0$, which implies that there is no selection. More accurate approximations for Δp also indicate very little selection when $\mu_o = \mu$ (Kimura 1981). Therefore, *when the mean phenotype in the population equals the optimum phenotype under stabilizing selection, then alleles with small effects on the trait are selectively nearly neutral.* Kimura (1981) has emphasized this finding in regard to the theoretical possibility of extensive neutral evolution of genes which affect traits subject to stabilizing phenotypic selection.

In the stabilizing selection in Figure 13, individuals whose phenotypes are between T_1 and T_2 have a fitness of one, and those outside of this range have a fitness of zero. A more usual way to model stabilizing selection is to suppose that the fitness of individuals decreases in proportion to the square of the deviation of their phenotype from the optimum. For examples of this approach, see Lande (1976, 1977, 1979), Felsenstein (1977), Kimura (1981), and Turelli (1985).

Genetic Architecture and Antagonistic Pleiotropy

Stabilizing selection is of great interest for the role it may play in the persistence of heritable genetic variation in multifactorial traits. Intuitively, it seems reasonable to suppose that observed levels of additive genetic variation may result from a balance between stabilizing selection, which tends to reduce genetic variation, and new mutations, which tend to increase it. Deceptively simple when stated verbally, the models become very complex when reduced to mathematics, and include such complications as the number of genes, the type of action of alleles and their interactions, linkage between genes, the type and intensity of selection, and the influence of selection on other traits which are related through pleiotropy. **Pleiotropy** refers to the widespread tendency of genes to affect several traits simultaneously, sometimes seemingly unrelated traits, which usually results from the fact that complex phenotypic traits are determined by the interactions of the products of many genes during development. Complex as the situation is, the hypothesis that additive genetic variation might be maintained by stabilizing selection balanced against mutation has been supported by mathematical approximations in the models studied by Lande (1975, 1977, 1979, 1980).

On the other hand, the approximations appear to be quite sensitive to the particular genetic and mathematical assumptions that are made. The simplest models ignore dominance and epistasis, and they assume a large number of alleles of each gene, the distribution of effects of which are approximated by a normal or *gaussian* distribution. Assuming a gaussian distribution of genetic effects is a very different thing from assuming a gaussian distribution of *phenotypes*, as the latter is open to empirical validation.

Turelli (1984) has argued that the gaussian distribution of genetic effects is to be expected only when the mutation rates of the relevant genes are unrealistically high, such as greater than 10^{-4} per gene per generation. Alternative approximations are found to break down when selection also occurs indirectly, through traits related by pleiotropy (Turelli 1985). Pleiotropy can even obscure the relation between the intensity of selection and the magnitude of the additive genetic variance. When pleiotropy occurs, an increase in the intensity of selection may decrease the additive genetic variance, or it may increase it, depending on the particular assumptions that are invoked—for example, whether stabilizing selection occurs on each of the pleiotropically related traits, or whether directional selection in antagonistic directions favors alternative sets of alleles.

The conclusion seems to follow that the simple mathematical models require so many artificial assumptions that predictions based on them cannot inspire great confidence (Turelli 1985). This difficulty is inherent in evaluating whether observed values of additive genetic variance and heritability can be explained by mutation–selection balance. All the more does it complicate the wider goal of attempting evolutionary explanations of the relative amounts of the variance in phenotype which are contributed by additive gene effects, dominance, interactions between genes, and environmental components. The variance structure of traits implied by the relative magnitude of the variance components is often called the **genetic architecture**.

The genetic architecture also includes genetic and environmental correlations among different traits, which result from pleiotropy and common environmental effects. Genetic variation can be maintained by selection acting alone, when the favorable effects of alleles in respect to some traits are offset by unfavorable pleiotropic effects on different traits. Negative pleiotropic effects which result in offsetting selection pressures are known as **antagonistic pleiotropy**. An important implication of antagonistic pleiotropy is that the action of natural selection cannot maximize every trait simultaneously, because favorable changes in some traits bring about unfavorable changes in others, and adaptive responses in some directions are nonadaptive in others. In animal and plant breeding, secondary responses to selection are called *correlated responses*, and they are discussed later in this chapter. When natural selection acts on traits which are negatively correlated, the best that can usually be achieved is an optimum mix of phenotypes, a sort of best choice of bad bargains. For example, in *Drosophila*, there is antagonistic pleiotropy in females between the mean rate of egg laying and longevity (Rose and Charlesworth 1981a), and selection for increased female longevity decreases the mean rate of egg laying (Rose and Charlesworth 1981b). Theoretical models of antagonistic pleiotropy have been studied by Rose (1982, 1985).

LIMITATIONS OF HERITABILITY

Equation 4.14 for heritability h^2 is important in showing that heritability says virtually nothing about the actual mode of inheritance of a quantitative trait, useful as the concept may be in predicting response to selection. The heritability of a trait represents the cumulative effect of all genes which affect the trait. Even for a single gene, heritability depends in a complex manner on the values of p, a, and d, and these individual components cannot be disentangled. (The values of p, a, and d are said to be *statistically confounded*.) With more than one gene, the heritability includes a summation of terms for each gene, and each term has its own particular values of p, q, a, and d. Here, precisely, is the problem: for a quantitative trait determined by, say, 10 genes, there would be 30 quantities involved in heritability—10 allele frequencies, 10 values of a, and 10 values of d. Heritability is but a single number that gives the combined effect of all 30 quantities. It says nothing about any one of them. With more genes affecting the trait or with multiple alleles, matters would be even worse. Also, we have assumed that all genes affecting the trait act independently of one another and are unlinked. In actual cases, genes often interact and can be linked. Moreover, we have assumed that environmental effects are the same for all genotypes, but this assumption can easily be violated, even in some well-designed breeding programs. All in all, while heritability, especially realized heritability, is an indispensable aid to plant and animal breeders, it lends itself to no easy interpretation in simple genetic terms. Exceptionally thoughtful discussions of the concept of heritability and its strengths and limitations are found in Kempthorne (1978) and Jacquard (1983). Problem 15 makes the connection between heritability and simple Mendelian inheritance.

PROBLEM 15 Show that the heritability of a trait determined by a single autosomal-recessive allele with frequency q equals $2q/(1 + q)$. Show that the heritability of a trait determined by a single autosomal-dominant allele with frequency q equals $2(1 - q)/(2 - q)$. Assume random mating and apply Equation 4.13.

ANSWER For convenience, let the phenotypes of AA, AA', and $A'A'$ be assigned phenotypic values 0, 0, and 1, respectively, so that the A' allele is recessive. Therefore, $\mu^* = 1/2$, $a = -1/2$, and $d = -1/2$. The numerator of Equation 4.13 is the additive genetic variance in phenotypic value, which equals $2pq^3$. Then the mean phenotype equals q^2, and the variance in phenotypic value equals $q^2 - (q^2)^2 = q^2(1 - q^2) = q^2(1 - q)(1 + q) = pq^2(1 + q)$. Therefore, the heritability of the autosomal-recessive trait is $h^2 = 2q/(1 + q)$. When the autosomal-recessive trait is rare, $q \simeq 0$, and the heritability is approximately equal to the frequency of heterozygous carriers. For the second half of the problem, since the A' allele is recessive, the A allele is dominant. In terms of the frequency of the dominant allele, therefore, the heritability is $h^2 = 2(1 - p)/(2 - p)$. Except for the use of p for the frequency of the dominant allele, instead of q, this is exactly what was to be shown.

Problem 15 demonstrates that the heritability even of simple Mendelian traits depends on the allele frequency. Another difficulty in interpreting heritability values is that they depend on the range of environments which occur. The denominator σ^2 in Equation 4.14 is the total variance in phenotypic value in the population. Because the total variance includes the variance resulting from environmental differences among individuals, increasing the variation in the environment decreases h^2.

The environment can also affect the value of heritability because the values of a and d depend on the environment. A hypothetical example is illustrated in Figure 14, which shows the norms of reaction of AA, AA', and $A'A'$ genotypes. The **norm of reaction** of any genotype is the distribution of phenotypes that result when the genotype undergoes growth and development in a range of environments. For example, if we were dealing with the range of environments denoted E_1 in Figure 14, A would be the favored allele and A would be nearly dominant to A'. If we were dealing with the range of environments denoted E_2, on the other hand, A' would be the favored allele and there would be essentially no dominance. Thus, switching a pop-

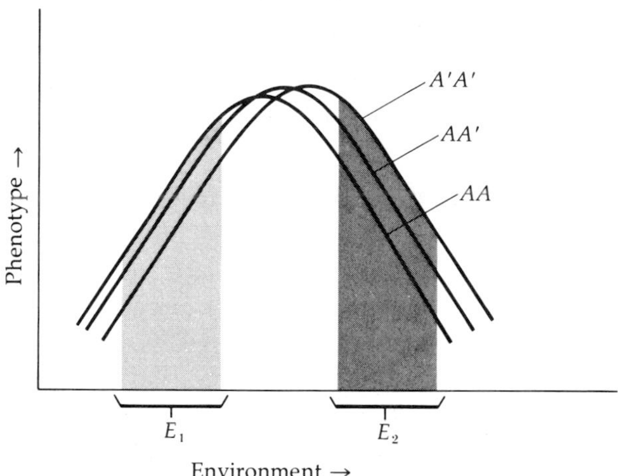

Figure 14. Hypothetical norms of reaction—relationships between phenotype and environment—for genotypes *AA*, *AA′* and *A′A′*. In the range of environments denoted E_1, *A* is very nearly dominant to *A′* (that is, *AA* and *AA′* have nearly the same phenotype). However, in the range E_2, *A* and *A′* are very nearly additive (no dominance). The heritability of the trait resulting from this gene differs according to whether the population is reared in E_1 environments or E_2 environments.

ulation from E_1 to E_2 would change the values of *a* and *d* and substantially alter the heritability of the trait, even though the total phenotypic variance of the population might remain the same.

PROBLEM 16

Dobzhansky and Spassky (1944) studied the norms of reaction with respect to temperature of the viability of various genotypes containing chromosomes extracted from natural populations of *Drosophila pseudoobscura*. For two such chromosomes (designated *A* and *B*), the following relative viabilities were obtained.

Temperature (°C)	Genotype		
	A/A	***A/B***	***B/B***
16.5	0.92	1.00	0.71
25.5	0.32	1.00	0.75

Estimate μ^*, *a*, and *d* for these genotypes in populations maintained at 16.5 and 25.5°C. Then, letting *p* represent the frequency of the *A* chromosome and *q* that of the *B* chromosome, assume $p = 0.3$, and estimate the additive genetic variance of viability resulting from these genotypes at both temperatures.

ANSWER At 16.5°C, $\langle \mu^* \rangle = (0.92 + 0.71)/2 = 0.815$; $\langle a \rangle = 0.92 - 0.815 = 0.105$; $\langle d \rangle = 1.00 - 0.815 = 0.185$. At 25.5°C, $\langle \mu^* \rangle = 0.535$; $\langle a \rangle = -0.215$; $\langle d \rangle = 0.465$. Additive genetic variance due to these genotypes is given by Equation 4.15, where

$p = 0.3$ and $q = 0.7$. At 16.5°C, $\langle\sigma_a^2\rangle = 0.01346$. At 25.5°C, $\langle\sigma_a^2\rangle = 0.00035$. Note that the additive genetic variance has been decreased by a factor of almost 40, yet all we did was raise the temperature!

The norm of reaction is important because the genetic architecture of traits depends on the norm of reaction. In practical breeding, the norm of reaction may affect the rate of response to selection and the performance of selected populations under different environmental conditions. With reference to natural populations, the norm of reaction may itself be under genetic control and subject to adaptive evolution by means of natural selection.

Selection for developmental stability is illustrated in diazonin-resistant populations of the Australian sheep blowfly, *Lucilia cuprina*, in which the insecticide resistance results from a single dominant gene. In the genetic background of the evolved natural population, the resistant genotypes show less sensitivity in their developmental response to environmental fluctuations than do the progeny of crosses in which the genetic background has been replaced with that of an unevolved laboratory population, as is shown by the finding of a smaller average degree of asymmetry between the left and right sides of the body (measured by the difference in bristle number between the two sides) (Clarke and McKenzie 1987; Jones 1987). The implication of this finding is that natural selection can favor genetic modifiers that influence the manner in which other genotypes respond to the environment.

As is evident in Problem 16, the heritability of a trait has reference only to a particular population at a particular time in a particular range of environments. On the other hand, practical breeders are usually interested in improving a particular population at a particular time in a particular range of environments, so the concept of heritability is of great utility when used for its intended purpose—the prediction of response to selection.

Heritability values are determined in part by gene frequencies. Because gene frequencies change during the course of selection, the heritability is also expected to change. In practice, however, the heritability changes sufficiently slowly that over the course of a few generations it can be regarded as a constant.

The approximate constancy of heritability has a twofold cause: (1) if a particular gene accounts for only a small proportion of the total phenotypic variance in a quantitative trait, then the gene frequency does not change very rapidly, and (2) the values of a and d remain nearly constant provided that the environment does not change drastically from one generation to the next. Thus, at least for the first ten generations or so, heritability usually remains approximately constant and can be used as a constant in the prediction equation (Equation 4.7). To be precise, suppose h^2 is constant and let μ_t

and S_t represent the mean of the population and the selection differential in the tth generation. Then, over the length of time during which h^2 is approximately constant,

$$\mu_t - \mu_0 = h^2(S_0 + S_1 + \ldots + S_{t-1}) \qquad \textbf{(4.23)}$$

The quantity $\mu_t - \mu_0$ is the **total response** to selection, and $S_0 + S_1 + \ldots + S_{t-1}$ is called the **cumulative selection differential**. During the time in which h^2 is approximately constant, therefore, a plot of μ_t against cumulative selection differential is expected to yield a straight line with slope equal to h^2, as illustrated for a case in mice in Figure 15.

PROBLEM 17

In the long-term experiment involving selection for high oil content in maize depicted in Figure 4, oil content increased linearly for 76 generations from an initial mean $(\langle\mu_0\rangle)$ of 4.8 percent to a final mean $(\langle\mu_{76}\rangle)$ of 18.8 percent. In this same experiment, the cumulative selection differential increased at an approximately constant rate of 1.1 per generation (Dudley 1977). Estimate the realized heritability for the high-oil line in Figure 4.

ANSWER Cumulative selection differential for 76 generations is (1.1)(76) = 83.6. Total response = $\mu_{76} - \mu_0$ = 14. Applying Equation 4.23, realized $\langle h^2\rangle$ = 14/83.6 = 16.7 percent. [Note: Selection in Figure 4 was for entire ears of kernels having high average oil content. Since kernels on the same ear are related as half-siblings (same mother plant, different pollen donors), the estimated heritability actually refers to the heritability of half-sib family means.]

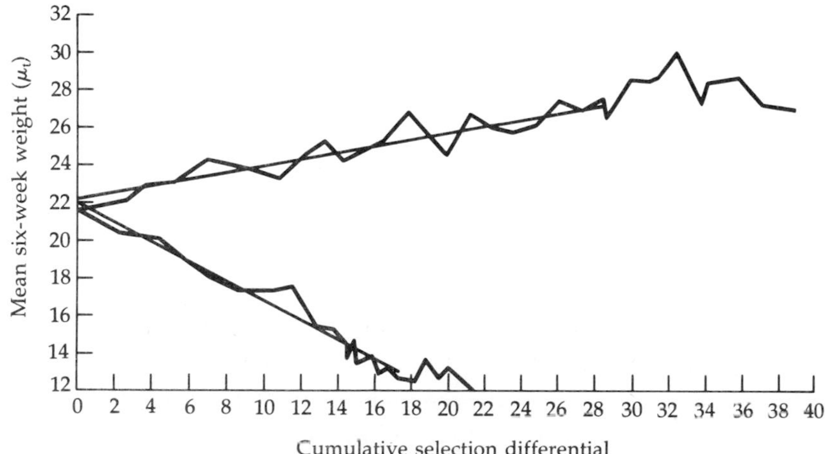

Figure 15. Linearity in response against cumulative selection differential for body weight in mice at age six weeks. Linearity in the *up* (high-weight) direction continues for about twice as long as it does in the *down* (low-weight) direction. (After Falconer 1955.)

If selection is continued for a sufficiently long time, of course, one does expect the value of heritability to change. Indeed, when a gene reaches an equilibrium under selection, the contribution of that gene to the additive genetic variance becomes zero. This principle can be illustrated with reference to overdominance. In Chapter 2 we used the symbols w_{11}, w_{12}, and w_{22} to refer to the relative fitnesses of *AA*, *Aa*, and *aa*, respectively. With w_{12} greater than w_{11} and w_{22}, there is an equilibrium when the allele frequency of *A* is $\hat{p} = (w_{12} - w_{22})/(2w_{12} - w_{11} - w_{22})$ (Equation 2.32). Converting to the symbolism of this chapter,

$$\begin{aligned} \mu^* &= (w_{11} + w_{22})/2 \\ a &= (1/2)(w_{11} - w_{22}) \\ d &= (1/2)(2w_{12} - w_{11} - w_{22}) \end{aligned} \qquad \textbf{(4.24)}$$

The equilibrium allele frequency of *A* is $\hat{p} = (a + d)/2d$, so $\hat{q} = (d - a)/2d$. Therefore, $[a + (q - p)d] = 0$ at equilibrium, and the contribution of this gene to the additive genetic variance (Equation 4.15) is zero also.

PROBLEM 18 In Problem 23 in Chapter 2, the relative fitnesses of the Standard (*ST*) and Arrowhead (*AR*) inversions in *Drosophila pseudoobscura* were estimated as 0.47, 1.0, and 0.62, respectively, for *ST/ST*, *ST/AR*, and *AR/AR*. Let *p* represent the frequency of the *ST* inversion. Estimate the additive genetic variance due to these inversions for $p = 0$, $p = 0.2$, $p = 0.4176$, $p = 0.8$, and $p = 1.0$.

ANSWER Let $q = 1 - p$ represent the frequency of the *AR* inversion. Because $\sigma_a^2 = 2pq[a + (q - p)d]^2$, then $\sigma_a^2 = 0$ for $p = 0$ and $p = 1.0$. For these genotypes, $\langle\mu^*\rangle = (0.47 + 0.62)/2 = 0.545$, $\langle a\rangle = 0.47 - 0.545 = -0.075$, and $\langle d\rangle = 1.0 - 0.545 = 0.455$. The equilibrium frequency of *ST* is $\hat{p} = (a + d)/2d = 0.4176$. For $p = 0.2$, $\langle\sigma_a^2\rangle = 0.01255$; for $p = 0.4176$ (the equilibrium), $\langle\sigma_a^2\rangle = 0$; for $p = 0.8$, $\langle\sigma_a^2\rangle = 0.03875$.

USE OF HERITABILITY IN THE DESIGN OF BREEDING PROGRAMS

A plant or animal breeder must choose among many alternatives in the design of breeding programs, and heritability is a useful aid in making these choices. For example, virtually all breeding programs attempt to improve performance for several traits simultaneously, and the decision as to whether an individual is saved for breeding depends upon its phenotypic value for each of the traits. Faced with the problem of simultaneous selection of, for example, two traits, a breeder has basically three options:

1. Selection could favor animals that are superior with respect to both traits simultaneously, which is called **independent culling**.

2. Selection could alternate between the traits, one trait being selected in one generation and the other trait in the next, which is known as **tandem selection**.
3. Selection could be practiced on both traits simultaneously, with the overall phenotypic value of an individual calculated according to a **selection index**, which is an arithmetic combination of the measurements of all relevant traits in the individual. Selection based on such a combination is known as **index selection**. In designing an optimal selection index, heritability is one of the components used to decide which traits should receive the most emphasis in the overall selection index.

Another use of heritability is in selection of traits which cannot be evaluated in certain individuals—for example, egg production in roosters, milk production in bulls, and litter size in boars. In such cases, the selection criterion for males must be based on the performance of their female relatives, often sisters, in which case the selection procedure is called **sib selection**. The heritability of a trait can be used to determine the number of sisters of each male that should be evaluated.

Heritability is also useful in deciding which mode of selection will best achieve progress. Many modes of selection are possible. Up to now we have discussed primarily **individual selection**, in which individuals are selected for breeding according to their own phenotypic value. Individual selection is most useful for traits with high heritabilities, for example $h^2 = 0.20$ or greater. For traits with lower heritability, other modes of selection are often used, particularly **family selection**, in which whole families are saved or discarded according to the value of the family mean. Family selection is useful for traits with low heritability when environmental effects on the trait are large but independent from individual to individual because, by averaging among members of a family, the environmental effects tend to cancel out. A breeder may in some cases apply **within-family selection**—a selection procedure in which individuals are saved or culled according to the deviation of their own phenotypic value from the family mean. Within-family selection is most useful when environmental effects on the trait are large but common to members of a family (preweaning weight in pigs and mice are examples of traits with a large but familial component of variation, which in these cases is due to the nurturing ability of the mother). In within-family selection, the selection of individuals with large deviations from their family means has the effect of eliminating the nongenetic familial effects because the selection is among individuals whose familial effects are held in common.

As with individual selection, prediction equations for family and within-family selection can be derived (see Falconer 1981 for details). Remarkably, the prediction equations have the same form as Equation 4.18 for individual selection. Specifically, for family selection, the prediction equation is simply

$$R = i\sigma_f h_f^2$$

where σ_f refers to the phenotypic standard deviation among family means and h_f^2 is the heritability of family means. Likewise, the prediction equation for within-family selection is

$$R = i\sigma_w h_w^2$$

where σ_w is the phenotypic standard deviation among within-family deviations and h_w^2 is the heritability of within-family deviations.

In natural populations, the most common mode of selection is undoubtedly individual selection—selection based on an individual's own fitness. If entire groups of progeny from the same mother tend to survive or perish as a unit, as may be the case in many parasitic organisms, then family selection becomes important. If there is intense competition within families, as again may occur in some parasites, then within-family selection becomes important. In the terminology of Chapter 2, family selection and within-family selection are special types of group selection.

Individual selection, family selection, and within-family selection are often used to improve the performance of a single *purebred* population. In commercial application, especially in plants, superior agricultural performance often involves crossbreeding, as in the case of hybrid corn. In such cases, selection is not based upon the phenotypic value of an individual but upon the individual's **combining ability** for the trait in question, measured as the average phenotypic value of the offspring formed by crossbreeding the individual with members of another population. Selection may be for **general combining ability** for the trait of interest, which refers to the average performance of crossbreds with a genetically heterogeneous *tester* population or with a series of genetically uniform testers; or selection may be for **specific combining ability** for the trait of interest, which refers to the performance of crossbreds with a single, usually genetically uniform, tester. Two principal modes of selection for combining ability are in widespread use. One mode is called **recurrent selection**, in which the selection criterion is the performance of the offspring obtained from matings between an individual and members of an unselected (and usually genetically uniform) tester strain; the other mode is called **reciprocal recurrent selection**, in which case two populations are selected simultaneously, and the selection criterion for each population is the performance of the offspring obtained from matings with members of the other population. Recurrent selection and reciprocal recurrent selection are particularly useful for traits in which a major part of the genetic variation is due to nonadditive gene action, such as overdominance.

These modes of selection represent only a sample of the many possibilities open to breeders, and decisions about the mode of selection to be applied in any individual case must be based not only on heritability but also on factors such as population size and economic and management considerations.

RESEMBLANCE BETWEEN RELATIVES

To carry out the artificial selection required for the estimation of realized heritability is in some cases impractical and, in humans, impossible. Alternative methods of estimating heritability are based on the similarity between relatives.

Parent–Offspring Resemblance

There are many degrees of relationship—for example, parent–offspring, sibling, half-sibling, and first cousin—but in genetics one of the most important relationships is that between parents and their offspring. Figure 16 shows a plot of the mean of male offspring for a quantitative trait (y values) against the phenotypic value of the father (x values). The line is the best fitting straight line, called the **regression line**, of offspring on parent. For reasons that will become clear in a moment, we are interested in the slope of this line, which is most easily expressed in terms of the covariance of x and y, as defined in the next section.

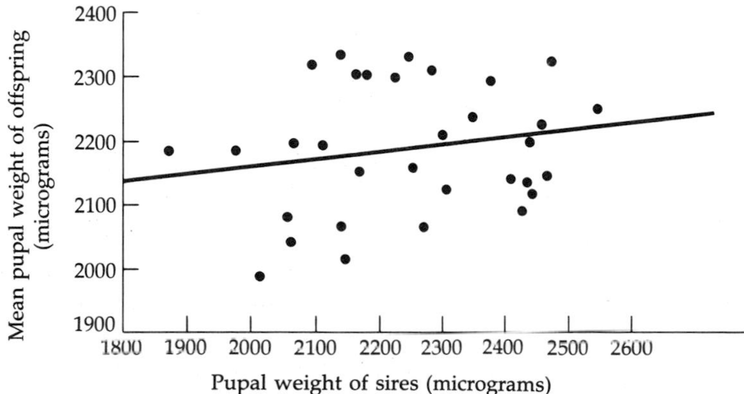

Figure 16. Mean weight of male pupae of the flour beetle *Tribolium castaneum*, against pupal weight of father (sire). Each point is the mean of about eight male offspring. The regression coefficient of male offspring weight on sire's weight is $\langle b \rangle = 0.11$, and h^2 is estimated as $2b$. (Courtesy of F. D. Enfield.)

Covariance and Regression

The covariance between two variables, for example the phenotypic values of parents and their offspring, is a measure of the degree to which the variables are associated. Suppose, to be specific, that the data consist of n pairs of measurements, where n_i is the frequency of pairs (x_i, y_i). The **covariance** of x and y, denoted σ_{xy}, is defined as the average of the product of the deviations of x and y from their respective means. The covariance of x and y is estimated as

$$\langle\sigma_{xy}\rangle = [\Sigma n_i(x_i - \overline{x})(y_i - \overline{y})]/(n - 1) \quad \textbf{(4.25)}$$

$$\langle\sigma_{xy}\rangle = [\Sigma n_i x_i y_i - n\overline{x}\,\overline{y}]/(n - 1) \quad \textbf{(4.26)}$$

The estimate in Equation 4.25 is faithful to the spirit of the definition of covariance in manifestly being related to the mean of the product of the deviations of x and y from their respective means. However, the estimate in Equation 4.26 is usually easier to use in computation. The $n - 1$ which appears in the denominator is a correction for sampling error.

Inspection of Equation 4.26 indicates that, when n is reasonably large, $\langle\sigma_{xy}\rangle$ approximately equals the mean of the product of x and y, minus the product of the means of x and y.

Since the covariance between any two variables measures their degree of association, the covariance may be positive or negative. Positive covariance means that values of x and y tend to increase or decrease together, negative covariance means that, as one variable increases, the other tends to decrease. The limiting values of the covariance are $-\sigma_x \sigma_y$ on the negative side, and $\sigma_x \sigma_y$ on the positive side. The limits are achieved only when the variables demonstrate a perfect linear relation with each other.

Returning now to Figure 16, if σ_{xy} represents the covariance between phenotypic values of fathers (sires) and those of their male offspring, and σ_x^2 represents the variance of phenotypic values of the fathers, then the slope of the regression line is simply σ_{xy}/σ_x^2. This can be seen as follows. Suppose that the equation of the line is represented as

$$y = c + bx$$

where c and b are constants, b being the slope. Taking means of both sides yields

$$\overline{y} = c + b\overline{x}$$

Subtracting the second equation from the first yields

$$y - \overline{y} = (c + bx) - (c + b\overline{x}) = b(x - \overline{x})$$

Now multiply through by $x - \bar{x}$ to obtain

$$(x - \bar{x})(y - \bar{y}) = b(x - \bar{x})^2$$

Taking means of both sides produces

$$\sigma_{xy} = b\sigma_x^2$$

In other words, the slope b of the regression line equals

$$b = \sigma_{xy}/\sigma_x^2 \quad \textbf{(4.27)}$$

This slope is called the **regression coefficient** of offspring on one parent.

A geometrical interpretation of regression is illustrated in Figure 17, which shows the distribution, in two dimensions, of the variables x and y. The variables may represent, for example, the phenotypic values of parents (x) and offspring (y). If there were no association between x and y, the surface would be completely symmetrical and resemble an inverted bowl. The ridge in Figure 17 results from the association. When pairs of (x, y) values are sampled from the two-dimensional distribution, a plot of the points yields the sort of scatter diagram shown in the inset in Figure 17, in which the straight line is the regression line of y on x. The slope of the regression line is given by Equation 4.27.

The procedure for estimating the regression coefficient of offspring on one

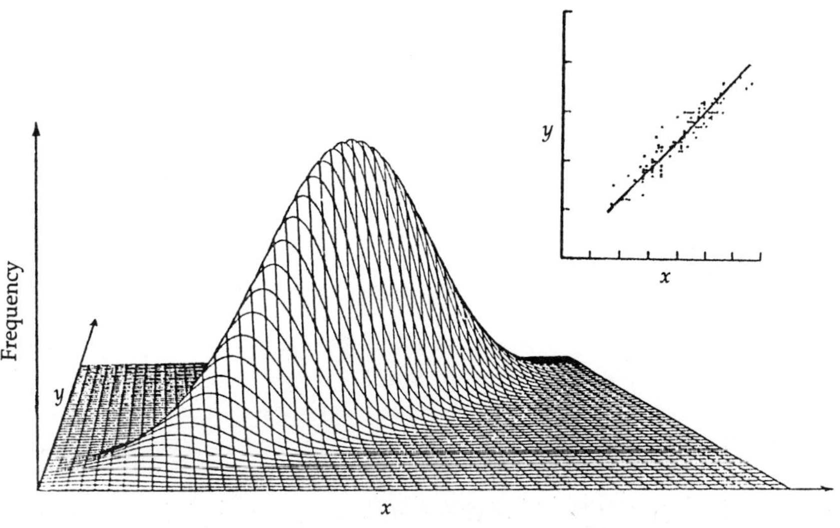

Figure 17. Two-dimensional normal distribution of correlated variables x and y, showing the ridge on the surface which results from the correlation. The inset shows pairs of (x, y) points sampled at random from the distribution and the resulting regression line. (After Sokal and Rohlf 1969).

parent is clearly related to the procedure for estimating realized heritability, because data pertaining to parents and their offspring are involved. The regression coefficient b of offspring on one parent can be calculated for any random-mating population, however, and the heritability (h^2) can be estimated from the relationship

$$b = h^2/2 \tag{4.28}$$

The 1/2 occurs in Equation 4.28 because the regression involves only a single parent (the father, in the case of Figure 16), and only half of the genes from any one parent are passed on to the offspring. In Figure 16, $\langle b \rangle = 0.11$, so $\langle h^2 \rangle = 0.22$. Notice the considerable scatter among the points in the figure, which represents data from 32 families. Because this sort of scatter is typical, heritability estimates tend to be quite imprecise unless based on data from several hundred families. One further point about Figure 16: the regression is better performed on the father's phenotype, rather than on the mother's, in order to avoid potential`bias in the estimate of heritability caused by such maternal effects as nurturing ability.

PROBLEM 19 The regression coefficient of offspring on the mean phenotypic value of the parents (often called the *midparent* value) can also be used to estimate heritability. If b is the regression coefficient of offspring on midparent, then $h^2 = b$. (Note that there is no factor of 1/2 in this case because both parents are involved.) Cook (1965) has studied shell breadth in 119 sibships of the snail *Arianta arbustorum*. For computational convenience, the data have been grouped into six categories. Estimate the heritability of shell breadth from these data.

Number of sibships	Midparent value (mm)	Offspring mean (mm)
22	16.25	17.73
31	18.75	19.15
48	21.25	20.73
11	23.75	22.84
4	26.25	23.75
3	28.75	25.42

ANSWER Let x refer to the midparent value and y refer to the offspring mean. Then, $\bar{x} = 20.2626$, $\bar{y} = 20.1786$, $\Sigma n_i x_i^2 = 49{,}823.4375$, $\Sigma n_i x_i y_i = 49{,}267.1875$, $\langle \sigma_{xy} \rangle = 5.1826$, $\langle \sigma_x^2 \rangle = 8.1801$, and $\langle b \rangle = \langle h^2 \rangle = 0.63$. (There is some loss of accuracy from grouping the data into categories. The regression coefficient for the ungrouped data is $b = 0.70$. It should be noted that there is substantial assortative mating for shell breadth, which makes the heritability estimate artificially large.)

ANALYSIS OF GENETIC VARIATION

The variance of a quantitative trait can be split into various components representing different causes of variation. Similarity between relatives is conveniently expressed in terms of the variance components, but variance partitioning is also of interest in its own right. Since the rate of change of a trait under selection depends on the amount of genetic variation affecting the trait, if there is no genetic variation, there is obviously no response to selection. What is not so obvious is that some components of genetic variation cannot be acted upon by some kinds of selection. In other words, certain populations have ample genetic variation, yet fail to respond to selection. The part of the genetic variation amenable to selection is clarified by partitioning the variance.

Table 4 shows the kind of reasoning involved in partitioning the variance. The phenotypic value of any individual is represented as a sum of three components: (1) the mean μ of the entire population, (2) a deviation from the population mean due to the specific genotype of the individual in question (symbolized as G_1, G_2, and G_3 for AA, AA', and $A'A'$ genotypes, re-

Table 4. Phenotypes of various genotypes as the sum of μ, G, and E.[a]

Genotype	Phenotypic value
AA	$\mu + G_1 + E_1$
AA	$\mu + G_1 + E_2$
AA	$\mu + G_1 + E_3$
AA'	$\mu + G_2 + E_4$
AA'	$\mu + G_2 + E_5$
AA'	$\mu + G_2 + E_6$
$A'A'$	$\mu + G_3 + E_7$
$A'A'$	$\mu + G_3 + E_8$
$A'A'$	$\mu + G_3 + E_9$

[a] μ is the population mean. G is a contribution due to genotype, different for each genotype. E is a contribution due to environment, different for each individual.

spectively), and (3) a deviation from the population mean due to the specific environment of the individual in question (the environmental deviations are unique to each individual and are represented as $E_1, E_2, \ldots, E_9$). It is important to note that the *G*s and *E*s are not directly observable. Nevertheless, as we shall see, the total variance in phenotypic value can be partitioned into a component due to variation among the *G*s and another component due to variation among the *E*s. The model in Table 4 can be summarized by writing

$$P = \mu + G + E \tag{4.29}$$

where *P* represents the phenotypic value of any individual and *G* and *E* are the genotypic and environmental deviations pertaining to that individual.

To connect the above symbols with actual numbers, we may use the earlier Problem 9 involving coat color in guinea pigs based on Table 2, in which we assumed an allele frequency for *A* of $p = 0.2$ in a hypothetical population. The mean of the population was calculated in Problem 7 as $\mu = 0.99368$. Thus, the respective G_1, G_2, and G_3 deviations for *AA*, *AA'*, and *A'A'* genotypes are

$$\begin{aligned} G_1 &= 1.202 - 0.99368 = 0.20832 \\ G_2 &= 1.059 - 0.99368 = 0.06532 \\ G_3 &= 0.948 - 0.99368 = -0.04568 \end{aligned}$$

For a particular animal of genotype *AA* whose actual coat color score is, for example, 1.312, the corresponding value of *E* for the animal would be calculated using Equation 4.29 from the expression $1.312 = 0.99368 + 0.20832 + E$; thus, for this animal, $E = 0.11$. Similarly, a particular animal of genotype *AA'* with an actual phenotype of $P = 1.009$ would have a value of *E* given by $1.009 = 0.99368 + 0.06532 + E$, or $E = -0.05$. Because the *E* values are defined as deviations from their mean, the average of *E*s for any genotype is 0. Likewise, since the *G*s are defined as deviations from their mean, the mean of the *G*s is 0, which can be verified in the guinea pig example because

$$(0.2)^2G_1 + 2(0.2)(0.8)G_2 + (0.8)^2G_3 = 0$$

Equation 4.29 is appropriate when the effects of genotype and environment are additive—that is, when the deviation of the phenotype of any particular individual from the population mean ($P - \mu$) can be written as the sum of an effect resulting from the genotype of that individual and a separate effect resulting from the environment of that individual. The additivity is true whenever the ratio of G_1:G_2:G_3 is the same in each of the relevant environments. For the genotypes in Figure 14, for example, if the actual range of environments is the range designated E_1, then the genetic and environmental effects are additive because the ratio G_1:G_2:G_3 is the same for any particular environment in E_1. For the same reason, the genetic and environmental effects are additive if the actual range of environments is E_2. However, if the actual range of environments includes both E_1 and E_2, then the ratio G_1:G_2:G_3 depends on the particular environment, and therefore the genetic and environmental effects are not additive. Nonadditivity of genetic and environmental effects is called **genotype–environment interaction**, and in writing Equation 4.29, we are assuming that there is no genotype–environment interaction.

When Equation 4.29 is valid, the total phenotypic variance σ_p^2 in the population equals the mean of $(P - \mu)^2$. However, Equation 4.29 says that $(P - \mu)^2$ equals $(\mu + G + E - \mu)^2$, which is

$$\sigma_p^2 = (G + E)^2 = G^2 + 2GE + E^2 \qquad \textbf{(4.30)}$$

Because G and E are already deviations from their means, the mean of G^2 is the phenotypic variance in the population resulting from differences in genotype, and the mean of E^2 is the phenotypic variance resulting from differences in environment. The mean of G^2 is called the **genotypic variance** and denoted σ_g^2. The mean of E^2 is called the **environmental variance** and denoted σ_e^2. The remaining term—the mean of $2GE$—is two times the **genotype–environment covariance**. If the genotypic and environmental deviations are independent—that is, if there is no systematic association between genotype and environment—then there is said to be no **genotype–environment association** and the mean of $2GE$ equals zero. When there is no genotype–environment association, therefore,

$$\sigma_p^2 = \sigma_g^2 + \sigma_e^2 \qquad \textbf{(4.31)}$$

Equation 4.31 is the theoretical foundation for Equation 4.17 used earlier in this chapter. The assumption that genotype–environment association is negligible is frequently a valid assumption in animal and plant breeding where, because breeders have a degree of control not available to, for example, human geneticists, experiments can be intentionally designed in such a way as to minimize genotype–environment association. However, genotype–environment association can occur even in animal and plant breeding. For example, dairy farmers routinely provide more feed supplements to cows that produce more milk; because milk-producing ability is partly due to genotype, this feed regimen will provide superior environments (i.e., better feed) to cows that have superior genotypes to begin with, so there will be a genotype–environment association.

The biological meaning of Equation 4.31 is shown for the alleles of one gene in Figure 18. The solid curves represent the phenotypic distributions in the genotypes AA, AA', and $A'A'$ with means denoted G_1, G_2, and G_3, respectively, and the dashed curve represents the phenotypic distribution in

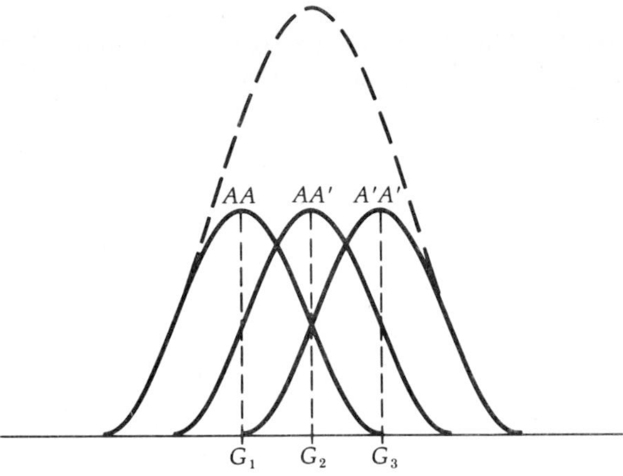

Figure 18. Phenotypic distribution (dashed curve) of a quantitative trait in a hypothetical population, showing distributions (solid curves) of three constituent genotypes for two alleles of a gene. The means of *AA*, *AA′* and *A′A′* genotypes are denoted G_1, G_2 and G_3, respectively.

the entire population. The total phenotypic variance σ_p^2 is the variance of the dashed distribution; the genotypic variance σ_g^2 is the variance among the *G*s (i.e., $\sigma_g^2 = p^2G_1^2 + 2pqG_2^2 + q^2G_3^2$, where p is the allele frequency of *A*); and the environmental variance σ_e^2 is obtained by subtraction: $\sigma_e^2 = \sigma_p^2 - \sigma_g^2$. Although the *G*s are not generally known, σ_g^2 must equal zero in a genetically uniform population. The observed variance of a randomly bred population, therefore, provides an estimate of $\sigma_g^2 + \sigma_e^2$, whereas the observed variance of a genetically uniform population provides an estimate of σ_e^2. The estimate of σ_g^2 is obtained by subtraction as noted previously in Equation 4.17. An example using thorax length in *Drosophila* is shown in Table 5. In this case,

Table 5. Calculation of genotypic variance (σ_g^2) and environmental variance (σ_e^2).[a]

	POPULATIONS	
Variance	Random-bred	Uniform
Theoretical	$\sigma_g^2 + \sigma_e^2$	σ_e^2
Observed	0.366	0.186

$\sigma_e^2 = 0.186$
$\sigma_g^2 = (\sigma_g^2 + \sigma_e^2) - \sigma_e^2 = 0.366 - 0.186 = 0.180$

[a] Trait is length of thorax in *Drosophila melanogaster* (in units of 10^{-2} mm). Data from Robertson (1957).

genetic variation among individuals in the randomly bred population accounts for about 0.180/0.366 = 49.2 percent of the phenotypic variance. Genetically uniform populations such as inbred lines or crosses between inbreds are not available in human populations, of course, but identical twins are often used instead because of the identical genotypes of the twins.

Components of Genotypic Variation

So far, the phenotypic variance has been partitioned into the genotypic variance and the environmental variance according to Equation 31. The genotypic variance can be partitioned further into terms that are particularly important for interpreting the resemblance between relatives. The appropriate model is shown in Table 6, where the means of AA, AA', and $A'A'$ genotypes are denoted in column 3 by $\mu^* + a$, $\mu^* + d$, and $\mu^* - a$, respectively, as they were earlier in Figure 8. To obtain the G values, the mean of each genotype must be expressed as a deviation from the population mean, which is $\mu = \mu^* + (p - q)a + 2pqd$, and the deviations are shown in column 4 of Table 6. The genotypic variance σ_g^2 works out to be

$$\begin{aligned}\sigma_g^2 &= p^2G_1^2 + 2pqG_2^2 + q^2G_3^2 \\ &= 2pq[a + (q - p)d]^2 + (2pqd)^2\end{aligned} \quad \textbf{(4.32)}$$

The first term in Equation 4.32 is the **additive genetic variance** σ_a^2 encountered earlier in Equation 4.15. The second term is a new quantity called the **dominance variance**, which is symbolized σ_d^2. From Equation 4.32, therefore,

$$\sigma_g^2 = \sigma_a^2 + \sigma_d^2 \quad \textbf{(4.33)}$$

which allows us to express the total phenotypic variance as the sum of three terms, namely

$$\sigma_p^2 = \sigma_a^2 + \sigma_d^2 + \sigma_e^2 \quad \textbf{(4.34)}$$

Table 6. Expressions for population mean and genotypic deviations.

Genotype	Frequency	Mean phenotype	Genotypic deviation from population mean (G)
AA	p^2	$\mu^* + a$	$G_1 = \mu^* + a - \mu = 2q[a + (q - p)d] - 2q^2d$
AA'	$2pq$	$\mu^* + d$	$G_2 = \mu^* + d - \mu = (q - p)[a + (q - p)d] + 2pqd$
$A'A'$	q^2	$\mu^* - a$	$G_3 = \mu^* - a - \mu = -2p[a + (q - p)d] - 2p^2d$

Population mean (μ) $= p^2(\mu^* + a) + 2pq(\mu^* + d) + q^2(\mu^* - a)$
$= (p^2 + 2pq + q^2)\mu^* + (p^2 - q^2)a + 2pqd$
$= (p + q)^2\mu^* + (p - q)(p + q)a + 2pqd$
$= \mu^* + (p - q)a + 2pqd$

PROBLEM 20 For the hypothetical guinea pig population in Problem 9, $a = 0.127$, $d = -0.016$, $p = 0.2$, $q = 0.8$, and $\sigma_p^2 = 0.0169$. Earlier we had calculated $G_1 = 0.20832$, $G_2 = 0.06532$, and $G_3 = -0.04568$. Show that $\sigma_g^2 = 0.00444$. Then calculate σ_a^2 and σ_d^2 and verify Equation 4.32. Finally, calculate σ_e^2 for the population.

ANSWER $\sigma_g^2 = (0.2)^2(0.20832)^2 + 2(0.2)(0.8)(0.06532)^2 + (0.8)^2(-0.04568)^2 = 0.00444$. Then, $\sigma_a^2 = 2(0.2)(0.8)[0.127 + (0.8 - 0.2)(-0.016)]^2 = 0.00441$; $\sigma_d^2 = [2(0.2)(0.8)(-0.016)]^2 = 0.00003$. Thus, $\sigma_g^2 = \sigma_a^2 + \sigma_d^2$. Finally, $\sigma_e^2 = \sigma_p^2 - \sigma_g^2 = 0.01246$.

When Equation 4.13 for heritability is written in terms of variance components rather than p, q, a and d, the equation implies that

$$h^2 = \sigma_a^2/\sigma_p^2 \qquad \textbf{(4.35)}$$

Equation 4.35 is an important result because it states that the heritability depends only on the additive genetic variance and not on the dominance variance. Therefore, if all the genetic variance in a population results from dominance variance (i.e., $\sigma_a^2 = 0$), then the population cannot respond to individual selection because h^2 equals zero. To say the same thing in another way, the dominance variance σ_d^2 represents that portion of the genetic variance that is not acted upon by individual selection.

PROBLEM 21 For the Standard and Arrowhead inversions in Problem 18, $\sigma_a^2 = 0.01255$ for $p = 0.2$ and $\sigma_a^2 = 0$ for $p = 0.4176$. Calculate the dominance variance and the total genotypic variance for these allele frequencies.

ANSWER $\langle d \rangle = 0.455$ was estimated in Problem 18. Thus, for $p = 0.2$, $\langle \sigma_d^2 \rangle = [2(0.2)(0.8)(0.455)]^2 = 0.02120$ and $\langle \sigma_g^2 \rangle = 0.01255 + 0.02120 = 0.03375$. For $p = 0.4176$, $\langle \sigma_d^2 \rangle = 0.04898$ and $\langle \sigma_g^2 \rangle = 0.04898$ also. The total genotypic variance is larger at $p = 0.4176$ than at $p = 0.2$, but the allele frequencies do not change with selection because all of the genotypic variance is in the form of dominance variance.

Equation 4.35 means that the heritability of a trait is the ratio of the additive genetic variance to the total phenotypic variance. Sometimes the word *heritability* is used in reference to a different variance ratio, namely the ratio of the total genotypic variance to the total phenotypic variance (i.e., σ_a^2/σ_p^2). To avoid confusion, quantitative geneticists distinguish the two types of heritability as follows:

1. The ratio σ_a^2/σ_p^2 is called **heritability in the narrow sense**. (This is the variance ratio we have been using all along.)
2. The ratio σ_g^2/σ_p^2 is called **heritability in the broad sense**.

Generally speaking, narrow-sense heritability is the more important with individual selection (or any mode of selection that capitalizes primarily on the additive genetic variance), whereas broad-sense heritability is the more important when selection is practiced among clones (a *clone* is a group of genetically identical individuals), inbred lines, or varieties. As previously in this book, we use the term *heritability* to mean narrow-sense heritability unless otherwise stated.

PROBLEM 22

Use the model in Problem 15 to show that the broad-sense heritability of a simple Mendelian trait equals 1.0.

ANSWER Problem 15 demonstrated that the additive genetic variance of a simple Mendelian trait is $\sigma_a^2 = 2pq^3$ and that the total variance in phenotypic value is $\sigma_p^2 = pq^2(1 + q)$. From Equation 4.32, the dominance variance is $\sigma_d^2 = p^2q^2$. Therefore, $\sigma_g^2 = \sigma_a^2 + \sigma_d^2 = pq^2(1 + q)$, and the broad-sense heritability σ_g^2/σ_p^2 equals 1.0.

As emphasized earlier, heritability has no transparent interpretation in simple genetic terms. The same is true of the variance components σ_a^2 and σ_d^2. Even for a single gene, the variance components depend on allele frequency and on the particular values of a and d (Figure 19). With many genes which act together, σ_a^2 represents a summation of the values of $2pq[a + (q - p)d]^2$ for each gene affecting the trait, and σ_d^2 represents a summation of the values of $(2pqd)^2$ for each gene. Indeed, when the trait is affected by multiple genes, the formula for σ_g^2 in Equation 4.33 must be extended to include an additional term that pertains to interaction among the genes. This interaction term is called the **interaction variance** or the **epistatic variance** and is symbolized σ_i^2. With the interaction variance included, Equation 4.33 becomes

$$\sigma_g^2 = \sigma_a^2 + \sigma_d^2 + \sigma_i^2 \qquad \textbf{(4.36)}$$

The important point to remember about the components of genotypic variance is that they represent the cumulative, statistical effect of all genes affecting the trait. Few inferences about the actual mode of inheritance of the trait are possible from the variance components, particularly concerning the number of genes involved and their individual effects.

Covariance Between Relatives

Components of genetic variation are important because they may be used to express the phenotypic covariance between relatives. The algebra involved in deriving the covariances is lengthy and tedious, but the most important results are summarized in Table 7. As can be seen, the additive genetic variance can be estimated directly either from parent–offspring covariance or from half-sib covariance. However, full-sib covariance includes a term re-

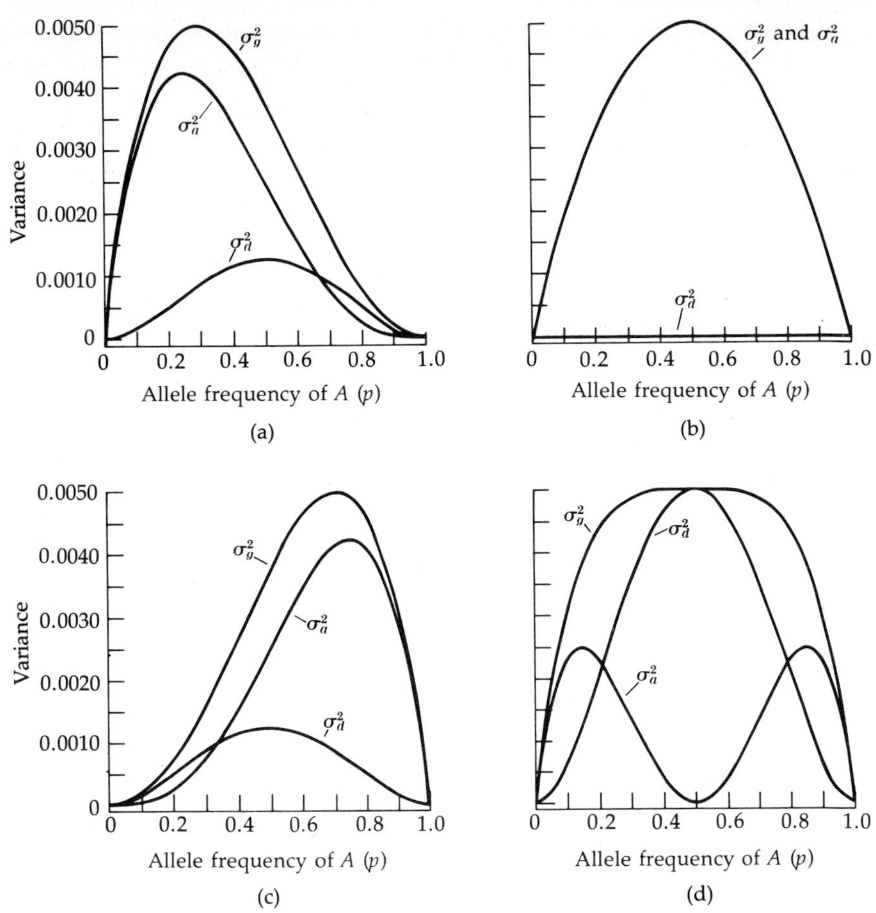

Table 7. Theoretical covariance in phenotype between relatives.[a]

Degree of relationship	Covariance
Offspring and one parent	$\sigma_a^2/2$
Offspring and average of parents (midparent)	$\sigma_a^2/2$
Half siblings	$\sigma_a^2/4$
Full siblings	$(\sigma_a^2/2) + (\sigma_d^2/4)$
Monozygotic twins	$\sigma_a^2 + \sigma_d^2$
Nephew and uncle	$\sigma_a^2/4$
First cousins[b]	$\sigma_a^2/8$
Double first cousins	$(\sigma_a^2/4) + (\sigma_d^2/16)$

[a] Variance terms due to interaction between loci (epistasis) have been ignored.
[b] First cousins are the offspring of matings between siblings and unrelated individuals; double first cousins are the offspring of matings between siblings from two different families.

◄ **Figure 19. Total genotypic variance σ_g^2, additive genetic variance σ_a^2, and dominance variance σ_d^2 for a gene with two alleles A and A', against allele frequency p of A. The mean phenotypes found in AA, AA', and $A'A'$ are denoted $\mu^* + a$, $\mu^* + d$, and $\mu^* - a$, respectively. In all cases $\sigma_a^2 = 2pq[a + (q - p)d]^2$, $\sigma_d^2 = (2pqd)^2$, and $\sigma_g^2 = \sigma_a^2 + \sigma_d^2$. (a) $a = d = 0.07071$ (A dominant to A'); (b) $a = 0.1$, $d = 0$ (no dominance); (c) $-a = d = 0.07071$ (A' dominant to A); (d) $a = 0$, $d = 0.14142$ (overdominance). For ease of comparison, the values of a and d have been chosen to make the maximum of σ_g^2 equal to 0.005 in each case.**

sulting from dominance. The expressions in Table 7 are correct as long as there are no complications such as genotype–environment associations or other nonrandom environmental effects such as full sibs sharing environmental factors common to the whole family but not shared by other families. Since the total variance in phenotypic value σ_p^2 can be estimated directly, once σ_a^2 is estimated from the covariance between relatives, the narrow-sense heritability can be estimated from Equation 4.35. The first three relationships in Table 7 are the most useful ones in quantitative genetics and are commonly used in animal and plant breeding. The other relationships are used mainly in human quantitative genetics.

Figure 20 presents the narrow-sense heritabilities of diverse quantitative traits in farm animals and one important crop plant as estimated from the correlation between relatives. The data are presented merely to show the values of heritability with which breeders typically must deal. It is important to keep in mind that the heritabilities in Figure 20 pertain to one population in one type of environment at one particular time. The same trait in a different population or in a different environment might well have a different heritability. Generally speaking, traits that are closely related to fitness (such as calving interval in cattle or eggs per hen in poultry) tend to have rather low heritabilities. Ignoring complications such as antagonistic pleiotropy, long-term natural selection is expected to gradually reduce the additive genetic variance until the effect is balanced against the input of new mutations.

For purposes of comparison, Figure 21 shows estimated broad-sense heritabilities of a number of quantitative traits in Caucasians. Broad-sense heritabilities vary widely for different traits, as they do in other species. Note the low heritability of fertility, a trait that is obviously closely related to fitness.

LONG-TERM ARTIFICIAL SELECTION

The long-term result of repeated artificial selection of a quantitative trait is determined by many factors, including (1) the number of segregating genes that affect the trait, (2) the allele frequencies in the base population, (3) the

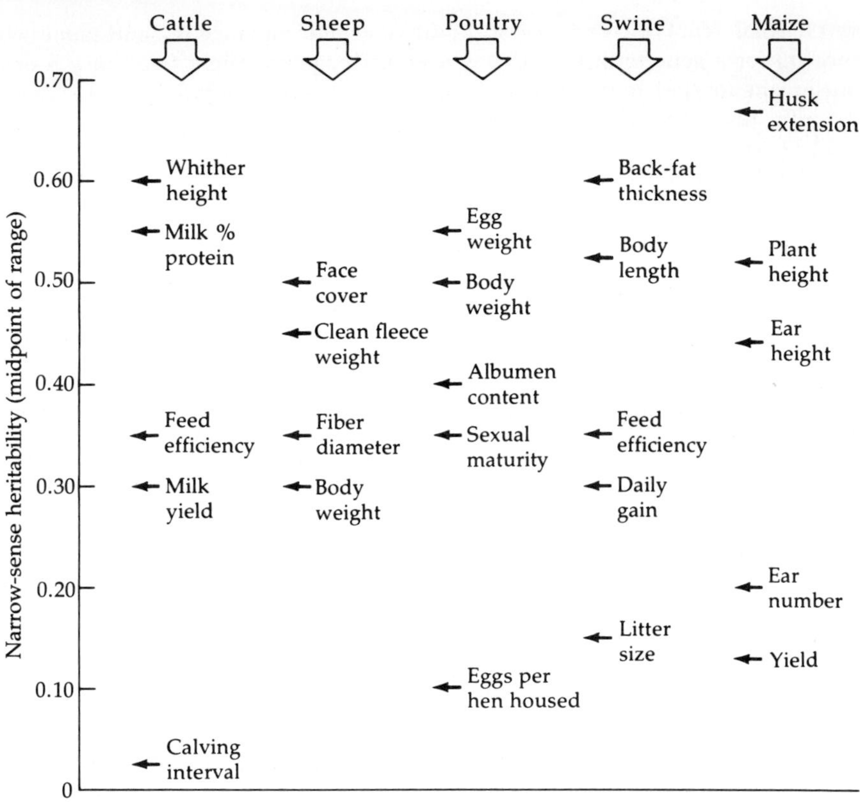

Figure 20. Narrow-sense heritabilities for representative traits in plants and animals. Traits closely related to fitness (calving interval, eggs per hen, litter size of swine, yield and ear number of corn) tend to have rather low heritabilities. (Animal data from Pirchner, 1969, who gives the range of heritabilities in various studies. The midpoint of the range is plotted here. Corn data from Robinson et al., 1949.)

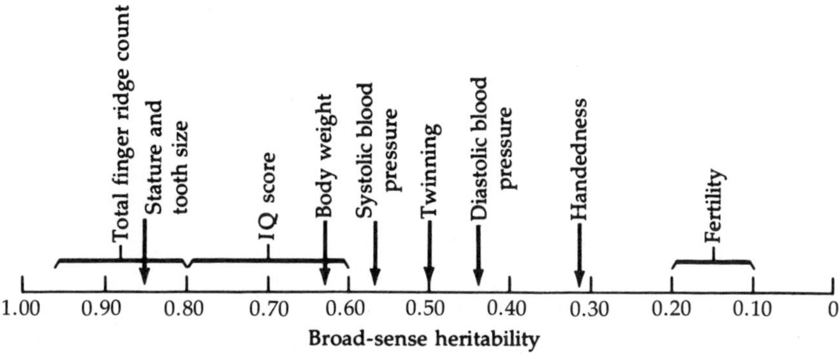

Figure 21. Broad-sense heritabilities and ranges of heritabilities of various traits in humans. (Data from Smith 1975.)

magnitude of the effects of the genes, (4) the magnitude and type of interaction among the genes, (5) genetic linkage between the genes, (6) the effective population number, (7) the intensity of selection, and (8) the relation between the quantitative trait and components of fitness. Several of these factors warrant a brief discussion.

Genetic Divergence in Small Populations

In many cases, the size of the population subjected to artificial selection must be limited for reasons of space or expense, especially in breeding programs involving large animals such as cattle, hogs, and sheep. Small population size implies that random genetic drift may become important, and alleles that are favorable for the trait might be lost from a small population by chance, as discussed in Chapter 2. Eventually, after long-continued artificial selection, a population may be expected to reach a **selection limit** or **plateau**—a phenotypic limit at which response to selection no longer occurs. A selection limit is reached whenever all alleles that affect the trait have become either fixed or lost, because in such cases the trait has no remaining genotypic variance. However, complete fixation or loss of all relevant alleles is not the most common reason that populations reach selection limits, except perhaps in very small populations. Other, more typical, reasons for selection limits are discussed in a few pages. Small populations undergo genetic differentiation resulting from random genetic drift even when they are also subjected to artificial selection, and small replicated subpopulations that have reached their selection limits may nevertheless differ in genotype, some favorable alleles having been fixed by selection in some subpopulations but lost by chance in others. Additional progress can sometimes be made by crossing the subpopulations and practicing selection in the population produced by the cross (or crosses), but the first offspring of such crosses may be phenotypically inferior to the parents. The temporary fallback is the price to be paid in order to obtain a selected population that may eventually surpass the original selection limit.

Inbreeding Depression

A second problem that arises in artificial selection because of small population size is a steady increase in the inbreeding coefficient in the selected population and a consequent reduction in fitness due to inbreeding depression (Charlesworth and Charlesworth in press). Traits that are closely related to fitness typically show the greatest inbreeding depression. Although there is considerable variation in the amount of inbreeding depression, there is often about a 5–10 percent reduction in phenotypic value for each 10 percent increase in the inbreeding coefficient (Falconer 1981). Because of inbreeding depression, breeders usually try to arrange matings to avoid inbreeding as much as possible.

Correlated Responses

Genes are **pleiotropic**—that is, every gene potentially affects every trait in the organism, either as a primary effect or as a secondary, indirect effect. Therefore, the alleles that are favorable for one quantitative trait may have unfavorable effects on another quantitative trait, and as these alleles are increased in frequency by artificial selection (thereby improving the phenotypic value with respect to the selected quantitative trait), the very same alleles may bring about a deterioration of some other aspect of performance. Pleiotropy is one cause of **correlated response**—a change in phenotypic value of one trait that accompanies response to selection of a different trait. A second possible cause of correlated responses is linkage disequilibrium (Chapter 1)—a favorable allele for one trait that increases in frequency under selection may drag along with it an allele of another, tightly linked gene that has a detrimental effect on an unselected trait.

Correlated responses are quite common in artificial selection and often, but not always, result in a deterioration in reproductive performance. In one case in Leghorn chickens, for example, 12 generations of selection for increased shank length reduced the egg hatchability by nearly half (Lerner 1958). In turkeys, to take another example, there was intense selection during the period 1944–1964 for growth rate, body conformation, and body size, but there was also a steady decline in some aspects of reproductive fitness such as fertility, egg production, and egg hatchability (Nordskog and Giesbrecht 1964). On the other hand, correlated responses can sometimes be useful. For example, selection for larger mature body size often increases litter size in mice and swine. If a trait has a low heritability or is difficult to measure, it is sometimes possible to practice selection for another, correlated trait, obtaining progress in the trait of interest by correlated response. Theoretically, the maximum response to artificial selection occurs when the criterion for selection is determined by a selection index which takes genetic correlations into account. However, the theoretical advantage of index selection is often overridden by practical difficulties in estimating the components of the index and implementing the selection procedure.

From a theoretical point of view, the covariance between two quantitative traits can be partitioned in a manner analogous to the partitioning of the variance for one trait outlined in an earlier section; the covariance can thus be partitioned into an additive covariance, a dominance covariance, an environmental covariance, and so on. The most important theoretical result is that the amount of correlated response with individual selection depends only on the additive covariance, much as the direct response to individual selection depends only on the additive variance. The components of covariance between traits can be estimated from the resemblance between relatives, but often it is preferable to estimate the correlated response by direct observation in a manner analogous to the determination of realized heritability (Falconer 1981).

Selection Limits

Progress under artificial selection cannot continue forever. The population eventually reaches a selection limit, or plateau, after which it no longer responds to selection. One of the obvious reasons why a population eventually reaches a plateau is exhaustion of additive genetic variance, when all alleles affecting the selected trait have become fixed, lost, or have attained a stable equilibrium. With no additive genetic variance, no progress under individual selection can be achieved. However, many experimental populations that have reached a selection limit readily respond to **reverse selection** (selection in the reverse direction of that originally applied), so additive genetic variance affecting the trait is still present. Indeed, in such populations, the phenotype may change in the direction of its original value if continuing artificial selection is simply suspended (**relaxed selection**). The consequences of relaxed selection for a case in *Drosophila* are illustrated in Figure 22.

One frequent reason for the occurrence of selection limits in populations

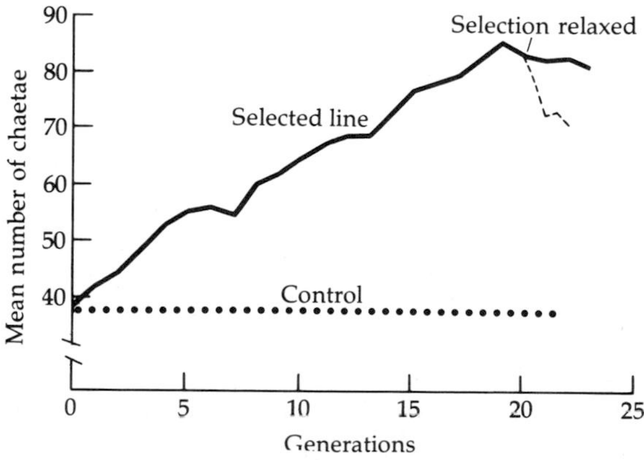

Figure 22. Response to selection for number of abdominal chaetae bristles in females of *Drosophila melanogaster* (solid line) and control (no selection applied, dotted line). Note the fallback in bristle number (dashed line) in the selected line upon relaxation of selection. (After Clayton et al. 1953.)

with considerable additive genetic variation is that artificial selection may be opposed by natural selection, an example of antagonistic pleiotropy. In mice, for example, selection for small body size ultimately ceases because small animals are less fertile than larger ones and the smallest animals are sterile (Falconer 1981). Selection for small body size gradually becomes less effective due to the opposing effects of natural selection until, eventually, no further progress is possible. When selection is relaxed, the natural selection is unopposed and results in a retrogression in the artificially selected trait. Some backward slippage with relaxed selection also results from diminution in the linkage disequilibrium which usually builds up during the course of long-term artificial selection.

A final point about selection limits: in most genetically heterogeneous populations, artificial selection can change phenotype well beyond the range of variation found in the original population. Pertinent data for populations of mice are presented in Table 8. As can be seen, a total selection response of three to five times the original phenotypic standard deviation is not unusual, and for selection to change a population of effective size N halfway to its selection limit typically requires about $0.5N$ generations.

In some cases the total response to artificial selection is very large. For example, in a long-term selection experiment for pupal weight in *Tribolium*, in which the base population consisted of the progeny of a cross between two inbred lines, 100 generations of selection resulted in a population in which the mean pupal weight in the selected population was 17 standard deviation units greater than the mean in the base population (Enfield 1980).

Table 8. Selection limits and duration of response for various traits in laboratory mice.[a]

Character selected	Direction of selection	Total response[b]	Half-life of response[c]
Weight (in strain N)	Up	$3.4\sigma_p$	$0.6N$
	Down	$5.6\sigma_p$	$0.6N$
Weight (in strain Q)	Up	$3.9\sigma_p$	$0.2N$
	Down	$3.6\sigma_p$	$0.4N$
Growth rate	Up	$2.0\sigma_p$	$0.3N$
	Down	$4.5\sigma_p$	$0.5N$
Litter size	Up	$1.2\sigma_p$	$0.5N$
	Down	$0.5\sigma_p$	$0.5N$

[a] From Falconer 1977.
[b] Total response is expressed as a multiple of the initial phenotypic standard deviation, σ_p.
[c] Half-life of response is the number of generations taken to progress halfway to the selection limit; here the half-life is expressed in multiples of effective population number (N).

In the classic long-term selection experiment of Clayton and Robertson (1956) for number of sternital bristles in *Drosophila melanogaster*, the short-term responses have been discussed in Problem 4. Below are shown the mean number of sternital bristles in females in the plateaued populations (plateaus were reached in 20–30 generations). The mean in females in the base population was 39.2, and the phenotypic standard deviation among females was 3.54. Complete the columns headed R_{total} (total response), R_{total}/σ_p (total response as a multiple of the initial phenotypic standard deviation), and R_{total}/R_1 (total response as a multiple of the response in the first generation). See Problem 4 for values of R_1. **PROBLEM 23**

	Plateau	$\langle \mathbf{R_{total}} \rangle$	$\langle \mathbf{R_{total}} \rangle / \langle \boldsymbol{\sigma_p} \rangle$	$\langle \mathbf{R_{total}} \rangle / \langle \mathbf{R_1} \rangle$
H1	83.6			
H2	70.4			
H3	70.3			
H4	88.9			
H5	75.4			

ANSWER $\langle R_{total} \rangle = \langle \text{mean at plateau} \rangle - \langle \mu_0 \rangle$, where $\langle \mu_0 \rangle = 39.2$ is given, estimates for H1 to H5 are 44.4, 31.2, 31.1, 49.7, and 36.2, respectively. Because $\langle \sigma_p \rangle = 3.54$, then $\langle R_{total} \rangle / \langle \sigma_p \rangle$ for H1 to H5 is 12.5, 8.8, 8.8, 14.0, and 10.2, respectively. From Problem 4, $\langle R_1 \rangle$ equals 2.6, 1.9, 2.3, 3.3, and 2.7 for H1 to H5, respectively. Thus, for H1 to H5, $\langle R_{total} \rangle / \langle R_1 \rangle$ equals 17.1, 16.4, 13.5, 15.1, and 13.4, respectively. (Note: Average $\langle R_{total} \rangle / \langle \sigma_p \rangle$ equals 10.9, that of $\langle R_{total} \rangle / \langle R_1 \rangle$ equals 15.1. You may be interested in knowing the half-life of the response: for H1 to H5, these are approximately 15, 7, 9, 10, and 5 generations, respectively, averaging 9.2 generations.)

THRESHOLD TRAITS

Some multifactorial traits do not exhibit continuous variation. Although, in these cases, individuals either express the trait or not, making the variation discontinuous, the trait is nevertheless influenced by multiple genetic factors and also by environment. Such traits are called **threshold traits**. A human example is diabetes, an abnormality in sugar metabolism that affects one or two percent of the Caucasian population. In a sense, diabetes is a continuous trait because the severity of the disease varies from nearly undetectable to extremely severe. On the other hand, diabetes can also be considered a threshold trait because all individuals may be classified according to whether or not they are so severely affected that clinical treatment is required. With such a classification, there are only two phenotypes, "affected" and "not affected," even though there is phenotypic variation within each category. That there is a genetic influence on the trait is evidenced by an enhanced risk of diabetes in relatives of affected individuals. However, environmental factors such as diet are also important in determining whether high-risk genotypes actually develop the disease. At one time, many threshold traits were "explained" by postulating a simple genetic mechanism (such as a single recessive allele in the case of diabetes) and invoking "incomplete penetrance" to account for the poor fit of pedigree data to a simple Mendelian hypothesis. Nowadays it is preferred, and probably more realistic, to consider threshold traits as bona fide polygenic traits and to calculate heritabilities as for any other quantitative trait. In most cases, however, it is simply not known whether the genetic influence on the trait results from one or a few major genes or is polygenic.

Heritability of Liability

The basic idea behind the current model of threshold traits is illustrated in Figure 23. The normal curve in panel (a) represents the (unobservable) distribution of a hypothetical **liability** (or risk) toward the threshold trait, measured on a scale such that the mean value is 0 and the variance is 1. It is assumed that individuals whose liability is above a certain threshold (T) actually express the trait. Thus, the shaded area in Figure 23(a) delimits the proportion of individuals in the population who are affected (B_p), and the mean liability among affected individuals is denoted μ_S. Figure 23(b) gives the (again unobservable) distribution of liability among the offspring of affected individuals. The offspring mean is denoted μ', and the proportion of offspring above the threshold is denoted B_o. The setup here is like that in the earlier section of this chapter in which we calculated the regression coefficient of offspring on one parent. In this model it can be shown that the regression coefficient b is given by $b = \mu'/\mu_S$, and the appropriate estimate of the **heritability of liability** is obtained from the relation

$$h^2 = 2b = 2\mu'/\mu_S \tag{4.37}$$

To illustrate the calculations, we will use data from Carter (1961) on the incidence of pyloric stenosis, which is an obstruction of the opening at the lower end of the stomach. The incidence of the condition differs in males and females, but we will use the incidence in fathers and their sons. For pyloric stenosis, the incidence among males in the general population is $B_p = 0.005$, and the incidence among sons of affected males is $B_o = 0.05$. From these two numbers, we wish to infer μ_S and μ' in order to calculate the heritability of liability.

The mean liability of fathers (μ_S) can be obtained immediately from Figure 12. For $B_p = 0.005$, Figure 12 gives $i = 2.89$. Since $S = \mu_S - \mu = i\sigma$ and we have assumed that $\mu = 0$ and $\sigma^2 = 1$, we have it that $\mu_S = 2.89$ also. The quantity B_p can also be used to obtain the threshold (T). Note that B_p repre-

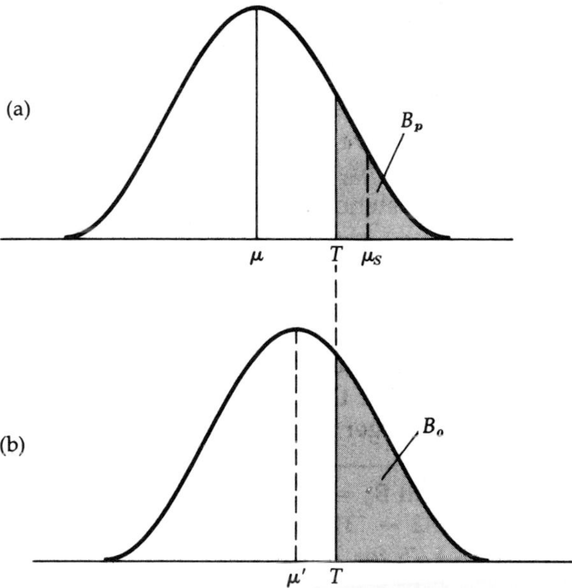

Figure 23. (a) Distribution of liability assumed for a threshold trait in a hypothetical population. The shaded area denotes individuals having liability above a critical threshold (T) and consequently affected with the trait. B_p is the frequency of affected individuals in the entire population, μ is the mean liability of individuals in the entire population, and μ_S the mean liability of affected individuals. (b) Distribution of liability among offspring who have one parent affected with the trait. μ' denotes the mean liability among offspring, and B_o is the proportion of affected offspring. Note similarity to Figure 5. In Figure 5, however, the quantitative trait could be measured directly in each individual. In this case, liability cannot be measured directly, and inferences about the distribution of liability must be made from B_p and B_o.

sents individuals with a positive deviation from the mean of an amount T or greater. Because of the symmetry of the distribution, B_p also represents individuals with a negative deviation of T or greater (although individuals with negative deviations are not affected). Overall, $2B_p$ is the proportion of the population having a positive or negative deviation of T or more units, and Figure 3 gives $T = 2.58$ for $2B_p = 0.010$.

From the value of T calculated above and the observed value of B_o, the mean liability of the offspring μ' can be obtained from Figure 3. Note in Figure 23(b) that the offspring distribution can be slid μ' units to the left to coincide with the parental distribution. The sliding procedure shifts the offspring threshold to the value $T - \mu'$. Among the offspring, therefore, $2B_o$ represents the proportion of the population having a positive or negative deviation of $T - \mu'$ or greater, and Figure 3 gives $T - \mu' = 1.64$ for $2B_o =$

0.10. From the paragraph above, we know that $T = 2.58$, so $\mu' = 2.58 - 1.64 = 0.94$. Finally, applying Equation 4.37, the estimate of heritability of liability of the condition pyloric stenosis based on father–son incidences is $\langle h^2 \rangle = 2\langle \mu' \rangle / \langle \mu_S \rangle = 2(0.94)/(2.89) = 65$ percent.

PROBLEM 24 The incidence of peptic ulcer in the general population is about 4 percent, but among the offspring of an affected parent the incidence is about 8.5 percent. Calculate the heritability of liability toward peptic ulcer. (Two hints: (1) use approximation I in Figure 12, which is very accurate for this case; and (2) in Figure 3, the probability of a deviation as large or larger than 1.37 is 0.17.)

ANSWER $\langle B_p \rangle = 0.04$ and $\langle B_o \rangle = 0.085$ are given. Using approximation I in Figure 12, $i = 1.132 - 0.3166 \ln(B_p) = 2.15$, and $\langle \mu_S \rangle = 2.15$. From Figure 3, $T = 1.75$ for $2B_p = 0.08$. Hint (2) implies that $T - \mu' = 1.37$ for $2B_o = 0.17$. Thus, $\langle \mu' \rangle = 1.75 - 1.37 = 0.38$, and the heritability of liability is estimated as $\langle h^2 \rangle = 2(0.38)/(2.15) = 35$ percent.

Application to Human Traits

The calculations in Problem 24 are a little tricky and require reference to charts or tables of the normal distribution. However, over a broad range of population incidences, the following approximation is sufficiently accurate for most purposes:

$$\log(B_r) \simeq \left(\frac{-0.274h^2}{1 + 0.742h^2}\right) + \left(\frac{1 - 0.579h^2}{1 + 0.371h^2}\right) \log(B_g) \qquad \textbf{(4.38)}$$

Equation 4.38 generalizes the symbolism somewhat in that B_g now represents the incidence of the trait in the general population (corresponds to B_P in Problem 24) and B_r represents the risk of the trait in first-degree relatives of affected individuals (corresponds to B_o in Problem 24). The logarithms are taken to the base 10. **First-degree relatives** share half their genes, and the most important first-degree relationships are parent–offspring and full siblings.

Equation 4.38 is a satisfactory approximation over the broad range of values of B_g from 0.00001 to 0.20. Therefore, within this range, given any values of B_g and h^2 (or B_r and h^2), the corresponding value of B_r (or B_g) can easily be estimated from Equation 4.38. In the example of peptic ulcer in Problem 24, $B_g = 0.04$ and $h^2 = 0.35$. Using Equation 4.38, B_r is estimated as $\langle B_r \rangle = 10^{-1.063} = 0.086$, which is close to the observed value.

Figure 24 shows the relation between the percentage affected among first-degree relatives of affected individuals B_r and the percentage of affected individuals in the general population B_g, for traits with population incidences

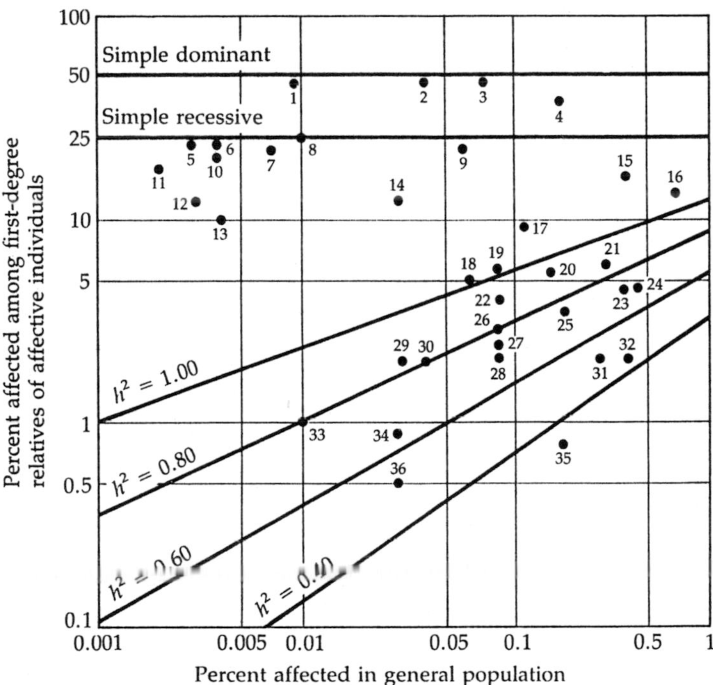

Figure 24. Risks of occurrence of common abnormalities in humans which are inherited or determined in part by inherited predispositions. Diagonal lines are the theoretical risks in first-degree relatives for threshold traits with narrow-sense heritabilities as indicated. Horizontal line for simple dominant indicates risk in offspring or siblings of affected individuals, horizontal line for simple recessive indicates risk in siblings. Numbered traits are: (1) achondroplasia, (2) target-cell anemia, (3) periodic paralysis, (4) otosclerotic deafness, (5) retinoblastoma, (6) hemophilia, (7) albinism, (8) retinitis pigmentosum, (9) cystic fibrosis, (10) phenylketonuria, (11) osteogenesis imperfecta, (12) microphthalmos, (13) Hirschsprung disease, (14) deafmutism, (15) manic-depressive psychosis, (16) mental deficiency, (17) schizophrenia, (18) congenital dislocated hip, (19) multiple sclerosis, (20) strabismus, (21) pyloric stenosis, (22) anencephaly, (23) diabetes, (24) rheumatic fever, (25) spina bifida aperta, (26) clubfoot, (27) patent ductus, (28) cleft lip, (29) celiac disease, (30) cleft palate, (31) congenital heart disease, (32) epilepsy, (33) situs inversus viscerum, (34) exomphalos, (35) hydrocephaly, (36) psoriasis. (After Freifelder et al. 1985, based on data of Newcombe 1964.)

smaller than one percent. In Figure 24, the diagonal lines for different values of narrow-sense heritability of liability are very nearly straight lines, as implied in Equation 4.38. The horizontal lines at 50 percent and 25 percent are the risks in first-degree relatives of individuals affected with dominant or recessive simple Mendelian disorders, respectively, although the risk for recessives refers only to the risk in full siblings, not parent–offspring.

Equation 4.38 is inconvenient to use in estimating the value of h^2 using B_r

and B_g, but the following approximation is sufficiently accurate for most purposes:

$$h^2 \simeq [-B - (B^2 - 4AC)^{1/2}]/2A \qquad \textbf{(4.39)}$$

where

$$\begin{aligned} A &= 0.102 + 0.275 \log(B_r) + 0.430 \log(B_g) \\ B &= 0.274 + 1.113 \log(B_r) - 0.163 \log(B_g) \\ C &= \log(B_r) - \log(B_g) \end{aligned} \qquad \textbf{(4.40)}$$

Equations 4.38 and 4.39 can also be adjusted for use with relatives other than those of the first degree. For the relationship between uncles/aunts and their nephews/nieces, replace h^2 in the equations by $h^2/2$, and for first cousins, replace h^2 by $h^2/4$. As before, B_g equals the frequency of the condition in the general population, and B_r equals the frequency in the specified relatives.

PROBLEM 25 Renal stone disease is a threshold trait which occurs at a frequency of 0.4 percent in the general population and approximately 2.5 percent among the offspring of affected individuals (Falconer 1965). Use Equation 4.39 to estimate the narrow-sense heritability of liability. Then use Equation 4.38 to estimate the frequency of the trait in brothers/sisters of affected individuals, nephews/nieces of affected individuals, and cousins of affected individuals.

ANSWER $\langle B_g \rangle = 0.004$ and $\langle B_r \rangle = 0.025$. From Equation set 4.40, $A = -1.369681$, $B = -1.118229$, and $C = 0.795880$. Substituting into Equation 4.39 yields $\langle h^2 \rangle = 0.46$. For the risk in brothers/sisters, use Equation 4.38 with $h^2 = 0.46$, for nephews/nieces use $0.46/2 = 0.23$ in place of h^2, and for first cousins use $0.46/4 = 0.115$ instead of h^2. Risks are 2.5 percent for brothers/sisters, 1.1 percent for nephews/nieces, and 0.67 percent for first cousins.

SUMMARY

Multifactorial traits are affected by multiple genes and usually by environmental factors as well. Some multifactorial traits, such as height or weight, are continuous in that they demonstrate a continuum of possible phenotypic values. With other multifactorial traits, known as meristic traits, the phenotypic value is determined by enumeration, for example by counting the number of bristles on a fruit fly sternite. Still other multifactorial traits feature an underlying continuum of liability or risk, and only certain individuals with liabilities above a threshold become affected. These three basic types of multifactorial traits are collectively called quantitative traits.

Many quantitative traits have a distribution of phenotypic values that is approximately normal. Traits that are not normally distributed sometimes become normal when measured on an appropriate scale, such as $\ln(x)$ or $\arcsin \sqrt{x}$, where x represents the original measurement of phenotype. The normal distribution is completely determined by two parameters—the mean (denoted μ) and the variance (denoted σ^2).

Truncation selection is a method of individual selection in which all individuals having a phenotype above a certain value (the truncation point T) are saved and bred to produce the next generation. If μ denotes the mean phenotype of the original population and μ_S denotes the mean phenotype among selected parents, then the mean phenotype among the progeny (μ') is given by $\mu' - \mu = h^2(\mu_S - \mu)$, where h^2 is the heritability of the trait. The quantity $\mu' - \mu$ is usually called the response to selection R and $\mu_S - \mu$ is called the selection differential S, so the prediction equation for individual selection can be written as $R = h^2S$. When h^2 is estimated from observed values of R and S, then h^2 is called the realized heritability. Otherwise h^2 can be estimated from the resemblance between relatives, and, theoretically, $h^2 = \Sigma 2pq[a + (q - p)d]^2/\sigma^2$ where summation is carried out across all genes affecting the trait and, for each gene, p is the frequency of the favorable allele, a is the effect of the gene (measured as the average difference between the homozygotes), and d is a measure of dominance (measured as the deviation of the heterozygote from the mean of the homozygotes). The quantity $\Sigma 2pq[a + (q - p)d]^2$ is known as the additive genetic variance, and, in the above formula for h^2, the symbol σ^2 represents the variance in phenotypic value in the population. Thus, the heritability is the ratio of additive genetic variance to total phenotypic variance.

The prediction equation for individual selection can be written equivalently as $R = i\sigma h^2$ where i is called the intensity of selection and $i = S/\sigma$. Intensity of selection is a useful quantity with which to compare diverse breeding programs because i depends only on the proportion of the population saved for breeding.

The fundamental theorem of natural selection applies the principles of quantitative genetics to the trait *fitness*. The fundamental theorem states that, for fitness, $R = \sigma_a^2/\overline{w}$, where σ_a^2 represents the additive genetic variance in fitness and $\overline{w}$ is the average fitness of the population. The intensity of selection for fitness can be shown to equal $\sigma/\overline{w}$ where σ is the phenotypic standard deviation in fitness. The quantity i^2 has been called the index of opportunity for selection because it defines an upper limit on the possible rate of increase in fitness. Stabilizing selection can be modeled as a double truncation selection against both tails of the distribution of phenotypic values. The selection coefficients of genes which affect traits subject to stabilizing selection are proportional to the deviation of the population mean from the optimum phenotype. When the population mean is close to the optimum phenotype, the magnitude of selective effects on the genes may be very small. The importance of stabilizing selection in determining the genetic architecture of traits in natural populations is uncertain. Genetic architecture refers to the relative magnitudes of the components of genetic variance and covariance. Antagonistic pleiotropy, in which alleles affect some traits favorably and others unfavorably, also influences the genetic architecture.

There are many alternatives to individual selection in the design of breeding programs, among them, family selection, within-family selection, and recurrent or reciprocal recurrent selection for combining ability. Each of these modes of selection is theoretically superior to individual selection under certain particular circumstances, and prediction equations for each mode of selection can be derived.

Heritability can be estimated from the resemblance between relatives, as exemplified by the regression coefficient of offspring on parent. If b denotes the regression coefficient of offspring on a single parent, then $h^2 = 2b$. If b denotes the regression coefficient of offspring on the mean of the parents (midparent), then $h^2 = b$.

Resemblance between relatives can also be used to partition the total phenotypic variance of a trait into components due to genotype (genotypic variance σ_g^2) and environment (environmental variance σ_e^2). The total genotypic variance can be further partitioned into the additive genetic variance σ_a^2, the dominance variance σ_d^2, and the variance due to interaction among nonallelic genes σ_i^2. In the absence of genotype–environment interaction (evidenced by parallel norms of reaction) and in the absence of genotype–environment association (evidenced by lack of correlation between genotype and environment), then the total phenotypic variance can be written as $\sigma_p^2 = \sigma_g^2 + \sigma_e^2 = \sigma_a^2 + \sigma_d^2 + \sigma_i^2 + \sigma_e^2$. The variance ratio σ_a^2/σ_p^2 is called heritability in the narrow sense, and this is the heritability that is important in individual selection. The variance ratio σ_g^2/σ_p^2 is called heritability in the broad sense, and it is important in certain other contexts.

After long-continued selection for a quantitative trait, response to selection eventually ceases and the population reaches a plateau. A selection limit may be reached even before all the additive genetic variance for the trait has been exhausted, and in many cases, the response to selection ceases because natural selection for fitness-related traits counterbalances the artificial selection for the trait in question. Nevertheless, artificial selection can produce a total response of five or ten or more phenotypic standard deviations, so the trait in the selected population may be well beyond the range of what it was in the original population.

Threshold traits are traits that are either present or absent in an individual, but their presence or absence is determined by the value of an underlying quantitative trait known as the liability. The heritability of liability can be calculated from established principles of quantitative genetics when one knows the incidence of the trait in the general population and also the incidence among offspring or other relatives of affected individuals.

FURTHER READINGS

Allard, R. W. 1960. *Principles of Plant Breeding*. John Wiley and Sons, New York.

Barker, J. S. (ed.). 1983. *Future Developments in the Genetic Improvement of Animals*. Academic Press, New York.

Bulmer, M. G. 1980. *The Mathematical Theory of Quantitative Genetics*. Clarendon, Oxford.

Chapman, A. B. (ed.). 1985. *General and Quantitative Genetics*. Elsevier, New York.

Falconer, D. S. 1986. *Introduction to Quantitative Genetics*, 2nd ed. Halsted Press, New York.

Falconer, D. S. 1986. *Problems of Quantitative Genetics*. Halsted Press, New York.

Haldane, J. B. S. et al. 1960. *Papers on Quantitative Genetics and Related Topics*. North Carolina State University Press, Raleigh.

Hartl, D. L. 1980. *Principles of Population Genetics*. Sinauer Associates, Sunderland, Massachusetts.

Jacquard, A. 1983. Heritability: One word, three concepts. *Biometrics* 39: 465–477.

Johansson, I. and J. Rendel. 1968. *Genetics and Animal Breeding*. (Trans. by M. Taylor). W. H. Freeman, San Francisco.

Kempthorne, O. 1969. *An Introduction to Genetic Statistics*. Iowa State University Press, Ames.

Kempthorne, O. 1978. Logical, epistemological and statistical aspects of nature–nurture data interpretation. *Biometrics* 34: 1–23.

Lasley, J. 1978. *Genetics of Livestock Improvement*, 3rd ed. Prentice-Hall, Englewood Cliffs, New Jersey.

Lerner, I. M. and H. P. Donald. 1966. *Modern Developments in Animal Breeding*. Academic Press, New York.

McClearn, G. E. and J. C. DeFries. 1973. *Introduction to Behavioral Genetics*. W. H. Freeman, San Francisco.

Mather, K. and J. L. Jinks. 1971. *Biometrical Genetics*. Cornell University Press, Ithaca, New York.

Morton, N. E. 1982. *Outline of Genetic Epidemiology*. Karger, New York.

Pirchner, F. 1969. *Population Genetics in Animal Breeding*. W. H. Freeman, San Francisco.

Pollak, E., O. Kempthorne and T. B. Bailey (eds.). 1977. *International Conference on Quantitative Genetics*. Iowa State University Press, Ames.

Rendel, J. M. 1967. *Canalization and Gene Control*. Academic Press, New York.

Robertson, A. 1967. The nature of quantitative genetic variation. *In* R. A. Brink (ed.). *Heritage from Mendel*. University of Wisconsin Press, Madison.

Robertson, A. (ed.). 1980. *Selection Experiments in Laboratory and Domestic Animals*, Commonwealth Agricultural Bureau, Slough.

Simmonds, N. W. 1979. *Principles of Crop Improvement*. Longman, New York.

Smith, C. 1975. Quantitative inheritance. *In* G. Fraser and O. Mayo (eds.). *Textbook of Human Genetics*. Blackwell, Oxford.

Thompson, J. M., Jr. and J. M. Thoday (eds.). 1979. *Quantitative Genetic Variation*. Academic Press, New York.

Turner, H. N. and S. S. Y. Young. 1969. *Quantitative Genetics in Sheep Breeding*. Cornell University Press, Ithaca, New York.

Walden, D. B. (ed.). 1978. *Maize Breeding and Genetics*. John Wiley and Sons, New York.

Warwick, E. J. and J. E. LeGates. 1979. *Breeding and Improvement of Farm Animals*, 7th ed. McGraw-Hill, New York.

Wricke, G. and E. Weber. 1986. *Quantitative Genetics and Selection in Plant Breeding*. De Gruyter, Hawthorne, New York.

Wright, S. 1977. *Evolution and the Genetics of Populations*. Vol. 3. *Evolutionary Results and Evolutionary Deductions*. University of Chicago Press, Chicago.

Wright, S. 1978. *Evolution and the Genetics of Populations*. Vol. 4. *Variability within and among Natural Populations*. University of Chicago Press, Chicago.

LITERATURE CITED

Adams, J. and R. H. Ward. 1973. Admixture studies and the detection of selection. *Science* 180: 1137–1143.

Alberts, B., D. Bray, J. Lewis, M. Raff, K. Roberts and J. D. Watson. 1983. *Molecular Biology of the Cell*. Garland, New York.

Allison, A. C. 1964. Polymorphism and natural selection in human populations. *Cold Spring Harbor Symp. Quant. Biol.* 29: 139–149.

Anderson, W. W., C. Oshima, T. Watanabe, Th. Dobzhansky and O. Pavlovsky. 1968. Genetics of natural populations. XXXIX. A test of the possible influence of two insecticides on the chromosomal polymorphisms in *Drosophila pseudoobscura*. *Genetics* 58: 423–434.

Antonovics, J., A. D. Bradshaw and R. G. Turner. 1971. Heavy metal tolerance in plants. *Adv. Ecol. Res.* 7: 1–85.

Aoki, K. 1981. Algebra of inclusive fitness. *Evolution* 35: 659–663.

Avise, J. C. 1986. Mitochondrial DNA and the evolutionary genetics of higher animals. *Phil. Trans. R. Soc. Lond. B.* 312: 325–342.

Avise, J. C., C. Giblin-Davidson, J. Laerm, J. C. Patton, and R. A. Lansman. 1979. Mitochondrial DNA clones and matriarchal phylogeny within and among geographical populations of the pocket gopher, *Geomys pinetis*. *Proc. Natl. Acad. Sci. USA* 76: 6694–6698.

Ayala, F. J. and M. L. Tracy. 1974. Genetic differentiation within and between species of the *Drosophila willistoni* group. *Proc. Natl. Acad. Sci. USA* 71: 999–1003.

Benveniste, R. E. 1985. The contributions of retroviruses to the study of mammalian evolution. pp. 359–417. *In* R. J. MacIntyre (ed.), *Molecular Evolutionary Genetics*. Plenum, New York.

Bingham, P. M., M. G. Kidwell and G. M. Rubin. 1982. The molecular basis of P-M hybrid dysgenesis: the role of the P element, a P-strain-specific transposon family. *Cell* 29: 995–1004.

Bishop, J. A. and L. M. Cook. 1975. Moths, melanism and clean air. *Sci. Amer.* 232: 90–99.

Bishop, J. M. 1981. Enemies within: The genesis of retrovirus oncogenes. *Cell* 23: 5–6.

Bishop, J. M. 1982. Oncogenes. *Sci. Amer.* 246 (March): 80–92.

Bonhomme, F. and R. K. Selander. 1978. Estimating total genic diversity in the house mouse. *Biochem. Genet.* 16: 287–298.

Brown, W. M. 1980. Polymorphism in mitochondrial DNA of humans as revealed by restriction endonuclease analysis. *Proc. Natl. Acad. Sci. USA* 77: 3605–3609.

Brown, W. M. 1983. Evolution of animal mitochondrial DNA. pp. 62–88. *In* M. Nei and R. K. Koehn (eds.), *Evolution of Genes and Proteins*, Sinauer Associates, Sunderland, Massachusetts.

Brown, W. M. 1985. The mitochondrial genome of animals. pp. 95–130. *In* R. J. MacIntyre (ed.), *Molecular Evolutionary Genetics*. Plenum, New York.

Bryan, G. J., J. W. Jacobson and D. L. Hartl. 1987. Heritable somatic excision of a *Drosophila* transposon. *Science* 235: 1636–1638.

Bucheton, A., R. Paro, H. M. Sang, A. Pelisson and D. J. Finnegan. 1984. The molecular basis of I-R hybrid dysgenesis in *Drosophila melanogaster*: identification, cloning and properties of the I factor. *Cell* 38: 153–163.

Buonagurio, D. A., S. Nakada, J. D. Parvin, M. Krystal, P. Palese and W. M. Fitch. 1986.

Evolution of human influenza A virus over 50 years: Rapid, uniform rate of change in NS gene. *Science* 232: 980–982.

Buri, P. 1956 Gene frequency in small populations of mutant *Drosophila*. *Evolution* 10: 367–402.

Burton, R. S. and A. R. Place. 1986. Evolution of selective neutrality: further considerations. *Genetics* 114: 1033–1036.

Carter, C. O. 1961. The inheritance of congenital pyloric stenosis. *Brit. Med. Bull.* 17: 251–254.

Cavalli-Sforza, L. L. and W. F. Bodmer. 1971. *The Genetics of Human Populations*. W. H. Freeman, San Francisco.

Cavalli-Sforza, L. L. and M. W. Feldman. 1978. Darwinian kin selection and "altruism." *Theoret. Pop. Biol.* 14: 268–280.

Charlesworth, B., C. H. Langley and W. Stephan. 1986. The evolution of restricted recombination and the accumulation of repeated DNA sequences. *Genetics* 112: 947–962.

Charlesworth, D. and B. Charlesworth. in press. Inbreeding depression. *Annu. Rev. Ecol. Syst.*

Christiansen, F. B. and O. Frydenberg. 1974. Geographical patterns of four polymorphisms in *Zoarces viviparus* as evidence of selection. *Genetics* 77: 765–770.

Clark, A. G. and M. W. Feldmen. 1986. A numerical simulation of the one-locus, multiple-allele fertility model. *Genetics* 113: 161–176.

Clarke, G. M. and J. A. McKenzie. 1987. Developmental stability of insecticide resistant phenotypes in blowfly: A result of canalizing natural selection. *Nature* 325: 345–346.

Clayton, G. A., G. R. Knight, J. A. Morris and A. Robertson. 1956a. An experimental check on quantitative genetical theory. III. Correlated responses. *J. Genet.* 55: 171–180.

Clayton, G. A., J. A. Morris and A. Robertson. 1953. Selection for abdominal chaetae in a large population of *Drosophila melanogaster*. pp. 7–15. *In Symposium on the Genetics of Population Structure*. Premiata Tipografia Successori Filli Fusi, Pavia, Italy.

Clayton, G. A., J. A. Morris and A. Robertson. 1956b. An experimental check on quantitative genetical theory. I. Short-term responses to selection. *J. Genet.* 55: 131–151.

Clayton, G. A. and A. Robertson. 1956. An experimental check on quantitative genetical theory. II. The long-term effects of selection. *J. Genet.* 55: 152–170.

Clegg, M. T., R. W. Allard and A. L. Kahler. 1972. Is the gene the unit of selection? Evidence from two experimental plant populations. *Proc. Natl. Acad. Sci. USA* 69: 2474–2478.

Cook, L. M. 1965. Inheritance of shell size in the snail *Arianta arbustorum*. *Evolution* 19: 86–94.

Cook, L. M., G. S. Mani and M. E. Varley. 1986. Postindustrial melanism in the peppered moth. *Science* 231: 622–623.

Cross, S. R. H. and A. J. Birley. 1986. Restriction endonuclease map variation in the *Adh* region in populations of *Drosophila melanogaster*. *Biochem. Genet.* 24: 415–433.

Crow, J. F. 1945 A chart of the χ^2 and t distributions. *J. Amer. Stat. Assn.* 40: 376.

Crow, J. F. 1966. The quality of people: Human evolutionary change. *BioScience* 16: 863–867.

Crow, J. F. 1985. The neutrality-selection controversy in the history of evolution and population genetics. pp. 1-18. *In* T. Ohta and K. Aoki (eds.), *Population Genetics and Molecular Evolution*. Japan Scientific Societies Press, Tokyo.

Crow, J. F. and K. Aoki. 1982. Group selection for a polygenic behavioral trait: A differential proliferation model. *Proc. Natl. Acad. Sci. USA* 79: 2628–2631.

Crow, J. F. and K. Aoki. 1984. Group selection for a polygenic behavioral trait: Estimating the degree of population subdivision. *Proc. Natl. Acad. Sci. USA* 81: 6073–6077.

Crow, J. F. and M. Kimura. 1970. *An Introduction to Population Genetics Theory*. Harper & Row, New York.

Crow, J. F. and A. P. Mange. 1965. Measurement of inbreeding from the frequency of marriages between persons of the same surname. *Eugen. Quart.* 12: 199–203.

Crow, J. F. and N. E. Morton. 1955 Measurement of gene frequency drift in small populations. *Evolution* 9: 202–214.

Curie-Cohen, M. 1982. Estimates of inbreeding in a natural population: A comparison of sampling properties. *Genetics* 100: 339–358.

Curtsinger, J. W. 1984. Evolutionary landscapes for complex selection. *Evolution* 38: 359–367.

Darwin, C. 1872. *The Origin of Species*, 6th ed. New American Library of World Literature, New York (1958).

Dice, L. R. and W. B. Howard. 1951. Distance of dispersal by prairie deer mice from birthplace to breeding sites. *Contr. Lab. Vert. Biol. Univ. Mich.* 50: 1–15.

Dobzhansky, Th. and B. Spassky. 1944 Genetics of natural populations. XI. Manifestation of genetic variants in *Drosophila pseudoobscura* in different environments. *Genetics* 20: 270–290.

Doolittle, F. W. and C. Sapienza. 1980. Selfish genes, the phenotype paradigm and genome evolution. *Nature* 284: 601–603.

Dover, G. 1982. Molecular drive: A cohesive mode of species evolution. *Nature* 299: 111–117.

Dudley, J. W. 1977. 76 generations of selection for oil and protein percentage in maize. pp. 459–473. *In* E. Pollak, O. Kempthorne and T. B. Bailey, Jr. (eds.), *International Conference on Quantitative Genetics*. Iowa State University Press, Ames.

DuMouchel, W. H. and W. W. Anderson. 1968. The analysis of selection in experimental populations. *Genetics* 58: 435–449.

Dykhuizen, D. E., A. M. Dean and D. L. Hartl. 1987. Metabolic flux and fitness. *Genetics* 115: 25–31.

Dykhuizen, D. and D. L. Hartl. 1980. Selective neutrality of 6PGD allozymes in *E. coli* and the effects of genetic background. *Genetics* 96: 801–817.

Dykhuizen, D. E. and D. L. Hartl. 1983. Selection in chemostats. *Microbiol. Rev.* 47: 150–168.

Easteal, S. 1985. The ecological genetics of introduced populations of the giant toad *Bufo marinus*. II. Effective population size. *Genetics* 110: 107–122.

Ehrman, L. 1970. The mating advantage of rare males in *Drosophila*. *Proc. Natl. Acad. Sci. USA* 515: 345–348.

Enfield, F. D. 1980. Long term effects of selection: The limits to response. pp. 69–86. *In* A. Robertson (ed.), *Selection Experiments in Laboratory and Domestic Animals*. Commonwealth Agricultural Bureau, Slough, England.

Engels, W. R. 1979. Hybrid dysgenesis in *Drosophila melanogaster*: rules of inheritance of female sterility. *Genet. Res.* 33: 219–223.

Engels, W. R. 1981. Estimating genetic divergence and genetic variability with restriction endonucleases. *Proc. Natl. Acad. Sci. USA* 78: 6329–6333.

Engels, W. R. 1983. The P family of transposable elements in *Drosophila*. *Annu. Rev. Genet.* 17: 315–344.

Ewens, W. J. 1979. *Mathematical Population Genetics*. Springer-Verlag, New York.

Falconer, D. S. 1955. Patterns of response in selection experiments with mice. *Cold Spring Harbor Symp. Quant. Biol.* 20: 178–196.

Falconer, D. S. 1965. The inheritance of liability to certain diseases, estimated from the incidence among relatives. *Ann. Hum. Genet.* 29: 51–76.

Falconer, D. S. 1977. Some results of the Edinburgh selection experiments with mice. pp. 101-115. *In* E. Pollak, O. Kempthorne, and T. J. Bailey, Jr. (eds.), *International Conference on Quantitative Genetics*. Iowa State Univ. Press, Ames.

Falconer, D. S. 1981. *Introduction to Quantitative Genetics*, 2nd ed. Longman, New York.

Falconer, D. S. 1985. A note on Fisher's "average effect" and "average excess." *Genet. Res.* 46: 337–347.

Felsenstein, J. 1977. Multivariate normal genetic models with a finite number of loci. pp. 227–245. *In* E. Pollak, O. Kempthorne and T. B. Bailey, Jr. (eds.), *International Conference on Quantitative Genetics*. Iowa State University Press, Ames.

Fisher, R. A. 1930. *The Genetical Theory of Natural Selection*. Clarendon, Oxford.

Flint, J., A. V. S. Hill, D. K. Bowden, S. J. Oppenheimer, P. R. Sill, S. W. Serjeantson, J. Bana-Koiri, K. Bhatia, M. P. Alpers, A. J. Boyce, D. J. Weatherall and J. B. Clegg. 1986. High frequencies of α-thalassaemia are the result of natural selection by malaria. *Nature* 321: 744–750.

Ford, E. B. and P. M. Sheppard. 1969. The *medionigra* polymorphism of *Panaxia dominula*. *Heredity* 24: 561–569.

Friefelder, D., L. A. Snyder and D. L. Hartl. 1985. *General Genetics*. Jones and Bartlett, Boston.

Fuerst, P. A., R. Chakraborty and M. Nei. 1977. Statistical studies on protein polymorphism in natural populations. I. Distribution of single locus heterozygosity. *Genetics* 86: 455–483.

Gilbert, W. 1986. The road not taken. *Nature* 320: 485–486.

Gillespie, J. H. 1978. A general model to account for enzyme variation in natural populations. V. The SAS-CFF model. *Theoret. Pop. Biol.* 14: 1–45.

Gillespie, J. H. 1986a. Natural selection and the molecular clock. *Mol. Biol. Evol.* 3: 138–155.

Gillespie, J. H. 1986b. Variability of evolutionary rates of DNA. *Genetics* 113: 1077–1091.

Gillespie, J. H. and C. H. Langley. 1974. A general model to account for enzyme variation in natural populations. *Genetics* 76: 837–848.

Gingerich, P. D. 1986. Temporal scaling of molecular evolution in primates and other mammals. *Mol. Biol. Evol.* 3: 205–221.

Golding, G. B., C. F. Aquadro and C. H. Langley. 1986. Sequence evolution within populations under multiple types of mutation. *Proc. Natl. Acad. Sci. USA* 83: 427–431.

Gray, P. W. and D. V. Goeddel. 1983. Cloning and expression of murine immune interferon cDNA. *Proc. Natl. Acad. Sci. USA* 80: 5842–5846.

Haldane, J. B. S. 1956. The estimation of viabilities. *J. Genet.* 54: 294–296.

Hamilton, W. D. 1964. The genetical evolution of social behaviour I. *J. Theoret. Biol.* 7: 1–16.

Hardies, S. C., S. L. Martin, C. F. Voliva, C. A. Hutchison III and M. H. Edgell. 1986. An analysis of replacement and synonymous changes in the rodent L1 repeat family. *Mol Biol. Evol.* 3: 109–125.

Harris, H. 1966. Enzyme polymorphisms in man. *Proc. R. Soc. Lond. B* 164: 298–310.

Harris, H. and D. A. Hopkinson. 1972. Average heterozygosity in man. *J. Hum. Genet.* 36: 9–20.

Harris, H., D. A. Hopkinson and Y. H. Edwards. 1977. Polymorphism and the subunit structure of enzymes. A contribution to the neutralist–selectionist controversy. *Proc. Natl. Acad. Sci. USA* 74: 698–701.

Hartl, D. L. 1975. Modifier theory and meiotic drive. *Theoret. Pop. Biol.* 7: 168–174.

Hartl, D. L. 1979. Four volume treatise on population biology. *BioScience* 29: 179–180.

Hartl, D. L. 1980. *Principles of Population Genetics*. Sinauer Associates, Sunderland, Massachusetts.

Hartl, D. L., A. M. Dean and D. E. Dykhuizen. 1986. The molecular biology of natural selection: Reply to Burton and Place. *Genetics* 114: 1037–1039.

Hartl, D. L. and D. E. Dykhuizen. 1981. Potential for selection among nearly neutral allozymes of 6-phosphogluconate dehydrogenase in *Escherichia coli*. *Proc. Natl. Acad. Sci. USA* 78: 6344–6348.

Hartl, D. L. and D. E. Dykhuizen. 1985. The neutral theory and the molecular basis of preadaptation. pp. 107–124. *In* T. Ohta and K. Aoki (eds.), *Population Genetics and Molecular Evolution*. Japan Scientific Societies Press, Tokyo.

Hartl, D. L., D. E. Dykhuizen and A. M. Dean. 1985. Limits of adaptation: The evolution of selective neutrality. *Genetics* 111: 655–674.

Hartl, D. L., D. E. Dykhuizen, R. D. Miller, L. Green and J. de Framond, 1983. Transposable element IS50 improves growth rate of *E. coli* cells without transposition. *Cell* 35: 503–510.

Hedrick, P. W. 1986. Genetic polymorphisms in heterogeneous environments: A decade later. *Annu. Rev. Ecol. Syst.* 17: 535–566.

Hedrick, P. W. and C. C. Cockerham. 1986. Partial inbreeding: Equilibrium heterozygosity and the heterozygosity paradox. *Evolution* 40: 856–861.

Hinds, K. R. and G. W. Litman. 1986. Major reorganization of immunoglobulin V_H segments during vertebrate evolution. *Nature* 320: 546–549.

Holland, J., K. Spindler, F. Horodyski, E. Grabau, S. Nichol and S. VandePol. 1982. Rapid evolution of RNA genomes. *Science* 215: 1577–1585.

Hood, L., M. Kronenberg and T. Hunkapiller. 1985. T cell antigen receptors and the immunoglobulin supergene family. *Cell* 40: 225–229.

Hunkapiller, T. and L. Hood. 1986. The growing immunoglobulin supergene family. *Nature* 323: 15–16.

Ikemura, T. 1985a. Codon usage and tRNA content in unicellular and multicellular organisms. *Mol. Biol. Evol.* 2: 13–34.

Ikemura, T. 1985b. Codon usage, tRNA content, and rate of synonymous substitution. pp. 385–406. *In* T. Ohta and K. Aoki (eds.), *Population Genetics and Molecular Evolution*. Springer-Verlag, New York.

Jacobson, J. W., M. M. Medhora, and D. L. Hartl. 1986. Molecular structure of a somatically unstable transposable element in *Drosophila*. *Proc. Natl. Acad. Sci. USA* 83: 8684–8688.

Jacquard, A. 1983. Heritability: One word, three concepts. *Biometrics* 39: 465–477.

Jain, S. K. and A. D. Bradshaw. 1966. Evolutionary divergence among adjacent plant populations. I. The evidence and its theoretical analysis. *Heredity* 21: 407–441.

Johannsen, W. 1903 *Über Erblichkeit in Populationen und in reinen Linien*. Gustav Fisher, Jena. [Translated in part in J. A. Peters (ed.), 1959. *Classic Papers in Genetics*, pp. 21–26. Prentice Hall, Englewood Cliffs, New Jersey.]

Jones, J. S. 1987. An asymmetrical view of fitness. *Nature* 325: 298–299.

Jukes, T. H. and C. R. Cantor. 1969. Evolution of protein molecules. pp. 21–132. *In* H. N. Munro (ed.), *Mammalian Protein Metabolism* III. Academic Press, New York.

Kacser, H. and J. A. Burns. 1973. The control of flux. *Symp. Soc. Exptl. Biol.* 32: 65–104.

Kaplan, N. L. and J. F. Y. Brookfield. 1983. Transposable elements in Mendelian populations. III. Statistical results. *Genetics* 104: 485–495.

Karlin, S. and J. McGregor. 1974. Towards a theory of the evolution of modifier genes. *Theoret. Pop. Biol.* 5: 59–103.

Kempthorne, O. 1978. Logical, epistemological and statistical aspects of nature-nurture data interpretation. *Biometrics* 34: 1–23.

Kettlewell, H. B. D. 1956. Further selection experiments on industrial melanism in the Lepidoptera. *Heredity* 10: 287–301.

Kettlewell, H. B. D. 1973. *The Evolution of Melanism: The Study of a Recurring Necessity*. Clarendon, Oxford.

Kidwell, M. G., J. F. Kidwell and J. A. Sved. 1977. Hybrid dysgenesis in *Drosophila melanogaster*: A syndrome of aberrant traits including mutation, sterility, and male recombination. *Genetics* 86: 813–833.

Kimura, M. 1955. Solution of a process of random genetic drift with a continuous model. *Proc. Natl. Acad. Sci. USA* 41: 144–150.

Kimura, M. 1964. *Diffusion Models in Population Genetics*. Methuen, London.

Kimura, M. 1968. Evolutionary rate at the molecular level. *Nature* 217: 624–626.

Kimura, M. 1981. Possibility of extensive neutral evolution under stabilizing selection with special reference to nonrandom usage of synonymous codons. *Proc. Natl. Acad. Sci. USA* 78: 5773–5777.

Kimura, M. 1983. *The Neutral Theory of Molecular Evolution*. Cambridge University Press, Cambridge, England.

Kimura, M. 1986. DNA and the neutral theory. *Phil. Trans. R. Soc. Lond. B* 312: 343–354.

Kimura, M. and J. F. Crow. 1978. Effect of overall phenotypic selection on genetic change at individual loci. *Proc. Natl. Acad. Sci. USA* 75: 6168–6171.

Kimura, M. and T. Ohta. 1971. *Theoretical Aspects of Population Genetics*. Princeton University Press, Princeton, New Jersey.

King, J. L. and T. H. Jukes. 1969. Non-Darwinian evolution: Random fixation of selectively neutral mutations. *Science* 164: 788–798.

Kiyasu, P. K. and M. G. Kidwell. 1984. Hybrid dysgenesis in *Drosophila melanogaster*: the evolution of mixed P and M populations maintained at high temperature. *Genet. Res.* 44: 251–259.

Kreitman, M. E. and M. Aguadé. 1986. Excess polymorphism at the *Adh* locus in *Drosophila melanogaster*. *Genetics* 114: 93–110.

Lande, R. 1975. The maintenance of genetic variability by mutation in a polygenic character with linked loci. *Genet. Res.* 26: 221–235.

Lande, R. 1976. Natural selection and random genetic drift in phenotypic evolution. *Evolution* 30: 314–334.

Lande, R. 1977. The influence of the mating system on the maintenance of genetic variability in quantitative characters. *Genetics* 86: 485–498.

Lande, R. 1979. Quantitative genetic analysis of multivariate evolution, applied to brain: body size allometry. *Evolution* 33: 402–416.

Lande, R. 1980. The genetic covariance between characters maintained by pleiotropic mutations. *Genetics* 94: 203–215.

Lande, R. 1981. The minimum number of genes contributing to quantitative variation between and within populations. *Genetics* 99: 541–553.

Langley, C. H., J. F. Y. Brookfield and N. Kaplan. 1983. Transposable elements in Mendelian populations. I. A theory. *Genetics* 104: 457–471.

Laski, F. A., D. C. Rio and G. M. Rubin. 1986. Tissue specificity of *Drosophila* P element transposition is regulated at the level of mRNA splicing. *Cell* 44: 7–19.

Leigh Brown, A. J. and C. H. Langley. 1979. Reevaluation of level of genic heterozygosity in natural populations of *Drosophila melanogaster* by two-dimensional electrophoresis. *Proc. Natl. Acad. Sci. USA* 76: 2381–2384.

Lerner, I. M. 1958. *The Genetic Basis of Selection*. Wiley, New York.

Levin, D. A. 1978. Genetic variation in annual *Phlox*: Self-compatible versus self-incompatible species. *Evolution* 32: 245–263.

Levin, D. A. and H. W. Kerster. 1968. Local gene dispersal in *Phlox*. *Evolution* 22: 130–139.

Li, W.-H., C.-C. Luo and C.-I. Wu. 1985. Evolution of DNA sequences. pp. 1–94. *In* R. J. MacIntyre (ed.), *Molecular Evolutionary Genetics*. Plenum, New York.

Li, W.-H., C.-I. Wu and C.-C. Luo. 1985. A new method for estimating synonymous and nonsynonymous rates of nucleotide substitution. *Mol. Biol. Evol.* 2: 150–174.

Mackay, T. 1984. Jumping genes meet abdominal bristles: Hybrid dysgenesis-induced quantitative genetic variation in *Drosophila melanogaster*. *Genet. Res.* 44: 231–237.

Marks, J., J.-P. Shaw, and C.-K. J. Shen. 1986. Sequence organization and genomic complexity of a primate θ1 gene, a novel α-globin-like gene. *Nature* 321: 785–788.

May, R. M. 1985. Evolution of pesticide resistance. *Nature* 315: 12–13.

McLellan, T. and L. S. Inouye. 1986. The sensitivity of isoelectric focusing and electrophoresis in the detection of sequence differences in proteins. *Biochem. Genet.* 24: 571–577.

McNeilly, T. 1968. Evolution in closely adjacent plant populations. III. *Agrostis tenuis* on a small copper mine. *Heredity* 23: 99–108.

Miklos, G. L. G. 1985. Localized highly repetitive DNA sequences in vertebrate and invertebrate genomes. pp. 241–321. *In* R. J. MacIntyre (ed.), *Molecular Evolutionary Genetics*. Plenum, New York.

Milkman, R. 1973. Electrophoretic variation in *E. coli* from natural sources. *Science* 182: 1024–1026.

Milkman, R. 1978. Selection differentials and selection coefficients. *Genetics* 88: 391–403.

Miller, R. D., D. E. Dykhuizen, L. Green and D. L. Hartl. 1984. Specific deletion occurring in the directed evolution of 6-phosphogluconate dehydrogenase in *Escherichia coli*. *Genetics* 108: 765–772.

Miyata, T. 1984. Evolution of DNA: Dynamically evolving eukaryotic genes. pp. 56–90. *In* M. Kimura (ed.), *Introduction to Molecular Evolutionary Study*. Baifukan, Tokyo. [in Japanese]

Montgomery, E. A. and C. H. Langley. 1983. Transposable elements in Mendelian populations. II. Distribution of three *copia*-like elements in a natural population of *Drosophila melanogaster*. *Genetics* 104: 473–483.

Mourant, A. E., A. C. Kopec and K. Domaniewska-Sobczak. 1976. *The Distribution of Human Blood Groups and Other Polymorphisms*, 2nd ed. Oxford University Press, New York.

Mukai, T., T. K. Watanabe and O. Yamaguchi. 1974. The genetic structure of natural populations of *Drosophila melanogaster*. XII. Linkage disequilibrium in a large local population. *Genetics* 77: 771–793.

Mukai, T. and O. Yamaguchi. 1974. The genetic structure of natural populations of *Drosophila melanogaster*. XI. Genetic variability in a local population. *Genetics* 76: 339–366.

Myrianthopoulos, N. C. and S. M. Aronson. 1966. Population dynamics of Tay-Sachs disease. I. Reproductive fitness and selection. *Amer. J. Hum. Genet.* 18: 313–327.

Nei, M. 1975. *Molecular Population Genetics and Evolution*. Elsevier, New York.

Nei, M. 1986. Definition and estimation of fixation indices. *Evolution* 40: 643–645.

Nei, M. and R. K. Chesser. 1983. Estimation of fixation indices and gene diversities. *Ann. Hum. Genet.* 47: 253–259.

Nei, M., P. A. Fuerst and R. Chakraborty. 1976. Testing the neutral mutation hypothesis by distribution of single locus heterozygosity. *Nature* 262: 491–493.

Nei, M. and A. K. Roychoudhury. 1982. Genetic relationship and evolution of human races. *Evol. Biol.* 14: 1–59.

Nei, M. and F. Tajima. 1985. Evolutionary change of restriction cleavage sites and phylogenetic inference for man and apes. *Mol. Biol. Evol.* 2: 189–205.

Nevo, E. 1978. Genetic variation in natural populations: Patterns and theory. *Theoret. Pop. Biol.* 13: 121–177.

Newcombe, H. B. 1964 Discussion. pp. 345–349. *In* M. Fishbein (ed.), *Papers and Discussions of the Second International Conference on Congenital Malformations*. The International Medical Congress, New York.

Nichols, B. P. and C. Yanofsky. 1979. Nucleotide sequences of *trpA* of *Salmonella typhimurium* and *Escherichia coli*: An evolutionary comparison. *Proc. Natl. Acad. Sci. USA* 76: 5244–5248.

Nilsson-Ehle, H. 1909. *Kreuzungsuntersuchungen an Hafer und Weizen. Lunds. Univ. Aarskr.* NF 5,2: 1–22.

Nordskog, A. W. and F. G. Giesbrecht. 1964. Regression in egg production in the domestic fowl when selection is relaxed. *Genetics* 50: 407–416.

Novick, A. 1955. Mutagens and antimutagens. *Brookhaven Symp. Biol.* 8: 201–215.

O'Brien, S. J., D. E. Wildt, M. Bush, T. M. Caro, C. FitzGibbon, I. Aggundey and R. E. Leakey. 1987. East African cheetahs: Evidence for two population bottlenecks. *Proc. Natl. Acad. Sci. USA* 84: 508–511.

Ohta, T. 1973. Slightly deleterious mutant substitutions in evolution. *Nature* 246: 96–98.

Ohta, T. 1974. Mutational pressure as the main cause of molecular evolution and polymorphisms. *Nature* 252: 351–354.

Ohta, T. 1982. Allelic and nonallelic homology of a supergene family. *Proc. Natl. Acad. Sci. USA* 79: 3251–3254.

Orgel, L. E. and F. H. C. Crick. 1980. Selfish DNA: the ultimate parasite. *Nature* 284: 604–607.

Palmer, J. D. 1985. Evolution of chloroplast and mitochondrial DNA in plants and algae. pp. 131-240. *In* R. J. MacIntyre (ed.), *Molecular Evolutionary Genetics*. Plenum, New York.

Pirchner, F. 1969. *Population Genetics in Animal Breeding*. W. H. Freeman, San Francisco.

Policansky, D. and E. Zouros. 1977. Gene differences between the sex ratio and standard gene arrangements of the X chromosome in *Drosophila persimilis*. *Genetics* 85: 507–511.

Prakash, S. 1977. Gene polymorphism in natural populations of *Drosophila persimilis*. *Genetics* 85: 513–520.

Provine, W. B. 1986. *Sewall Wright and Evolutionary Biology*. University of Chicago Press, Chicago.

Race, R. R. and R. Sanger. 1975. *Blood Groups in Man*, 6th ed. J. B. Lippincott, Philadelphia.

Reed, T. E. and J. V. Neel. 1959. Huntington's chorea in Michigan. *Amer. J. Hum. Genet.* 11: 107–136.

Robertson, F. W. 1957. Studies in quantitative inheritance. XI. Genetic and environmental correlation between body size and egg production in *Drosophila melanogaster*. *J. Genet.* 55: 428–443.

Robinson, H. F., R. E. Comstock and P. H. Harvey. 1949. Estimates of heritability and degree of dominance in corn. *Agron. J.* 41: 353–359.

Rose, M. R. 1982. Antagonistic pleiotropy, dominance, and genetic variation. *Heredity* 48: 63–78.

Rose, M. R. 1985. Life history evolution with antagonistic pleiotropy and overlapping generations. *Theoret. Pop. Biol.* 28: 342–358.

Rose, M. R. and B. Charlesworth. 1981a. Genetics of life history in *Drosophila melanogaster*. I. Sib analysis of adult females. *Genetics* 97: 173–186.

Rose, M. R. and B. Charlesworth. 1981b. Genetics of life history in *Drosophila melanogaster*. II. Exploratory selection experiments. *Genetics* 97: 187–196.

Rubin, G. M. 1983. Dispersed repetitive DNAs in *Drosophila*. pp. 329–361. *In* J. A. Shapiro (ed.), *Mobile Genetic Elements*. Academic Press, New York.

Ruvolo, M. and T. F. Smith. 1986. Phylogeny and DNA-DNA hybridization. *Mol. Biol. Evol.* 3: 285–289.

Saitou, N. 1986. On the delta Q-test of Templeton. *Mol. Biol. Evol.* 3: 282–284.

Sawyer, S. A., D. E. Dykhuizen, R. F. DuBose, L. Green, T. Mutangadura-Mhlanga, D. F. Wolczyk and D. L. Hartl. 1987. Distribution and abundance of insertion sequences among natural isolates of *Escherichia coli*. *Genetics* 115: 51–63.

Sawyer, S. A., D. E. Dykhuizen and D. L. Hartl. in press. Confidence interval for the number of selectively neutral amino acid polymorphisms. *Proc. Natl. Acad. Sci. USA*.

Sawyer, S. A. and D. L. Hartl. 1986. Distribution of transposable elements in prokaryotes. *Theoret. Pop. Biol.* 30: 1–17.

Scavarda, N. J. and D. L. Hartl. 1987. Germ line abnormalities in *Drosophila simulans* transfected with the transposable P element. *J. Genet.* 66: 1–15.

Schaal, B. A. 1980. Measurement of gene flow in *Lupinus texensis*. *Nature* 284: 450–451.

Schaal, B. A. and D. A. Levin. 1978. Morphological differentiation and neighborhood size in *Liatris cylindracea*. *Amer. J. Bot.* 65: 923–928.

Schmid, C. W. and C.-K. J. Shen. 1985. The evolution of interspersed repetitive DNA sequences in mammals and other vertebrates. pp. 323-358. *In* R. J. MacIntyre (ed.), *Molecular Evolutionary Genetics*. Plenum, New York.

Schultz, S. C. and J. H. Richards. 1986. Site-saturation studies of β- lactamase: Production and characterization of mutant β-lactamases with all possible amino acid substitutions at residue 71. *Proc. Natl. Acad. Sci. USA* 83: 1588–1592.

Selander, R. K. and B. R. Levin. 1980. Genetic diversity and structure in *Escherichia coli* populations. *Science* 210: 545–547.

Selander, R. K. and S. Y. Yang. 1969. Protein polymorphism and genic heterozygosity in a wild population of the house mouse (*Mus musculus*). *Genetics* 63: 653–667.

Shen, S., J. L. Slightom and O. Smithies. 1981. A history of the fetal globin gene duplication. *Cell* 26: 191–203.

Sibley, C. G. and J. E. Ahlquist. 1984. The phylogeny of the hominoid primates, as indicated by DNA-DNA hybridization. *J. Mol. Evol.* 20: 2–15.

Sibley, C. G. and J. E. Ahlquist. 1986. Reconstructing bird phylogeny by comparing DNAs. *Sci. Amer.* 254 (February): 82–92.

Simmonds, N. W. 1977. Approximations for *i*, intensity of selection. *Heredity* 38: 413–414.

Simmons, M. F. and J. F. Crow. 1977. Mutations affecting fitness in *Drosophila* populations. *Annu. Rev. Genet.* 11: 49–78.

Simon, M., J. Zieg, M. Silverman, G. Mandel and R. Doolittle. 1980. Phase variation: Evolution of a controlling element. *Science* 209: 1370–1374.

Singer, M. F. 1982. SINEs and LINEs: Highly repeated short and long interspersed sequences in mammalian genomes. *Cell* 28: 433–434.

Singh, L., C. Phillips and K. W. Jones. 1984. The conserved nucleotide sequences of Bkm, which define *Sxr* in the mouse, are transcribed. *Cell* 36: 111–120.

Singh, R. S., R. C. Lewontin and A. A. Felton. 1976. Genetic heterogeneity within electrophoretic "alleles" of xanthine dehydrogenase in *Drosophila pseudoobscura*. *Genetics* 84: 609–629.

Slatkin, M. 1985a. Rare alleles as indicators of gene flow. *Evolution* 39: 53–65.

Slatkin, M. 1985b. Gene flow in natural populations. *Annu. Rev. Ecol. Syst.* 16: 393–430.

Smith, C. 1975. Quantitative inheritance. pp. 382–441. *In* G. Fraser and O. Mayo (eds.), *Textbook of Human Genetics*. Blackwell, Oxford.

Sokal, R. R. and F. J. Rohlf. 1969. *Biometry*. W. H. Freeman, San Francisco.

Southern, E. M. 1975. Detection of specific sequences among DNA fragments separated by gel electrophoresis. *J. Mol. Biol.* 98: 503–517.

Sprague, G. F. 1978. Introductory remarks to the session on the history of hybrid corn. pp. 11–12. *In* D. B. Walden (ed.), *Maize Breeding and Genetics*. Wiley, New York.

Stebbins, G. L. 1976. Chromosome, DNA and plant evolution. *Evol. Biol.* 9: 1–34.

Stocker, B. A. D. 1949. Measurements of rate of mutation of flagellar antigenic phase in *Salmonella typhimurium*. *J. Hygiene* 47: 398–412.

Sved, J. A. 1975. Fitness of third chromosome homozygotes in *Drosophila melanogaster*. *Genet. Res.* 25: 197–200.

Sved, J. A. and B. D. H. Latter. 1977. Migration and mutation in stochastic models of gene frequency change. *J. Math. Biol.* 5: 61–73.

Teissier, G. 1942. Persistence d'un gène léthal dans une population de *Drosophiles*. *Compt. Rend. Acad. Sci.* 214: 327–330.

Templeton, A. R. 1983. Phylogenetic inference from restriction endonuclease cleavage site maps with particular reference to the evolution of humans and the apes. *Evolution* 37: 221–244.

Templeton, A. R. 1985. The phylogeny of hominid primates: A statistical analysis of the DNA-DNA hybridization data. *Mol. Biol. Evol.* 2: 420–433.

Templeton, A. R. 1986. Further comments on the statistical analysis of DNA-DNA hybridization data. *Mol. Biol. Evol.* 3: 290–295.

Turelli, M. 1984. Heritable genetic variation via mutation-selection balance: Lerch's zeta meets the abdominal bristle. *Theoret. Pop. Biol.* 25: 138–193.

Turelli, M. 1985. Effects of pleiotropy on predictions concerning mutation-selection balance for polygenic traits. *Genetics* 111: 165–191.

Uyenoyama, M. 1985. Quantitative models of hybrid dysgenesis: Rapid evolution under transposition, extrachromosomal inheritance, and fertility selection. *Theoret. Pop. Biol.* 27: 176–201.

Uyenoyama, M. and M. W. Feldman. 1980. Theories of kin and group selection: A population genetics perspective. *Theoret. Pop. Biol.* 17: 380–414.

Vogel, F. and A. G. Motulsky. 1986. *Human Genetics*, 2nd ed. Springer-Verlag, New. York.

Walsh, J. B. 1985. How many processed pseudogenes are accumulated in a gene family? *Genetics* 110: 345–364.

Watanabe, S., S. Kondo and E. Matsunaga (eds.). 1975. *Human Adaptability*, Vol. 2: *Anthropological and Genetic Studies on the Japanese*. University of Tokyo Press, Tokyo.

Weir, B. S. and C. C. Cockerham. 1984. Estimating *F*-statistics for the analysis of population structure. *Evolution* 38: 1358–1370.

Whittam, T. S., H. Ochman and R. K. Selander. 1983. Multilocus genetic structure in natural populations of *Escherichia coli*. *Proc. Natl. Acad. Sci. USA* 80: 1751–1755.

Widegren, B., Ú. Árnason and G. Akusjärvi. 1985. Characteristics of a conserved 1,579-bp highly repetitive component in the killer whale, *Orcinus orca*. *Mol. Biol. Evol.* 2: 411–419.

Wilson, D. S. 1983. The group selection controversy: History and current status. *Annu. Rev. Ecol. Syst.* 14: 159–187.

Wright, S. 1968. *Evolution and the Genetics of Populations*, Vol. 1: *Genetic and Biometric Foundations*. University of Chicago Press, Chicago.

Wright, S. 1977. *Evolution and the Genetics of Populations*, Vol. 3: *Experimental Results and Evolutionary Deductions*. University of Chicago Press, Chicago.

Wright, S. 1978. *Evolution and the Genetics of Populations*, Vol. 4: *Variability Within and Among Natural Populations*. University of Chicago Press, Chicago.

INDEX

ABOUT THE BOOK
The book and cover design are by Joe Vesely. The typeface is Linotron 202 Palatino with Helvetica display type and was generated by DEKR Corporation. R. R. Donnelley & Sons printed and bound the book.